TO MY WIFE

who has been most patient whilst
this book has been written

Electrical Circuits

INCLUDING MACHINES

A. DRAPER

M.B.E., B.Sc.(Eng.), C.Eng., F.I.E.E.

Formerly Dean of the Faculty of Engineering,
Lanchester Polytechnic, Coventry and Rugby

LONGMAN

LONGMAN GROUP LIMITED
London

Associated companies, branches and representatives
throughout the world

© *A. Draper 1964*

This Edition © Longman Group Limited 1972

First published 1964
Second edition 1972

ISBN 0 582 44487 X cased
0 582 44486 1 paper

Set by Eta Services (Typesetters) Ltd., Beccles
and printed in Great Britain by
J. W. Arrowsmith Ltd., Winterstoke Road, Bristol

PREFACE TO
THE FIRST
EDITION

Most textbooks on Electrical Circuits have been written from the point of view of the light current or communications engineer. This volume is intended for the student whose primary interest is in the subject Electrical Power, Machines and Control.

It is expected to be used in conjunction with the author's previous textbook, *Electrical Machines,* and the two cover most topics in the heavy current subjects of the Diploma in Technology Examinations, Part III Examinations of the Institution of Electrical Engineers and the final degree Examinations of London University.

The treatment is chiefly by the application of matrix methods, but the mathematical ability of the student is not expected to be beyond Advanced Level of the General Certificate of Education, or Ordinary National Certificate.

Chapter 3 contains all that is necessary in explanation of matrix presentation.

The first fourteen chapters deal with problems of General Circuit Theory of interest to all electrical engineers but especially the applications engineer concerned with machines. With this background the work proceeds naturally to the Unified Theory of Electrical Machines which is seen to be a series of specialised examples of Electrical Circuits.

The M.K.S. system of units (for both mechanical and electrical quantities) and the symbols of B.S. 1991, 1954, have been used throughout.

Special attention is drawn to the need for uniform conventions relating to the directions of flux, current and voltage in machine circuits (as first pointed out by J. G. Fleming, H.M.I.) and to the differences between generalised impedance and two-port theory. A full understanding of these conventions eliminates many of the difficulties encountered by students in problems relating to machines which may be either motors or generators and where ambiguity in the direction of power flow may exist.

The term 'phasor' diagram is used in preference to 'vector' diagram in the previous volume and a full explanation is given respecting the change.

The author is fully aware of the crowding of modern syllabuses and the immense burden on the present-day student trying to cope with so many of the recent advances in technology.

This textbook is planned to help the student towards streamlining his efforts and to provide a straighter road to his goal.

The author gratefully acknowledges the assistance of many colleagues and friends who have contributed by discussion and helpful criticism. Particularly to Mr. H. E. Dance, for his most thorough and detailed criticism of the first draft; to the Electrical Research Association for Table 1, Chapter 13; to Mr. B. H. Rudall for assistance with the computation of the tables in the Appendix and to Miss M. K. Stewart for typing the manuscript.

Rugby, May 1963   A. DRAPER

PREFACE TO THE SECOND EDITION

In order to give a 'new look' to the accepted teaching of Electrical Machine Theory the cult of the Generalised Machine is now with us. Manufacturers are now producing these machines specifically for use in University and Technical College laboratories. They have been accepted by progressive teachers and their value is being exploited in many different grades of course. Colleges are pleased to use this new flexible work horse which enables many different tests to be carried out on the same unit and if several similar machines have been installed a whole class has the opportunity of doing the same thing at the same time.

All this helps to show clearly that electrical machines of all the various types which have been developed are members of a family and not the subjects of separate technologies. Good teaching emphasises these similarities, at the same time pointing out the special characteristics which enable an induction machine for example to be more appropriate for a given purpose than a synchronous machine or vice versa.

This is not, however, the only reason why Unified Theory must be taught, Unified Theory also presents a method by which transient conditions can also be analysed as well as steady-state conditions. Transient problems are much more difficult and at the present time many courses tend to stick too closely to steady-state phenomena with the result that many a student well versed in phasor-diagram analysis applicable to constant speed, constant load conditions, is unprepared for dealing with variable-load or variable-voltage states. It is easy to assume that because these problems are difficult and require advanced mathematical treatment they should be relegated to postgraduate courses or mention of them at least left to the final year of an undergraduate course.

Unified Theory is difficult but these difficulties are minimised if a student is well versed in unified electrical circuit theory in the first place. *Electrical Circuits* (*including Machines*) has been written with this point in mind and it is intended that the two books, *Electrical Circuits* and *Electrical Machines* should be studied in parallel. Whilst a student is studying Transformers, Windings, D.C. Machines and the elementary theory of the Induction Motor, he should also be concerned with fundamental circuit principles and the methods of circuit analysis.

Matrix methods and Laplace transform theory are essential nowadays and the chapters in *Electrical Circuits* devoted to these topics cover all the necessary groundwork. The first half of *Electrical Circuits* is concerned with the principles of generalised impedance theory applied to static circuits and introduces the principles of transform calculus. These elementary principles are well within the capacity of first and second year students on Ordinary and Honours Degree Courses. The second part of the book extends these principles to include the special problems of rotating machines. This is the difficult step, but if a student is properly prepared he is well able to take it and the remainder of the book shows the extent which can reasonably be attempted by a student of average ability.

Additional material has been added to chapters on Matrices, Mesh and Nodal Analysis. The chapter on the Primitive Machine has been rewritten and the chapter on 2-Port Networks now includes a discussion of Transmission lines. The tables in the appendix have also been extended by a section on the *A B C D* constants of a transmission line.

A valuable section of the appendix describes Kron's Method of Tearing but the most significant change is the introduction of chapter 22 on Magnetic Circuits.

In this chapter a detailed comparison is made of the various types of field of interest to the electrical engineer, conduction, electric and magnetic, and since electric fields may be reduced to circuits, the corresponding treatment is applied to magnetic fields. By the use of the 'inverted analogue' studies are made of the interaction between electric and magnetic circuits in order to clarify conditions in such examples as the three- and five-limb three-phase transformer and the direct and quadrature-axis inductances of a synchronous machine.

SI units are used throughout.

Coventry, 1972   A. DRAPER

CONTENTS

ACKNOWLEDGEMENTS

We are grateful to the following for permission to use copyright material:

The Electrical Research Association for material used in the table on Typical values of Z_1, Z_2, Z_0, and The Institution of Electrical Engineers for the Table of Laplace Transforms.

SYMBOLS

Italic Symbols

$A, B,$ Two-port network con-
C, D stants.
      Constants of integration.

A Area.

B Magnetic flux density.
Number of branches in a network.
Viscous resistance.

C Capacitance.

D Compliance.
Distance.
Diameter.
Electric flux density.

E Electro-motive force, e.m.f.
Driving voltage.
Voltage gradient.

E_1 Loop e.m.f. (loop 1).

E_g Generator voltage.

E_s Source voltage.

F Magneto-motive force, m.m.f.

G Conductance.
Coefficient of generated voltage.

H Magnetising force.
Height.

I Current

I_1 Loop current (loop 1).

I_b Branch current.

I_g Generator current.

I_s Source current.

$I_2{}^\dagger$ Current at an output port.

I_d Direct-axis current.

I_q Quadrature-axis current.

J Moment of inertia.
Current density.

L Inductance.
Length.

M Mutual inductance.
Mass.
Transmission constant corrector.
Number of loops in a network.
Polynomial.

N Number of turns.
Transmission constant corrector.
Number of nodes in a network.
Polynomial.

P Active power.
Force.

Q Charge.
Reactive power.

R Resistance.

S Magnetic reluctance.
Modulus of complex power.

T Time constant.

V Voltage.
Potential difference, p.d.

V_b Branch voltage.

V_A Nodal voltage (node A).

$V(t)$ Voltage as a function of time.

$V(s)$ Voltage as a function of s.

V_0 Thevenin voltage.

V_{q0} Quadrature-axis voltage at time zero.

W Energy.

X Reactance.

X_m Magnetising reactance.

Y Admittance.

Y_g Generator admittance.

Y_A Nodal admittance (connected to node A).

Z Impedance.

Z_0 Thevenin impedance. Zero-sequence impedance.

Z_1 Positive-sequence impedance.

Z_2 Negative-sequence impedance.

Z_g Generator impedance.

Z_i Image impedance.

Z_b Branch impedance.

$a \ldots n$ Coefficients.

b Susceptance.

c Inductance coefficient due to short-circuited secondary winding.

d Diameter. Distance.

e e.m.f.

f Torque.

g Number of coils.

i Current.

j Number.

k Coefficient of coupling. Turns ratio. Multiplier. Number.

l Length.

m Number of phases.

n Number of turns. Number.

p Power. Number of ports in a network. Force.

s Fractional slip (induction motor). Distance.

t Time. Thickness.

v Velocity. Voltage, potential difference, p.d. Voltage drop.

w Energy.

x'_d Direct-axis transient reactance.

x''_d Direct-axis sub-transient reactance.

x''_q Quadrature-axis sub-transient reactance.

x Distance (along x axis).

y Distance (along y axis).

z Distance (along z axis).

Bold Symbols

A, B, C, D Transmission constants (complex numbers).

I Current phasor.

S Complex power ($P - jQ$).

S^* Conjugate complex power ($P + jQ$).

V Voltage phasor.

Y Admittance complex operator.

Z Impedance complex operator.

A^* Conjugate of A (if $A = a + jb$ $A^* = a - jb$)

$V_2^\dagger \ I_2^\dagger$ Output voltage and current phasors (two-port network conventions).

h Sequence operator.

s A complex number.

y Complex admittance per unit length.

z Complex impedance per unit length.

Matrix Symbols

$[A]$ The matrix A.

$[A_t]$,   The transpose of matrix A.

$[A]$ The matrix A (complex elements).

$[A_t{}^*]$ The conjugate transpose of matrix A.

$[A]^{-1}$ The inverse of matrix A.

$|A|$ or Δ The determinant of matrix A.

$[G]$ Torque coefficients matrix.

λ Eigenvalue.

$[X]$ Eigenvector.

$[A]$ A diagonal matrix of eigenvalues.

$[Q]$ A matrix of eigenvectors.

$[A]$ The incidence matrix (nodal analysis).

$[E_s]$ Symmetrical components of e.m.f.

$[I_s]$ Symmetrical components of current.

$[M]$ The minor of a matrix element.

$[c]$ The connection matrix.

$[c]$ Connection matrix (complex numbers).

$[h]$ The symmetrical component transform.

$[Z]$ The impedance matrix of a primitive system.

$[Z']$ The derived impedance matrix of a network.

$[H]$ The hybrid matrix.

$a\,b\,c\,..\,j\,k\,..\,n$ Elements of a matrix.

a_{jk} Element of matrix a in row j column k.

Roman Symbols

a, b, c, etc.   Nodes in a network. Points on a diagram.

$F(x)$, $f(x)$, $\phi(x)$ Function of x.

j   Complex operator.

dy/dx Differential coefficient of y with respect to x.

$\int y.dx$ Integral of y with respect to x (indefinite).

$\int_a^b y.dx$ Integral of y with respect to x from $x = a$ to $x = b$.

$\oint y.dx$ Integral of y with respect to x over a closed contour.

N   North pole.

S   South pole.

pu   Per unit.

k   Winding on primitive machine.

d   Direct axis.

q   Quadrature axis.

r   Rotor.

s   Stator.

f   Field.

p   Differential operator with respect to time (p.i. = di/dt).

T   Three-terminal network.

$Lf(t)$, $F(s)$ Laplace transform of $f(t)$.

Re V Real part of V.

Im V Imaginary part of V.

Greek Symbols

α alpha   The real part of the complex number s.
Load angle.
Axis.

β beta   Angle.
Coefficient.
Axis.

γ gamma   Angle.
Coefficient.
Axis.

$\varDelta$ delta   The determinant of a matrix.

δ delta   Increment of.

ϵ epsilon   Permittivity.

ϵ_0 epsilon   Permittivity of free space.

θ theta   Angular displacement.
Angle.

$\varLambda$ lambda   Permeance.

λ lambda   Permeance.
Eigenvalue.
Phase angle.

μ mu   Permeability.

μ_0 mu   Permeability of free space.

$\varPi$ pi   Three terminal network.

π pi   Ratio circumference/diameter of circle.

ρ rho   Resistivity.

σ sigma   Conductivity.

$\varPhi$ phi   Magnetic flux.

ϕ phi   Magnetic flux.
Phase angle.

ω omega   Angular velocity.
Frequency (radians per second).
Imaginary part of complex number s.

General

The recommendations of B.S. 1991 are used throughout.

Italic symbols for physical quantities viz:

V voltage, I current, R resistance.

Roman type for abbreviations of units viz:

V volts, A amperes, Ω ohms.

Clarendon type for complex numbers (e.g. phasors) $V = a + jb$.

The solidus is used where possible $(a+b)/(c+d)$ in preference to

$$\frac{a+b}{c+d}.$$

The exponential prefix is used, $\exp(-at)$ in preference to e^{-at}.

CHAPTER 1

The fundamentals of circuit theory

1.1 Introduction

It is not easy to define the field of operation of an electrical engineer. His activities are so widespread that it is easier to compile a list of what he does not undertake. He does not build skyscrapers, bridges or cable-railways, but he is responsible for the construction of aerial masts and pylons and he strings steel-cored aluminium lines across valleys and rivers. He does not provide the necessary propulsion to eject a satellite into space: his task is the design and manufacture of the complex measuring equipment and means of communication with the space explorer, to say nothing of the control system ensuring the stability of the launching.

His mission is not the repair and maintenance of the human body, though his contribution to medical science ranges from electron microscopes, X-ray equipment to ingenious transistorised pills for exploring the alimentary tract; from high intensity lighting and colour television aids in the operating theatre to the simple comfort of the electric blanket.

His electrical machines provide cities with heat, ligh and transport, and drive countless mechanical devices from shavers to steel rolling mills.

.If one attempts to sum up, he is concerned with the generation, transmission, utilisation and control of power. To this end he uses electrical circuits, closed loops in which currents of electricity flow.

Electrical circuits vary in complexity from the simple circuit of an electric torch to the windings of electrical machines, the wiring of a digital computer and the electrical power network of a continent.

The systematic understanding of electrical circuits is, therefore, fundamental to any course of study for a potential electrical engineer, at all levels.

1.2 Simple circuit

The simplest electrical circuit (Fig. 1.1) consists of three parts:
1. A source of electrical energy, a cell or battery of cells;

2. A load or sink of energy such as a lamp or heater; and
3. Two connecting wires which should be good conductors of electricity.

   The purpose of the circuit is to transfer energy from the source to the load, and this is accomplished by the passage of electrons around the circuit.

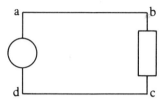

FIG. 1.1 A simple electric circuit

In order to explain what is happening, we define the following terms related to the circuit:

1.2.1 Current

The current flow at any point in a circuit is the rate at which charge is passing across a cross-section of the conductor at that point.*

$$i_{bc} \quad = \quad dq/dt$$
$$\text{(amperes)} \quad \text{(coulombs/sec.)} \tag{1.1}$$

and the total charge passed in time t seconds

$$q = \int_0^t i_{bc} \, dt \tag{1.2}$$

1.2.2 Potential difference

As each unit of charge passes through a load, it is said to fall in potential, and so give out energy.

   A p.d. of one *volt* exists between two points (say **bc** in Fig. 1.1), if one coulomb passing from **b** to **c** delivers one *joule*. Thus the energy w delivered in time t (seconds)

$$w = \int_0^t v_{bc} i_{bc} \, dt \text{ (joules)} \tag{1.3}$$

1.2.3 Power

Power is the rate of doing work. Then

$$p = dw/dt = v_{bc} i_{bc} \text{ (watts)} \tag{1.4}$$

if v and i are volts and amperes respectively.

---

* For the use of double subscript notation to ensure absence of ambiguity in direction—see A. Draper, *Electrical Machines*. Longmans, Chap. 1.

1.2.4 Electromotive force

If electric charge is flowing round the circuit (Fig. 1.1) and there is a fall of potential from **b** to **c**, it is obvious that there must be a corresponding rise as charge moves through the source from **d** to **a**. Indeed this is how the source puts energy into the system.

Each unit of charge is raised in potential as it moves through the source, acquiring energy which is released again, as it falls through the load, thus accomplishing the required transfer of energy from the source to the load.

In order to give prominence to this fundamental property of a source, (to raise the potential of charge passing through it) , the source is said to possess *Electromotive Force* (e.m.f.), which is caused by the chemical, thermal or magnetic changes occurring within the device, and is defined in terms of the voltage produced by it. Thus in Fig. 1.1, where a source is connected to a load by resistanceless conductors, if the e.m.f. of the source acting in the direction from **d** to **a** is e_{da},

then
$$e_{\mathrm{da}} = v_{\mathrm{bc}} \qquad (1.5)$$

there being no voltage drop along **ab** or **cd**, i.e. $v_{\mathrm{ab}} = 0$ and $v_{\mathrm{cd}} = 0$.

The reader's attention is drawn to the fact that this terminology is just convention. There is, of course, only one voltage, namely the potential difference between lines **ab** and **cd**. This one voltage can be described either as a rise through the source or a fall through the load. Some authorities take violent exception to the use of two different symbols to mean the same thing. The author can sympathise with this point of view, and will use the symbol v as much as possible, when it is clear that v refers to 'the volt drop across a load'. On the other hand, if a circuit contains more than one source, there is merit in using e for the e.m.f. or 'driving voltage' of each source and equating volt rises in sources with voltage drop in the loads in the manner of Eq. (1.5).

Multiplying both sides of Eq. (1.5) by the current in the circuit gives the power equation

$$e_{\mathrm{da}}i_{\mathrm{da}} = v_{\mathrm{bc}}i_{\mathrm{bc}} \qquad (1.6)$$

power input by source = power output from load.

1.3 An alternative viewpoint

At this point it is worth noting, in passing, that the circuit approach is not the only way of considering the flow of energy from the source to the load. In Fig. 1.2(*a*), which is the same circuit as Fig. 1.1, the magnetic field due to the current is shown. The field is at right angles to the plane of the paper, as shown by the crosses in Fig. 1.2(*b*). At the same time electric field exists due to the potential difference between conductors **ab** and **cd** and this is depicted by the dotted lines in Fig. 1.2(c). Finally, the direction of energy flow from source to load is shown in Fig. 1.2(d).

From these diagrams it is seen that the electric field, the magnetic field and the direction of energy flow are mutually at right angles. This notion is the basis of Poynting's Law.

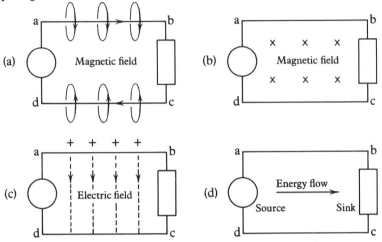

FIG. 1.2 The electric and magnetic fields associated with a simple electric circuit

1.3.1 Poynting's law

When the same physical system produces electrical field and magnetic field at right angles to each other, then energy is being transferred in the direction at right angles to both fields. These directions are shown again in Fig. 1.3. This view-

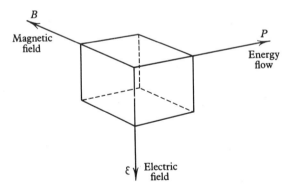

FIG. 1.3 The direction of energy flow when electric and magnetic fields exist at right angles to each other

point concentrates attention on the fields and the relation of energy flow to the field parameters. With this treatment, energy flow is seen as a field phenomenon and the circuit is merely the boundary of the field.

1.4 Stimuli and responses

Ideas of cause and effect are fundamental in electrical problems. How will a machine respond to certain conditions of applied voltage and load? What will be the value of its speed and how will it change with changes of load? What current will flow in a transmission line connecting two power stations under particular conditions of bus-bar voltage? What will be the value of the current under fault conditions? Will the circuit breaker be large enough to clear the worst possible fault? What happens when a circuit breaker is closed on a system and voltage is suddenly applied? What happens when lightning strikes? Is the output voltage of an amplifier proportional to the input for all frequencies?

This type of question, with many others, is the sort of problem the electrical engineer is required to solve. In all these, he is required to discover what the response of the apparatus system or circuit will be when a particular stimulus

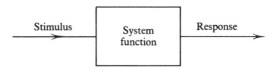

FIG. 1.4 Block diagram representing system function

FIG. 1.4a Block diagrams representing (a) resistance, (b) inductance, (c) capacitance

is applied. With an electrical circuit the stimulus is usually an applied voltage, and a resulting current somewhere in the system is regarded as the response. Both stimulus and response are functions of time and the relationship between them, or the way in which the application modifies the stimulus to produce the response, is known as the system function. Block diagrams such as Fig. 1.4 are drawn to illustrate this principle.

1.5 System function of an electric circuit

If the voltage applied to a circuit is time variant, that is to say unless we are merely considering the effects of a steady-state d.c. voltage, the resulting current depends not only on resistance, but also on the inductance and capacitance of the system. In a purely electrical network with no moving parts these three items alone determine the system function.

All capacitance, resistance and inductance is distributed in space, but often the physical form of the component makes it possible to concentrate attention on the terminal connections and to treat the circuit between them as R, L or C or a combination of them. Where the components are not so arranged and the

R, L and C are distributed continuously, it is convenient to imagine them to be concentrated. For example, a magnet coil has both inductance and resistance, each turn having an appropriate fraction of the total. Capacitance exists between the turns. To study the behaviour of this coil, it may be replaced by the idealised circuit of Fig. 1.5, where the inductance is shown concentrated in one part of the

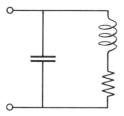

FIG. 1.5 The lumped parameters of a simple magnet coil

circuit and resistance in another part, the two being connected in series and the system capacitance represented by a single capacitor across the terminals.

Concentrating and separating R, L and C in this manner, is known as 'lumping'. This procedure necessarily introduces error which might be serious. Considerable skill is required to select a suitable equivalent circuit of reasonable accuracy.

1.6 Circuit elements

The individual effects of resistance, inductance and capacitance must now be studied.

1.6.1 Resistance

Resistance is that property of a circuit which determines the heating effect of current. For a metallic conductor at constant temperature the graph showing

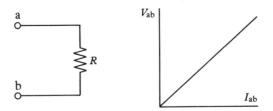

FIG. 1.6 Current and voltage are linear for a constant resistor

the relationship between v_{ab} and i_{ab} the steady state d.c. values of voltage and current is a straight line passing through the origin (Fig. 1.6).
Thus

$$v_{ab}/i_{ab} = R \qquad (1.7)$$

or

$$v_{ab} = Ri_{ab} \qquad (1.8)$$

where R is a constant and known as the resistance of the device.
The rate at which energy is converted to heat is given by

$$p = vi = i^2R \qquad (1.9)$$

The resistance of a component usually varies with temperature and in some materials is affected by other factors. Resistance may be held nearly constant by controlling the temperature or by choosing for the component a combination of materials for which the resulting variation with temperature is small.

In many circuits the resistances are either nearly constant or they change slowly in comparison with other changes. Mathematical work is simplified if they are treated as constant resistances since on this assumption the relation between voltage and current is linear.

In strain gauges the resistance is not constant because the shape of the component is changed by the strain and this effect can be used for the determination of strain at a point in a structure to which the gauge is attached.

For materials such as semi-conductors the volt/ampere relationship is not linear and, consequently, the ratio V/I is not constant. The use of the term resistance in these circumstances is somewhat misleading.

When alternating current flows in a conductor it is often noticed that the power loss is greater than the loss due to a direct current of the same r.m.s. value.

The difference is due to the alternating current not being uniformly distributed through the cross-section of the conductor* and not due to any change in the

* This non-uniform distribution of current can be shown if we consider a circular conductor of a transmission line divided into a parallel system of coaxial cylinders each having the same resistance. The inner cylinder has a slightly greater inductance than the outer cylinder and,

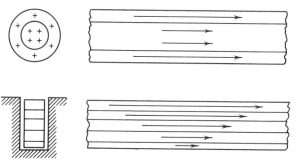

FIG. 1.13 Non-uniform current distribution in (a) a cylindrical conductor, (b) a deep conductor in an armature slot

consequently, under a.c. conditions its impedance will be greater. Since the cylinders are in parallel, the current will be divided in inverse proportion to the impedances and the current will be greatest at the outside and will taper towards the centre (Fig. 1.13).

The non-uniform distribution can be represented by assuming an additional current flowing

resistivity of the material. It is convenient to speak of and to use an imaginary
resistance R_{ac} (greater than R) such that

$$(I_{r.m.s.})^2 R_{ac} = P \qquad (1.10)$$

It should be noticed that R_{ac} varies with frequency but for power system
problems frequency is usually constant.

Equation (1.7) indicates that the system function of an idealised circuit
element comprising resistance only is a simple number equal to the value of R.

1.6.2 Inductance

Considering now the magnetic properties of a circuit, and neglecting its resistance, some relationship exists between the flux ϕ linking the circuit and the
current i producing it, determined by the geometry of the circuit, and the
magnetic properties of the medium of the field. Unless the field is ferromagnetic,
ϕ will be proportional to i. If these flux links are concentrated in the part of
the circuit **a** to **b**, Fig. 1.7, and if the current is time variant, the flux will also be
time variant.

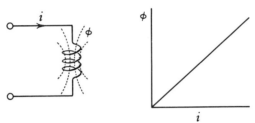

FIG. 1.7 Current and flux are linear for a constant inductor

A voltage will be required across **ab** to produce the changes. By Faraday's
Law

$$v_{ab} = d\phi/dt \qquad (1.11)$$

thus

$$v_{ab} = \frac{d\phi}{di} \cdot \frac{di}{dt} \qquad (1.12)$$

If ϕ is proportional to i, $d\phi/di$ is a constant parameter for the magnetic field
hence

$$v_{ab} = L\,di/dt \qquad (1.13)$$

---

along the outer conductor and back along the centre cylinder superimposed on the normal
uniform distribution.

This additional current is the cause of additional loss on a.c. working.

The same effect is very pronounced if a deep conductor is used in an armature slot in a
machine. Considering this conductor divided into strips, the inductance and impedance of the
bottom strip is much greater than that of the top strip and the current distribution will be as
shown in the figure. For this reason, deep conductors are always stranded and the strands
are not connected in parallel until they have been sufficiently transposed to equalise the
impedances.

where

$$L = d\phi/di \tag{1.14}$$

and is known as the inductance of the circuit.

Equation (1.13) indicates that the system function (i.e. the relationship between stimulus and response) of an idealised circuit element, comprising inductance only, is not merely the number L but includes the process of differentiation.

1.6.3 Capacitance

If a charge q is taken from one insulated conductor to another, an electric field is set up between them and a potential difference is produced. In Fig. 1.8,

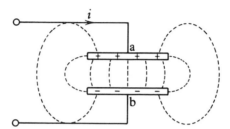

FIG. 1.8 The electric field between parallel plates

charge $+q$ is taken from conductor **b** and placed on conductor **a** (thus leaving charge $-q$ on **b**) and a p.d. v_{ab} is set up. The relationship between v_{ab} and q depends on the geometry of the electric field and the material of the dielectric medium. For most practical materials and with the geometry unchanged

$$v_{ab} = q/C \tag{1.15}$$

where C is a constant for the field and is known as the capacitance of the system.

$$q = \int_{-\infty}^{t} i \, dt \tag{1.16}$$

thus

$$v_{ab} = (1/C) \int_{-\infty}^{t} i \, dt \tag{1.17}$$

and

$$i_{ab} = C \, dv/dt \tag{1.18}$$

The system function of a capacitance element includes the process of integration as well as the coefficient $1/C$.

1.6.4 Potential source

Two other circuit elements are needed to provide the stimuli and these are potential sources and current sources. A potential source is an idealised device devoid of resistance, inductance or capacitance, but possessing e.m.f. which is a function of time only.

Figure 1.9(*a*) shows a potential source *e*(*t*) connected between terminals **a**

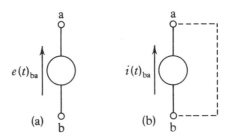

FIG. 1.9 Symbols for (a) a voltage source, (b) a current source

and **b**. At all times v_{ab} is equal to *e*(*t*) and is independent of the value of current flowing.

1.6.5 Current source

Similarly Fig. 1.9(*b*) shows a current source. This also is a perfect generator which maintains a current *i*(*t*), a given function of time, flowing from **b** to **a** independently of the p.d. which happens to be across them.

A potential source cannot be short-circuited or infinite current would flow, and similarly, a current source must not be open-circuited or the p.d. across it would be infinite.

This statement is a reminder that we are dealing with idealised circuit elements.

1.7 Energy considerations

A source delivers energy to a circuit (Eq. (1.6)) if the product $e_{da}i_{da}$ is positive. If at a given time the e.m.f. and current are in opposite directions within the source, the product $e_{da}i_{da}$ will be negative and the source will be 'delivering negative power' or, in other words, absorbing energy or acting as a load.

On the other hand a load or sink removes energy from the circuit, and again referring to equation (1.6), the power removed is $v_{bc}i_{bc}$.

In the case of a resistor *R*, the power $p = vi = i^2R$ and this is always positive whatever the direction of current, hence a resistor is always a sink or load.

We now examine the amount of energy associated with an inductance. To establish magnetic field, energy is required, and so to establish a current *I* in an inductance *L*, a voltage *v* must be applied for a given length of time during which

$$v = L \, di/dt \tag{1.13}$$

The energy supplied by the source will be

$$W = \int_{i=0}^{i=I} vi\,dt \tag{1.19}$$

$$= L \int_0^I i\,di$$

$$= \tfrac{1}{2}LI^2 \tag{1.20}$$

An inductor is thus an energy store. Similarly to charge a capacitor C to a voltage V, energy once more is needed.

If the capacitor is charged from a current source i the voltage will rise at a rate such that

$$i = C\,dv/dt \tag{1.18}$$

The energy supplied

$$W = \int_{v=0}^{v=V} vi\,dt \tag{1.21}$$

$$= \int_0^V Cv\,dv$$

$$= \tfrac{1}{2}C V^2 \tag{1.22}$$

A capacitor is also an energy store. When a store is receiving energy it behaves as a load (viz. current increasing in an inductor or voltage increasing in a capacitor), but when a store is returning energy to the system (current decreasing in an inductor, voltage decreasing in a capacitor) it behaves temporarily as a source.

1.8 Steady-state and transient conditions

If the applied stimulus to a system is constant, as for example when a steady d.c. voltage is applied to a circuit, a state of equilibrium will ultimately be reached when the response (i.e. the current) will also become constant and the power input to the system at any instant will be equal to the power dissipated in the sinks. The voltages across any capacitor in the circuit will cease to change and the voltages across the inductors will be zero. Energy will be neither entering nor leaving the stores. The system is now operating in what is termed steady-state conditions, and the final current value is known as a steady-state direct current.

But whilst the currents and voltages have been growing and energy has been entering or leaving the stores, the distribution of power will have been quite different, indeed oscillations may have been set up between the stores. This condition is temporary and exists until the state of equilibrium is achieved. Whenever an applied stimulus is changed from one steady-state value to another (the most common example is a sudden change from zero to normal value), a time interval exists before the output settles down to its new steady-state value. The term *transient* is the name given to this intermediate phase. Only as the transient phase dies away are the final steady-state values assumed.

1.8.1 *R.C. circuit transient*

A simple example of transient conditions occurs when a d.c. voltage is suddenly applied to a circuit which has both resistance and capacitance. Figure 1.10 shows such a circuit with a resistor R connected between points **a** and **b** and a capacitor between **b** and **c**. The switch connecting the source is shown open and the capacitor is initially uncharged. The applied voltage is zero until the switch is closed,

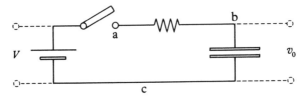

FIG. 1.10 **Switching a series** *R.C.* **circuit across a constant voltage source**

but after closure the battery voltage is applied between **a** and **c**. Let the switch be closed at the time $t = 0$.

The graph of the applied voltage is the one shown in Fig. 1.11(*a*) and the instantaneous value is described by the expression

$$v(t) = 0\Big]_{-\infty}^{0} + V\Big]_{0}^{\infty} \qquad (1.23)$$

We will designate v_{bc} the capacitor voltage (or output voltage v_0) to be the response of the system due to the applied stimulus $v(t)$ and determine an expression for its magnitude.

Immediately before the switch is closed the voltage across the capacitor is zero, and this voltage cannot change instantaneously unless infinite current flows into the capacitor. At the instant of closing the switch the voltage V_{ab} across the resistance R is equal to V, the voltage V_{bc} being zero, and the instantaneous current is given by

$$i_{(t=0)} = V/R \qquad (1.24)$$

This current flows into the capacitor, consequently the voltage across the capacitor begins to rise and as it does, the voltage across R is reduced and with it the current.

At a time t

$$V = v_{ab} + v_{bc}$$
$$= iR + v_0$$

where

$$i = dq/dt = C\,dv_0/dt$$

Thus

$$V = CR \, dv_0/dt + v_0 \qquad (1.25)$$

This is the differential equation governing the behaviour of the circuit. It can be solved for v_0 by the following simple mathematical procedure.

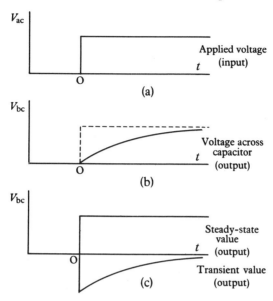

FIG. 1.11 Applying a step-function voltage to an *R.C.* circuit showing how the response compares with the stimulus

Re-arranging the terms in equation (1.25)

$$V - v_0 = CR \, dv_0/dt$$

$$dv_0/(V - v_0) = dt/CR \qquad (1.26)$$

Integrating

$$-\log_e (V - v_0) = t/CR + D \qquad (1.27)$$

where D is a constant of integration.
At time $t = 0$, $v_0 = 0$, hence substituting these values in Eq. (1.27)

$$-\log_e V = D$$

Thus

$$\log_e \frac{V - v_0}{V} = -\frac{t}{CR} \qquad (1.28)$$

and

$$(V - v_0)/V = \exp(-t/CR)$$

thus

$$v_0 = V - V \exp(-t/CR) \qquad (1.29)$$

The graph of v_0 is shown in Fig. 1.11(b). We see that v_0 is the sum of two components V and $-V \exp(-t/CR)$. The first component V is the steady-state response and is equal to the input voltage The second component $-V \exp(-t/RC)$ is the transient term in the output voltage. It is equal to $-V$ when $t = 0$ but as t increases this term diminishes and ultimately becomes negligible.

It is the presence of this transient which prevents v_0 from responding immediately and following exactly the sudden change of V. The transient term dies away because of the negative sign attributed to (t/RC) and the rapidity of the decay depends inversely on the product RC.

We shall usually find that transient terms relating to electrical systems decay in an exponential manner, but when they do not, it means that steady-state (equilibrium) conditions are never achieved, the system being unstable and the final response unrelated to the magnitude of the input.

At any instant

$$\begin{bmatrix} \text{instantaneous} \\ \text{value} \end{bmatrix} = \begin{bmatrix} \text{steady-state} \\ \text{response} \end{bmatrix} + \begin{bmatrix} \text{transient} \\ \text{response} \end{bmatrix} \tag{1.30}$$

1.8.2 Steady-state a.c. conditions

The term steady-state can also be applied to a periodic function. When alternating current flows in a circuit, although the current is continually changing, it does so in accordance with a prescribed series of events. Although the waveform may be complex, if each cycle is an exact replica of the previous one, steady-state conditions are said to have been achieved.

Once again, the total energy per cycle obtained from the sources is equal to that dissipated in the sinks, although now additional energy flows to and fro between the stores, in a manner which has bearing on the overall relationship between the stimulus and the output response.

A good example of this is the steady swing of the pendulum of a clock. Energy continually flows from the kinetic energy of the bob which is a maximum at the centre of the swing to potential energy which is a maximum when the bob comes momentarily to rest at the top of the swing. The small amount of energy lost in each cycle is supplied by the escapement mechanism and so the pendulum maintains steady-state.

If the periodic stimulus applied to a system is changed either in magnitude, frequency, or waveform to another steady-state value, once again transient conditions exist until the circuit response achieves steady-state values corresponding to the new stimulus.

Transient conditions are always introduced by change in the applied stimulus and the final steady-state response is not regarded as being fully established until the transient response has become completely negligible.

The determination of the transient response of a system, as well as of its steady-state behaviour, is therefore of immense importance.

1.9 Circuit differential equations

When all three elements R, L and C exist together in a circuit, the equilibrium equations can be defined. These are differential/integral equations of the following type. For example, in the series circuit of Fig. 1.12.

$$v_{ad} = v_{ab} + v_{bc} + v_{cd} \qquad (1.31)$$

whence from Eqs. (1.13, 1.8, 1.17)

$$e(t) = L\, di/dt + Ri + (1/C) \int_{-\infty}^{t} i\, dt \qquad (1.32)$$

If the voltage of the source $e(t)$ can be expressed analytically, as a function of time, this equation can be solved for $i(t)$ by the classical methods of solution of

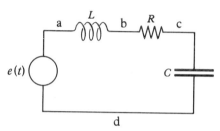

FIG. 1.12 A simple L.R.C. circuit with lumped parameters

differential equations, and thus the response of this particular circuit to the applied stimulus can be determined.

The solution of Eq. (1.32) proceeds as follows: first substitute for $i(t)$ since

$$i(t) = dq/dt \qquad (1.33)$$

and Eq. (1.32) becomes

$$L\, d^2q/dt^2 + R\, dq/dt + (1/C)q = e(t) \qquad (1.34)$$

which is of the form

$$a\, d^2x/dt^2 + b\, dx/dt + cx = f(t) \qquad (1.35)$$

This is a familiar linear differential equation of the second order with constant coefficients (see footnote).*

---

* **Linearity** A linear differential equation is one of the form

$$P\, d^n x/dt^n \ldots + Q\, dx/dt + Rx = S \qquad (1.36)$$

where P, Q, R and S are independent of x. (They may be functions of t or vary in other ways.) When P, Q and R are constants, this condition is clearly satisfied.

One of the properties of a linear equation is that if $x = u_1$, is the solution of

$$P\, d^2x/dt^2 + Q\, dx/dt + Rx = S_1 \qquad (1.37)$$

and if
$x = u_2$, is the solution of

$$P\, d^2x/dt^2 + Q\, dx/dt + Rx = S_2 \qquad (1.38)$$

It is well known that the solution of this equation is obtained in two parts, giving

$$x = F_1(t) + F_2(t) \tag{1.40}$$

$F_1(t)$ is a particular integral which will satisfy the complete equation (1.35). $F_1(t)$ obviously depends to a large extent on the type of function $f(t)$ on the right hand side of the equation. This part of the solution is the steady-state response.

$F_2(t)$ is known as the complementary function and satisfies the reduced equation

$$a \, d^2x/dt^2 + b \, dx/dt + cx = 0 \tag{1.41}$$

This is the transient response of the circuit and the type of function resulting is obviously in no way dependent on $f(t)$. It is important to note that the complementary function depends only on the constants a, b and c, the parameters of the circuit.

Mathematics textbooks show that the reduced equation is satisfied by

$$F_2(t) = A \exp(\alpha t) + B \exp(\beta t) \tag{1.42}$$

where α and β are the roots of the auxiliary equation

$$am^2 + bm + c = 0 \tag{1.43}$$

If a, b and c are all positive as they must be for an electric circuit, α and β will be either real and both negative, or complex conjugate with a negative real part. In either case it will be seen that $F_2(t)$ is subject to exponential decay.

It is thus appreciated that when the transients have become negligible after sufficient lapse of time, the charge on the capacitor and the current will assume steady-state values.

More complicated circuits are represented by simultaneous differential equations including many terms, although all of them are similar to those of Eq. (1.34). Classical methods of solution, when applied to such problems, are too laborious and consequently more powerful, though less general, methods have been developed.

These devices are discussed in Chap. 7.

1.9.1 Operational nomenclature

A simplified nomenclature will be used in later chapters. This is obtained by writing

$$pf(t) \quad \text{for} \quad df(t)/dt \tag{1.44}$$

Then

$x = u_1 + u_2$, is the solution of

$$P \, d^2x/dt^2 + Q \, dx/dt + Rx = S_1 + S_2 \tag{1.39}$$

(the proof is obtained by substitution).

This is the basis of the principle of superposition in electrical circuits, which is to be discussed later (Chap. 10).

and

$$(1/p)f(t) \quad \text{for} \quad \int_{-\infty}^{t} f(t)\, dt \tag{1.45}$$

With this notation, Eq. (1.32) becomes

$$e(t) = Lpi(t) + Ri(t) + (1/Cp)i(t) \tag{1.46}$$

or

$$e(t) = (Lp + R + 1/Cp)i(t) \tag{1.47}$$

The expression $(Lp + R + 1/Cp)$ is known as a *differential operator* not to be confused with an algebraic multiplier. The equation indicates that the function $i(t)$ has to be operated upon in various ways, including differentiation and integration, in order to obtain the function $e(t)$. It describes precisely and concisely how, given the current $i(t)$, the volt drop across each part of the circuit can be determined and how these drops are then added to find the applied voltage.

Given $i(t)$, the value of $e(t)$ can be obtained relatively easily, but the opposite (given $e(t)$, find $i(t)$) requires the procedure of solving differential equations.

If we re-write equation (1.47) in the form

$$i(t) = \frac{1}{Lp + R + 1/Cp}\, e(t) \tag{1.48}$$

we see that the process of solving the equation becomes implicit in the under-standing of what is meant by the inverse operator

$$\frac{1}{Lp + R + 1/Cp}$$

and how this operator, acting on $e(t)$, the stimulus, produces the response $i(t)$. Again it is emphasised that $1/(Lp + R + 1/Cp)$ is not an algebraical multiplier.

CHAPTER 2

Network
terminology

2.1 Series and parallel circuits

The reader will already be familiar with simple d.c. circuits such as are shown in Figs. 2.1(a) and 2.1(b) and with analytical techniques which enable the currents in each part of the circuit to be determined, given the applied voltage and all values of resistance.

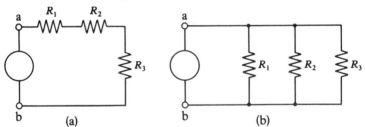

FIG. 2.1 **Three resistors (a) in series, (b) in parallel**

In the circuit of Fig. 2.1(a), which is a series circuit

$$I = V/(R_1 + R_2 + R_3) \tag{2.1}$$

and in the parallel circuit of Fig. 2.1(b)

$$I = V(G_1 + G_2 + G_3) \tag{2.2}$$

where

$$G_1 = 1/R_1 \quad G_2 = 1/R_2 \quad G_3 = 1/R_3 \tag{2.3}$$

The simple rules are
(1) in a series circuit the equivalent resistance is the sum of the individual resistances;
(2) in parallel circuit the equivalent conductance is the sum of the individual conductances.

In Fig. 2.2 we have a more complicated circuit consisting of resistors in series and parallel. The method of systematic reduction is applied, using the laws for series and parallel circuits. First it is recognised that R_2 and R_3 are in series and so they can be replaced by a single resistor equal in magnitude to $(R_2 + R_3)$. R_4 and R_5 are treated similarly, and the circuit is reduced to that of Fig. 2.2(b).

We now see that there are two resistors in parallel between points **b** and **c**. These are replaced by their equivalent resistance, first converting the resistance

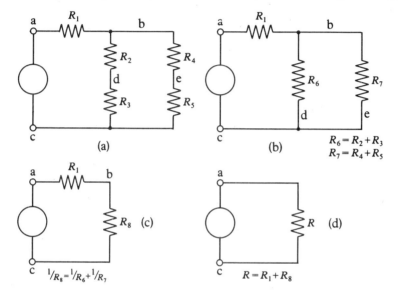

$$R_6 = R_2 + R_3$$
$$R_7 = R_4 + R_5$$

$$1/R_8 = 1/R_6 + 1/R_7$$

$$R = R_1 + R_8$$

FIG. 2.2 The systematic reduction of a series parallel circuit

to conductance and adding the conductances according to the laws of the parallel circuit. The circuit then becomes that of Fig. 2.2.(c).

The final reduction to Fig. 2.2(d) is obtained by adding the resistances in Fig. 2.2(c), giving the total equivalent resistance R. Since the voltage across this resistance is the applied voltage, the current is determined from Ohm's Law

$$I = V_{ac}/R \tag{2.4}$$

Knowing I_{ac}, the currents and voltages applied to the individual resistors of the original circuit can be traced back.
From Fig. 2.2(c)

$$I_{ab} = I_{bc} = I \tag{2.5}$$

Therefore

$$V_{bc} = IR_8 \tag{2.6}$$

In Fig. 2.2(b)

$$I_{bdc} = V_{bc}/R_6 \tag{2.7}$$

and

$$I_{\text{bec}} = V_{\text{bc}}/R_7 \qquad\qquad (2.8)$$

2.2 Networks

This simple technique will no longer apply if the circuit of Fig. 2.2(*a*) is modified to that of Fig. 2.3. The introduction of R_9 between points **d** and **e** destroys the elementary series parallel arrangement. No longer are R_2 and R_3 in series, nor can we find any parallel sections to begin the process of simplification.

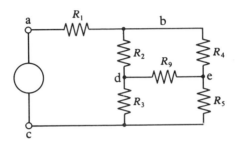

FIG. 2.3 An elementary circuit with three loops

The system is now described as a network and a more elaborate method is needed for its analysis. The method must be sufficiently general in application so as to include the effects of inductance and capacitance in addition to resistance.

The following terms used extensively in network theory are now defined:

2.2.1 Circuit elements—the constituent components of a network

These have already been discussed in Chap. 1, and are given again below

Active elements	Voltage source	$e(t)$
	Current source	$i(t)$
Passive elements	Resistor	R
	Inductor	L
	Capacitor	C

Elements are termed linear if a linear differential equation applies to the response to all types of stimulus, and bilateral if the terminals can be interchanged without affecting external results.

2.2.2 Branch

A branch is a single element or group of elements connected in either series or parallel to form a device with two terminals only, these terminals forming the only connections to other branches. A branch may include combinations of R, L and C.

2.2.3 Node (junction)

This is a point in a network which forms the connection between three or more branches. In a circuit diagram of the network a node may be a line representing a busbar or impedanceless connection, but in the graph of a network (Fig. 2.5) the nodes are designated by small circles.

In Fig. 2.3, points **b, c, e, d**, are nodes. Point **a** might be described as a node if we wish to consider the voltage source and R_1 separately, but in accordance with the definition **cab** is a single active branch and the network has only four nodes.

2.2.4 Loop

Any path through two or more branches which forms a closed circuit is called a loop or mesh.

2.2.5 Network

The term network is applied to any circuit which has two or more loops.

2.2.6 Graph of a network

A network may be depicted by a circuit diagram in which the elements are shown by usual symbols, as for example in Fig. 2.4. Inspection shows that this

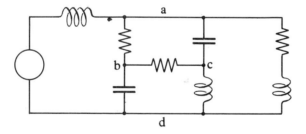

FIG. 2.4 A network with four nodes

network has four nodes, **a, b, c** and **d**. In order to see more clearly the connections of the network, a map may be made commencing with the nodes and drawing lines between them representing the branches.
The geometrical figure produced is referred to as the graph of the network.

It is relatively easy to determine the number of nodes by inspection, but to find the number of independent loops in the system is less simple. Out of the twelve possible loops in Fig. 2.5, only four are independent.

The number of independent loops is the *minimum* number of loops or closed paths through the network which when put together constitute the whole network.

As the number of independent loops is needed to establish the equilibrium equations of the system, a routine method should be used, and for this two new terms are introduced, *network trees* and *tree links*.

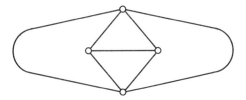

FIG. 2.5 The graph corresponding to the network of Fig. 2.4

2.2.7 Network trees and tree links

A network tree is any arrangement of branches which connects *all* the nodes together without forming any loops.

Three possible trees corresponding to the network of Fig. 2.5 are given in Fig. 2.6.

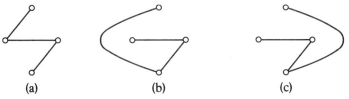

(a)   (b)   (c)

FIG. 2.6 Three possible trees for the network of Fig. 2.4

It will be noticed that if a network has N nodes, the number of branches forming a tree is

$$N-1 \tag{2.9}$$

The remaining branches are known as tree links, and hence the number of tree links in a network with B branches is

$$B-(N-1). \tag{2.10}$$

Figure 2.7 shows the tree links in dotted lines corresponding to the trees of Fig. 2.6.

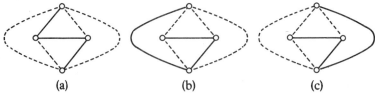

(a)   (b)   (c)

FIG. 2.7 The three trees of Fig. 2.6 with their tree links

Now it will be seen that if the tree links are drawn in, one at a time, each connection adds another loop to the system, each loop consisting of one-tree link closed by branches which are part of the tree.

Each tree link is thus associated with only one loop and since when all the links have been connected, the network is complete, it is clear that the number of independent loops is equal to the number of tree links.

If M is the number of loops then

$$M = B - N + 1. \tag{2.11}$$

2.3 Ports

We have seen that a network is a system constructed of branches, some of which may be active and others passive. It is often convenient to concentrate the sources on the perimeter of the network and also any other loops of particular

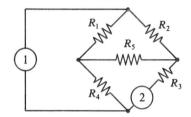

FIG. 2.8 A conventional method of representing a circuit

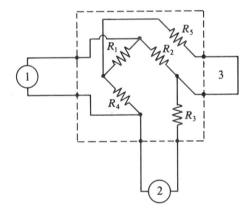

FIG. 2.9 The circuit of Fig. 2.8 re-drawn as a passive network with three ports

interest, and regard these as points of entry or *ports* to an otherwise passive network. This is of particular importance in certain types of power-system problems where the interest is in the load sharing between generators and when it is unnecessary to determine the current in all parts of the network.

For example the network of Fig. 2.8 has two active branches, and the branch R_5 may be of particular importance. This circuit is redrawn in Fig. 2.9 as a 3-port network with the two sources and a loop in branch R_5 brought out through the dotted lines enclosing the network.

The following terms apply to a network with P ports.

2.3.1 Driving-point impedance

The term *driving-point impedance* is applied to the ratio between the voltage and the current *at a given port*. The condition or termination of the other ports must be stated.

This is illustrated in Fig. 2.10 where all ports are open-circuited, the voltage V_1 is applied and the current I_1 flows.

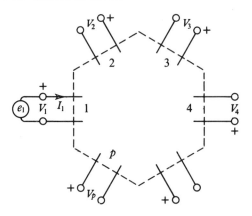

FIG. 2.10 A p-port network with ports open-circuited

The open-circuit driving point impedance at port **1**

$$= V_1/I_1 \qquad (2.12)$$

and the open-circuit driving-point admittance at port **1**

$$= I_1/V_1 \qquad (2.13)$$

Similarly in Fig. 2.11 where all ports are short-circuited except the first, the short-circuit driving-point impedance at port **1**

$$= V_1/I_1 \qquad (2.14)$$

and the short-circuit driving-point admittance at port **1**

$$= I_1/V_1 \qquad (2.15)$$

The term *immittance* is sometimes used as a general term to mean either admittance or impedance whichever is the easier to specify for a given condition.

2.3.2 Transfer impedance or admittance

It is often convenient to relate the voltage at one port to the current at a second port always remembering to specify the terminations at all the ports. Thus we have in Fig. 2.10

$V_2/I_1 =$ the open-circuit transfer impedance between port **2** and port **1**   (2.16)

It should be noted that the ratio V_1/I_2, the open-circuit transfer impedance between port **1** and port **2**, is not the same as V_2/I_1.

In Fig. 2.11

$$I_2/V_1 = \text{short-circuit transfer admittance between port } \mathbf{2} \text{ and port } \mathbf{1} \quad (2.17)$$

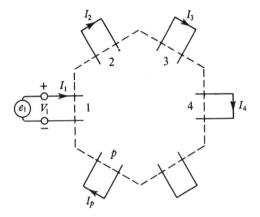

FIG. 2.11 A p-port network with ports short-circuited

2.3.3 Voltage and current ratios

The relationship between voltages for a pair of ports expressed as a ratio is often required and similarly the ratio of the currents.

Thus in Fig. 2.10

$$V_2/V_1 = \text{open-circuit voltage ratio between port } \mathbf{2} \text{ and port } \mathbf{1} \quad (2.18)$$

and in Fig. 2.11

$$I_2/I_1 = \text{short-circuit current ratio between port } \mathbf{2} \text{ and port } \mathbf{1} \quad (2.19)$$

The termination of all ports must be specified and also which port is used as the source.

2.3.4 Source impedance

A practical generator is not a perfect constant-voltage or constant-current source since it usually includes conductors which possess impedance. It may be regarded as either (*a*) a constant-voltage source with series impedance (Fig. 2.12(*a*)) or (*b*) a constant-current generator with shunt admittance (Fig. 2.12(*b*)).

Such impedance (or admittance) is known as the *source impedance* (or *admittance*).

2.3.5 Input impedance

If often happens that a network can be divided into two zones with a pair of terminals (**a** and **b** in Fig. 2.13) connecting them. If the sources are all in the first zone and the second zone is entirely passive the terminals **a** and **b** form an output port of zone **A** connected to the input port of zone **B**. It will be shown in Chap. 10 that zone **B** can be reduced to a single effective impedance Z_2 looked

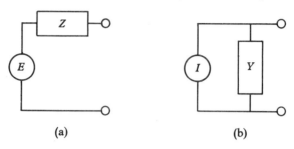

(a)   (b)

FIG. 2.12 A practical generator is either: (a) a constant voltage source with series impedance or (b) a constant voltage source with shunt admittance

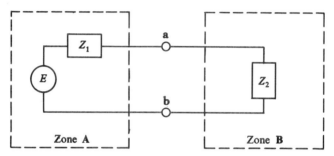

FIG. 2.13 A network divided into two zones; zone A is active, zone B is passive

at from the terminals **a** and **b** and this impedance is known as the *input impedance* of zone **B**. It is connected as a *load impedance* to zone **A**.

In a similar way source **A** can be reduced to an effective source voltage in series with a source impedance Z_1.

2.3.6 Image impedance

This term is applied to a passive network with two or more ports. At each of the ports an external impedance can be connected (Z_{i1}, Z_{i2} and Z_{i3}, Fig. 2.14)

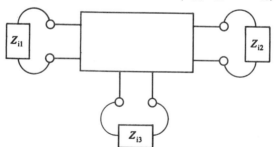

FIG. 2.14 A network with three ports

so that with all impedances in position, the impedance at any port looking into the network (the input impedance at that port) is equal to the external impedance connected there. These external impedances are known as the *image impedances* corresponding to the ports.

Example

If a two-port network consists of three impedances Z_1, Z_2 and Z_3 in 'tee' formation the image impedances Z_{i1} and Z_{i2} will be such that

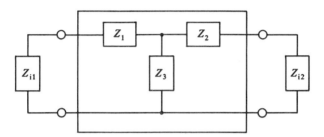

FIG. 2.15 An example of image impedance

$$Z_{i1} = Z_1 + \frac{Z_3(Z_{i2}+Z_2)}{Z_{i2}+Z_2+Z_3} \qquad (2.20)$$

$$Z_{i2} = Z_2 + \frac{Z_3(Z_{i1}+Z_1)}{Z_{i1}+Z_1+Z_3} \qquad (2.21)$$

from which

$$Z_{i1}(Z_{i2}+Z_2+Z_3) = Z_1(Z_{i2}+Z_2+Z_3)+Z_3Z_{i2}+Z_2Z_3$$
$$= (Z_1Z_2+Z_2Z_3+Z_3Z_1)+Z_{i2}(Z_1+Z_3)$$

and

$$Z_{i2}(Z_{i1}+Z_1+Z_3) = (Z_1Z_2+Z_2Z_3+Z_3Z_1)+Z_{i1}(Z_2+Z_3)$$

hence

$$Z_{i1}^2 = \frac{(Z_1+Z_3)}{(Z_2+Z_3)} \cdot (Z_1Z_2+Z_2Z_3+Z_3Z_1) \qquad (2.22)$$

and

$$Z_{i2}^2 = \frac{(Z_2+Z_3)}{(Z_1+Z_3)} \cdot (Z_1Z_2+Z_2Z_3+Z_3Z_1) \qquad (2.23)$$

2.3.7 Iterative impedance

The term iterative impedance is applied to 2-port networks. When a 2-port network is terminated by a load impedance, the input impedance at the first port is dependent on the value of the load.

The particular value of the load impedance which makes the two impedances equal is known as the iterative impedance of the network.

FIG. 2.16 An example of iterative impedance when the input impedance and load impedance are alike

It could also be defined as the input impedance of an infinite chain of such networks.

CHAPTER 3

Matrix
nomenclature

3.1 Terminology

A matrix is an orderly arrangement of numbers. Let us consider different ways of displaying one million numbers. They might be placed in a horizontal line or row, or alternatively, in a vertical column.

Each of these methods is a matrix arrangement. Each number is now identified by its position in the row or column.

Another suggestion would be to use one thousand rows with one thousand numbers in each row, placing all numbers equally spaced in each row and the rows immediately above each other. Each number has now two co-ordinates, since each exists both in a row and in a column. The matrix would be known as a square matrix, 1000×1000.

Single row or column matrices are termed one-way matrices and mathematicians also refer to them as row vectors or column vectors respectively. A rectangular matrix which has both rows and columns is known as a 2-way matrix.

It is possible to think of matrices with a still greater number of ways. For example, the million numbers could be arranged in 10 rows, each of 10 numbers on 10 pages of 10 books on 10 shelves of 10 bookcases.

Each number has now six co-ordinates to define its position and the display is then a 6-way matrix.

We shall show that matrices are valuable ways of displaying numbers representing physical quantities. Only 1-way and 2-way matrices will be discussed in this chapter; when the term matrix alone is used, it will be understood to mean a 2-way matrix either rectangular or square.

The purpose of these orderly displays is not only to enable individual numbers to be easily identified, but to enable a group of numbers to be treated as a whole or as a single expression. Subject to certain precautions, the normal mathematical processes of addition, subtraction, multiplication and division can be carried out on these matrix units, this being the basis of a special method of computation, known as matrix algebra.

3.2 Types of matrices

3.2.1 Row, column, rectangular and square matrices

Examples of row, column and 2-way matrices, rectangular, and square, are shown below. It is usual to display the matrix by enclosing the numbers in square sided brackets or to arrange a framework around them. In the text we use [A] when we wish to refer to 'a matrix A', or sometimes when formulae and equations tend to become cumbersome, A in **bold** type is used, the brackets being omitted.

$$[a] = [a_1 \quad a_2 \quad a_3 \quad a_4] \quad \text{or} \quad \boxed{a_1 \mid a_2 \mid a_3 \mid a_4} \tag{3.1}$$

a row matrix (or row vector).

$$[b] = \begin{bmatrix} b_1 \\ b_2 \\ b_3 \\ b_4 \end{bmatrix} \quad \text{or} \quad \boxed{\begin{array}{c} b_1 \\ b_2 \\ b_3 \\ b_4 \end{array}} \quad \begin{array}{l} \text{a column matrix} \\ \text{(or column vector)} \end{array} \tag{3.2}$$

$$[c] = \begin{bmatrix} c_{11} & c_{12} & c_{13} \\ c_{21} & c_{22} & c_{23} \\ c_{31} & c_{32} & c_{33} \end{bmatrix} \quad \text{or} \quad \boxed{\begin{array}{c|c|c} c_{11} & c_{12} & c_{13} \\ c_{21} & c_{22} & c_{23} \\ c_{31} & c_{32} & c_{33} \end{array}} \quad \begin{array}{l} \text{a square} \\ \text{matrix} \end{array}$$

$$[d] = \begin{bmatrix} d_{11} & d_{12} & d_{13} & d_{14} \\ d_{21} & d_{22} & d_{23} & d_{24} \end{bmatrix} \tag{3.3}$$

a rectangular matrix

$$\text{or} \quad \boxed{\begin{array}{c|c|c|c} d_{11} & d_{12} & d_{13} & d_{14} \\ d_{21} & d_{22} & d_{23} & d_{24} \end{array}} \tag{3.4}$$

Notice how a double subscript notation is used with rectangular and square matrices. The position of each matrix element is indicated by the subscripts.

The first number relates to the row in which the element is situated, reading from top to bottom, and the second number refers to the column reading from left to right (example: c_{23} is in the second row and third column).

3.2.2 Diagonal elements

In a square matrix, special attention should be drawn to the terms which have the same number for both suffixes.

$$\begin{bmatrix} Z_{11} & 0 & 0 & 0 \\ 0 & Z_{22} & 0 & 0 \\ 0 & 0 & Z_{33} & 0 \\ 0 & 0 & 0 & Z_{44} \end{bmatrix}$$

These lie in the diagonal top left to bottom right.

If all the other terms of the matrix are zero, the matrix is known as a diagonal matrix.

3.2.3 Symmetrical matrix

If all non-diagonal terms of $[Z]$ are such that $Z_{mn} = Z_{nm}$ the matrix is said to be symmetrical.
Example:

$$\begin{bmatrix} 1 & 5 & 6 & 7 \\ 5 & 2 & 8 & 9 \\ 6 & 8 & 3 & 10 \\ 7 & 9 & 10 & 4 \end{bmatrix}$$

3.2.4 Transpose of a matrix

This is a second matrix, having the same elements as the first but with rows and columns interchanged. For example

$$[A] = \begin{bmatrix} 1 & 2 \\ 3 & 4 \\ 5 & 6 \end{bmatrix} \quad [A_t] = \begin{bmatrix} 1 & 3 & 5 \\ 2 & 4 & 6 \end{bmatrix}$$

3.3 Matrix operations

3.3.1 Addition and subtraction

Only matrices of the same dimension can be added. The addition is performed by adding corresponding elements. For example,

$$\begin{bmatrix} 2 & 3 \\ 4 & 5 \\ 6 & 7 \end{bmatrix} + \begin{bmatrix} 2 & 1 \\ 3 & 4 \\ 6 & 5 \end{bmatrix} = \begin{bmatrix} 2+2 & 3+1 \\ 4+3 & 5+4 \\ 6+6 & 7+5 \end{bmatrix} = \begin{bmatrix} 4 & 4 \\ 7 & 9 \\ 12 & 12 \end{bmatrix} \tag{3.5}$$

Subtraction is carried out in a similar way.

3.3.2 Multiplication

3.3.2.1 *Single number.* Any matrix can be multiplied by a single number. Each element of the matrix is multiplied by the given number.
Example:

$$[B] = [24 \quad 7 \quad -12]$$

$$2[B] = [2 \times 24 \quad 2 \times 7 \quad 2(-12)] = [48 \quad 14 \quad -24] \tag{3.6}$$

3.3.2.2 *Row by column.* These matrices can be multiplied together only if there are the same number of elements in each matrix. The product is obtained by multiplying corresponding elements and adding the result, thus forming a single number.
If

$$[a] = [a_1 \quad a_2 \quad a_3] \qquad [b] = \begin{bmatrix} b_1 \\ b_2 \\ b_3 \end{bmatrix}$$

$$[a] \cdot [b] = [a_1 \quad a_2 \quad a_3] \cdot \begin{bmatrix} b_1 \\ b_2 \\ b_3 \end{bmatrix} = a_1b_1 + a_2b_2 + a_3b_3 \tag{3.7}$$

Example:

$$[c] = [2 \quad 3 \quad -5] \qquad [d] = \begin{bmatrix} 4 \\ 3 \\ 1 \end{bmatrix}$$

$$[c] \cdot [d] = [2 \quad 3 \quad -5] \cdot \begin{bmatrix} 4 \\ 3 \\ 1 \end{bmatrix} = 2 \times 4 + 3 \times 3 - 5 \times 1 = 8 + 9 - 5$$
$$= 12 \tag{3.8}$$

The arrows assist in identifying the individual numbers which form the products.

3.3.2.3 *Square by column.* When a square matrix is multiplied by a column, each row is multiplied in turn by the column as above, and the result is then displayed as a column matrix

$$\begin{bmatrix} a_1 & a_2 \\ b_1 & b_2 \end{bmatrix} \cdot \begin{bmatrix} c_1 \\ c_2 \end{bmatrix} = \begin{bmatrix} a_1c_1 + a_2c_2 \\ b_1c_1 + b_2c_2 \end{bmatrix} \tag{3.9}$$

The row and column must both have the same number of elements.

3.3.2.4 *Rectangle by rectangle*

$$\begin{bmatrix} a_1 & a_2 & a_3 \\ b_1 & b_2 & b_3 \end{bmatrix} \cdot \begin{bmatrix} c_1 & c_2 \\ d_1 & d_2 \\ e_1 & e_2 \end{bmatrix} = \begin{bmatrix} a_1c_1 + a_2d_1 + a_3e_1 & a_1c_2 + a_2d_2 + a_3e_2 \\ b_1c_1 + b_2d_1 + b_3e_1 & b_1c_2 + b_2d_2 + b_3e_2 \end{bmatrix} \tag{3.10}$$

There must be the same number of elements in each matrix, and the number of columns in the second matrix must be equal to the number of rows in the first.

The first matrix is multiplied by each of the columns of the second (as in 3.3.2.3) in turn and the resulting columns are displayed side by side.

It will be observed that the product of two rectangular matrices is a square matrix.

3.3.2.5 *Order of multiplication*. It is important to observe that matrix multiplication does not obey the commutation rule applicable to ordinary numbers. Simple evaluation demonstrates that $[A][B]$ is *not* equal to $[B][A]$ but is equal to the transpose of $[B_t][A_t]$

That is to say

$$[A] \cdot [B] \neq [B] \cdot [A] \tag{3.11}$$

but

$$[A] \cdot [B] = ([B_t] \cdot [A_t])_t \tag{3.12}$$

3.3.2.6 *Unit matrix*. The unit matrix is any square matrix in which all elements along the diagonal are unity, and all other elements zero.

$$[1] = \begin{bmatrix} 1 & 0 & 0 \\ 0 & 1 & 0 \\ 0 & 0 & 1 \end{bmatrix} \tag{3.13}$$

It will be observed that if any matrix is multiplied by the appropriate unit matrix, the matrix is unchanged

$$[1] \cdot [A] = [A] \cdot [1] = [A] \tag{3.14}$$

Example:

$$\begin{bmatrix} 1 & 2 \\ 3 & 4 \\ 5 & 6 \end{bmatrix} \cdot \begin{bmatrix} 1 & 0 \\ 0 & 1 \end{bmatrix} = \begin{bmatrix} 1 & 2 \\ 3 & 4 \\ 5 & 6 \end{bmatrix}$$

3.4 Representation of simultaneous equations

Consider a number of simultaneous equations, for example

$$\begin{aligned} V_1 &= Z_{11}I_1 + Z_{12}I_2 + Z_{13}I_3 \\ V_2 &= Z_{21}I_1 + Z_{22}I_2 + Z_{23}I_3 \\ V_3 &= Z_{31}I_1 + Z_{32}I_2 + Z_{33}I_3 \end{aligned} \tag{3.15}$$

where I_1, I_2 and I_3 are unknown currents, other quantities being known. It will be observed that the numbers on the left hand side of the equation form a column matrix.

On the right hand side, the coefficients form a 3×3 matrix which has been multiplied by a column matrix of I_1, I_2 and I_3. Thus the matrix way of writing down the above equations is as follows:

$$\begin{bmatrix} V_1 \\ V_2 \\ V_3 \end{bmatrix} = \begin{bmatrix} Z_{11} & Z_{12} & Z_{13} \\ Z_{21} & Z_{22} & Z_{23} \\ Z_{31} & Z_{32} & Z_{33} \end{bmatrix} \cdot \begin{bmatrix} I_1 \\ I_2 \\ I_3 \end{bmatrix} \tag{3.16}$$

This presentation requires I_1, I_2 or I_3 to be written down once only and gives prominence to the coefficients Z. In this form the individual voltages, currents and impedances are directly displayed—moreover, the whole set of equations is reduced to the statement

$$[V] = [Z] \cdot [I] \tag{3.17}$$

where

$$[V] = \begin{bmatrix} V_1 \\ V_2 \\ V_3 \end{bmatrix} \quad [Z] = \begin{bmatrix} Z_{11} & Z_{12} & Z_{13} \\ Z_{21} & Z_{22} & Z_{23} \\ Z_{31} & Z_{32} & Z_{33} \end{bmatrix} \quad \text{and} \quad [I] = \begin{bmatrix} I_1 \\ I_2 \\ I_3 \end{bmatrix} \tag{3.18}$$

Considerable importance is attached to this simplified presentation. It will be used in later chapters whenever simultaneous equations emerge.

3.5 The determinant of a matrix

The determinant of matrix $(n \times n)$ is the name given to the algebraic sum of all the possible products of n terms obtained by taking one term from each row, no two terms being from the same column. Positive or negative signs are attributed to each product according to whether, when the terms of the product are arranged in the natural order of the rows, the number of inversions in the column order is even or odd. The symbol Δ is used for a determinant or straight lines are used in place of the side brackets of the matrix notation. For example in the matrix

$$[A] = \begin{bmatrix} a_1 & a_2 \\ b_1 & b_2 \end{bmatrix} \quad \Delta = |A| = \begin{vmatrix} a_1 & a_2 \\ b_1 & b_2 \end{vmatrix} \tag{3.19}$$

The products are

$$a_1 b_2 \text{ and } a_2 b_1$$

The order 1, 2, has no inversion therefore the sign is positive and the order 2, 1 has one inversion therefore the sign is negative.

Thence

$$\Delta = a_1 b_2 - a_2 b_1 \tag{3.20}$$

Again if

$$\Delta = \begin{vmatrix} a_1 & a_2 & a_3 \\ b_1 & b_2 & b_3 \\ c_1 & c_2 & c_3 \end{vmatrix}$$

the products are

$$a_1 b_2 c_3 \quad \text{no inversion} \quad +$$
$$a_2 b_3 c_1 \quad \text{no inversion} \quad +$$
$$a_3 b_1 c_2 \quad \text{no inversion} \quad +$$
$$a_2 b_1 c_3 \quad \text{one inversion} \quad -$$
$$a_3 b_2 c_1 \quad \text{one inversion} \quad -$$
$$a_1 b_3 c_2 \quad \text{one inversion} \quad -$$

and the determinant

$$\Delta = a_1 b_2 c_3 + a_2 b_3 c_1 + a_3 b_1 c_2$$
$$- (a_2 b_1 c_3 + a_3 b_2 c_1 + a_1 b_3 c_2) \tag{3.21}$$

It should be noted that the determinant of a matrix is a single number. The determinant of a 4×4 matrix involves 24 products and the amount of computation required for higher order matrices increases considerably. Appropriate methods for evaluating such determinants are described in mathematical textbooks of advanced algebra.*

3.5.1 The minor of an element in a matrix

This is the name given to the determinant of the matrix that remains when the row and column containing the given element are crossed out. For example in the matrix

$$\begin{bmatrix} a_{11} & a_{12} & a_{13} & a_{14} \\ a_{21} & a_{22} & a_{23} & a_{24} \\ a_{31} & a_{32} & a_{33} & a_{34} \\ a_{41} & a_{42} & a_{43} & a_{44} \end{bmatrix}$$

the minor M_{11} corresponding to a_{11} is obtained by crossing out the first row and first column and finding the determinant of the remainder

thus

$$\begin{bmatrix} - & - & - & - \\ - & a_{22} & a_{23} & a_{24} \\ - & a_{32} & a_{33} & a_{34} \\ - & a_{42} & a_{43} & a_{44} \end{bmatrix} \qquad M_{11} = \begin{vmatrix} a_{22} & a_{23} & a_{24} \\ a_{32} & a_{33} & a_{34} \\ a_{42} & a_{43} & a_{44} \end{vmatrix} \tag{3.22}$$

Also the minor M_{23} corresponding to a_{23} is obtained by removing the second row and third column, viz.

$$\begin{bmatrix} a_{11} & a_{12} & - & a_{14} \\ - & - & - & - \\ a_{31} & a_{32} & - & a_{34} \\ a_{41} & a_{42} & - & a_{44} \end{bmatrix} \qquad M_{23} = \begin{vmatrix} a_{11} & a_{12} & a_{14} \\ a_{31} & a_{32} & a_{34} \\ a_{41} & a_{42} & a_{44} \end{vmatrix} \tag{3.23}$$

* S. A. Stigant, *The Elements of Determinants, Matrices and Tensors*, Macdonald, 1959.

3.6 The inverse of a matrix

If the product of two matrices is the unit matrix, the second matrix is said to be the inverse of the first.
If

$$[A] . [B] = [1]$$

then $[B]$ is the inverse of $[A]$ and is designated $[A]^{-1}$.

This nomenclature compares with usual algebraic rules since if

$$x . y = 1$$
$$y = 1/x = x^{-1}$$

The procedure of determining the inverse of a matrix is the basis of the solution of simultaneous equations.

Considering the equations

$$V_1 = Z_{11}I_1 + Z_{12}I_2 + Z_{13}I_3$$
$$V_2 = Z_{21}I_1 + Z_{22}I_2 + Z_{23}I_3 \qquad (3.15)$$
$$V_3 = Z_{31}I_1 + Z_{32}I_2 + Z_{33}I_3$$

in which I_1, I_2 and I_3 are the unknown currents, voltages and impedances being known.

We have seen that the equations can be rewritten in matrix form

$$[V] = [Z][I] \qquad (3.17)$$

multiplying by $[Z]^{-1}$ we have

$$[Z]^{-1}[V] = [Z]^{-1}[Z][I]$$
$$= [1][I]$$
$$= [I] \qquad (3.24)$$

This is interpreted

$$\begin{bmatrix} I_1 \\ I_2 \\ I_3 \end{bmatrix} = [Z]^{-1} . \begin{bmatrix} V_1 \\ V_2 \\ V_3 \end{bmatrix} \qquad (3.25)$$

where

$$[Z]^{-1} = \begin{bmatrix} Y_{11} & Y_{12} & Y_{13} \\ Y_{21} & Y_{22} & Y_{23} \\ Y_{31} & Y_{32} & Y_{33} \end{bmatrix} \qquad (3.26)$$

or

$$I_1 = Y_{11}V_1 + Y_{12}V_2 + Y_{13}V_3$$
$$I_2 = Y_{21}V_1 + Y_{22}V_2 + Y_{23}V_3 \qquad (3.27)$$
$$I_3 = Y_{31}V_1 + Y_{32}V_2 + Y_{33}V_3$$

Thus the unknown currents are determined if the elements of $[Z]^{-1}$ can be obtained from the elements of $[Z]$.

3.6.1 Rules for determining the inverse of a matrix

The following rules refer to the method of determining the elements of an inverse matrix. For proof the reader is once again referred to a suitable mathematical textbook.

The inverse of a matrix $[A]$ is determined in five steps:

1. Find the determinant $|A|$
2. Write down the transpose $[A_t]$
3. Replace each element in $[A_t]$ by its minor
4. Multiply each of the minors in (3) by $(+1)$ or (-1) according to the following pattern

$$\begin{bmatrix} + & - & + & - & . & . & . \\ - & + & - & . & . & . & . \\ + & - & + & . & . & . & . \\ . & . & . & . & . & . & . \\ . & . & . & . & . & . & . \\ . & . & . & . & . & . & . \end{bmatrix}$$

5. Divide all terms by the determinant $|A|$

Thus if the minors of $[A_t]$ are

$$\begin{bmatrix} M_{11} & M_{12} & M_{13} & . & . & . \\ M_{21} & M_{22} & M_{23} & . & . & . \\ M_{31} & M_{32} & M_{33} & . & . & . \\ . & . & . & . & . & . \\ . & . & . & . & . & . \\ . & . & . & . & . & . \end{bmatrix}$$

$$[A]^{-1} = \frac{1}{|A|} \begin{bmatrix} M_{11} & -M_{12} & M_{13} & . & . & . \\ -M_{21} & M_{22} & -M_{23} & . & . & . \\ M_{31} & -M_{32} & M_{33} & . & . & . \\ . & . & . & . & . & . \\ . & . & . & . & . & . \end{bmatrix}$$

Example:

$$[A] = \begin{bmatrix} 1 & 2 & 0 \\ 3 & 5 & 4 \\ 5 & 6 & 7 \end{bmatrix}$$

$$[A_t] = \begin{bmatrix} 1 & 3 & 5 \\ 2 & 5 & 6 \\ 0 & 4 & 7 \end{bmatrix}$$

$$|A| = 1 \times 5 \times 7 + 3 \times 6 \times 0 + 5 \times 2 \times 4 - 0 \times 5 \times 5 - 4 \times 6 \times 1 - 7 \times 2 \times 3$$
$$= 35 + 0 + 40 - 0 - 24 - 42$$
$$= 75 - 66$$
$$= 9$$

$$[A]^{-1} = \frac{1}{\Delta}[M] = \frac{1}{9}\begin{bmatrix} +\begin{vmatrix} 5 & 6 \\ 4 & 7 \end{vmatrix} & -\begin{vmatrix} 2 & 6 \\ 0 & 7 \end{vmatrix} & +\begin{vmatrix} 2 & 5 \\ 0 & 4 \end{vmatrix} \\[12pt] -\begin{vmatrix} 3 & 5 \\ 4 & 7 \end{vmatrix} & +\begin{vmatrix} 1 & 5 \\ 0 & 7 \end{vmatrix} & -\begin{vmatrix} 1 & 3 \\ 0 & 4 \end{vmatrix} \\[12pt] +\begin{vmatrix} 3 & 5 \\ 5 & 6 \end{vmatrix} & -\begin{vmatrix} 1 & 5 \\ 2 & 6 \end{vmatrix} & +\begin{vmatrix} 1 & 3 \\ 2 & 5 \end{vmatrix} \end{bmatrix}$$

$$= \frac{1}{9}\begin{bmatrix} 35-24 & -14+0 & 8-0 \\ -21+20 & 7-0 & -4+0 \\ -25+18 & 10-6 & 5-6 \end{bmatrix}$$

$$= \frac{1}{9}\begin{bmatrix} 11 & -14 & 8 \\ -1 & 7 & -4 \\ -7 & 4 & -1 \end{bmatrix}$$

The numerical complexity involved in the inversion of higher order matrices increases rapidly and, consequently, many methods have been devised for systematising the arithmetic and reducing the labour involved.

Even with these methods, however, the time required is much too long to be worthwhile.

In recent years, the advent of the high-speed digital computer has completely changed the position. Sub-routine programmes for matrix inversion are now available and the computer time required for the inversion of, say, a 20×20 matrix is reduced to less than one minute.

This remarkable reduction in computational time, therefore, allows engineers to use matrix methods with confidence, knowing that if a problem produces a high order matrix requiring inversion, the machine can be relied upon to complete the arithmetical work with speed and accuracy.

No longer has the engineer to indulge in complicated 'short-cutting' procedures, but of course, ready access to the computer is essential.

3.7 Complex numbers in matrices

For steady-state solutions of a.c. problems, voltages, currents and impedance are represented by complex numbers and we have equations such as

$$V = Z \cdot I \tag{3.28}$$

$$Z = R + jX \tag{3.29}$$

where
$$V = V_d + jV_q \tag{3.30}$$

$$I = I_d + jI_q \tag{3.31}$$

When a set of such equations is expressed in matrix form

viz.
$$\begin{bmatrix} V_1 \\ V_2 \end{bmatrix} = \begin{bmatrix} R_{11} + jX_{11} & R_{12} + jX_{12} \\ R_{21} + jX_{21} & R_{22} + jX_{22} \end{bmatrix} \cdot \begin{bmatrix} I_1 \\ I_2 \end{bmatrix} \tag{3.32}$$

the normal rules of matrix algebra are applicable, but all the individual multi-plication and addition of elements involve complex numbers.

On the other hand, expanding equation

$$V = ZI \tag{3.28}$$

$$V_d + jV_q = (R + jX)(I_d + jI_q)$$
$$= RI_d - XI_q + jXI_d + jRI_q \tag{3.33}$$

and so, equating real and imaginary terms

$$V_d = RI_d - XI_q \tag{3.34}$$

$$V_q = XI_d + RI_q \tag{3.35}$$

which can be expressed as the matrix equation

$$\begin{bmatrix} V_d \\ V_q \end{bmatrix} = \begin{bmatrix} R & -X \\ X & R \end{bmatrix} \cdot \begin{bmatrix} I_d \\ I_q \end{bmatrix} \tag{3.36}$$

This means then, instead of using complex numbers in a matrix equation such as (3.32) we can replace each Z element by a sub-matrix

$$\begin{bmatrix} R & -X \\ X & R \end{bmatrix} \tag{3.37}$$

Equation (3.32) thus becomes

$$\begin{bmatrix} V_{1d} \\ V_{1q} \\ V_{2d} \\ V_{2q} \end{bmatrix} = \begin{bmatrix} R_{11} & -X_{11} & R_{12} & -X_{12} \\ X_{11} & R_{11} & X_{12} & R_{12} \\ R_{21} & -X_{21} & R_{22} & -X_{22} \\ X_{21} & R_{21} & X_{22} & R_{22} \end{bmatrix} \cdot \begin{bmatrix} I_{1d} \\ I_{1q} \\ I_{2d} \\ I_{2q} \end{bmatrix} \tag{3.38}$$

Although the matrix is of a higher order, all elements are simple numbers and the presentation enables a computer solution to be obtained, using the standard matrix inversion sub-routine.*

---

* The use of 'j' as an operator automatically separates a system of numbers into two groups, those that are operated on by 'j' and those that are not.

In Sec. 3.7 the separation is achieved by classifying the numbers as direct and quadrature.

3.8 Partitioning a matrix

It is often convenient to consider a high order matrix to be subdivided into smaller sections, each of which is in itself a matrix. Such a division is termed partitioning, and the original matrix is termed a compound matrix. For example

$$
\begin{bmatrix} Z_{11} & \cdot & \cdot & \cdot & \cdot \\ \cdot & \cdot & \cdot & \cdot & \cdot \\ \cdot & \cdot & \cdot & \cdot & \cdot \\ \cdot & \cdot & \cdot & \cdot & \cdot \\ \cdot & \cdot & \cdot & \cdot & Z_{55} \end{bmatrix} = \begin{bmatrix} \begin{bmatrix} \cdot & \cdot \\ Z_1 \\ \cdot & \cdot \end{bmatrix} & \begin{bmatrix} \cdot & \cdot & \cdot \\ Z_2 \\ \cdot & \cdot & \cdot \end{bmatrix} \\ \begin{bmatrix} \cdot & \cdot \\ \cdot Z_3 \cdot \\ \cdot & \cdot \end{bmatrix} & \begin{bmatrix} \cdot & \cdot & \cdot \\ \cdot Z_4 \cdot \\ \cdot & \cdot & \cdot \end{bmatrix} \end{bmatrix} \text{ or } \begin{bmatrix} \begin{bmatrix} \cdot & \cdot & \cdot \\ Z_5 \\ \cdot & \cdot & \cdot \end{bmatrix} & \begin{bmatrix} \cdot & \cdot \\ Z_6 \\ \cdot & \cdot \end{bmatrix} \\ \begin{bmatrix} \cdot & \cdot & \cdot \\ \cdot Z_7 \cdot \\ \cdot & \cdot & \cdot \end{bmatrix} & \begin{bmatrix} \cdot & \cdot \\ Z_8 \\ \cdot & \cdot \end{bmatrix} \end{bmatrix}
$$

(3.39)

This is of particular value if we are only interested in results arising from only one section of a complicated matrix, when it is advisable to eliminate the unwanted sections in the following manner.

Consider a higher order Z matrix partitioned into four sections

$$
\begin{bmatrix} [V_1] \\ [V_2] \end{bmatrix} = \begin{bmatrix} [Z_1] & [Z_2] \\ [Z_3] & [Z_4] \end{bmatrix} \cdot \begin{bmatrix} [I_1] \\ [I_2] \end{bmatrix}
$$

(3.40)

$[V_1]$ and $[V_2]$ now represent sets of voltages and $[I_1]$ and $[I_2]$ sets of currents. Let us suppose that we are only interested in the set of currents represented by $[I_1]$. Then to eliminate $[I_2]$, since

$$[V_1] = [Z_1][I_1] + [Z_2][I_2] \tag{3.41}$$

$$[V_2] = [Z_3][I_1] + [Z_4][I_2] \tag{3.42}$$

by normal algebraical rules, but taking care not to alter any orders of multiplication, we have from (3.42)

$$[I_2] = [Z_4]^{-1}([V_2] - [Z_3][I_1]) \tag{3.43}$$

and substituting in (3.41)

$$[V_1] = [Z_1][I_1] + [Z_2][Z_4]^{-1}([V_2] - [Z_3][I_1]) \tag{3.44}$$

or

$$[V_1] - [Z_2][Z_4]^{-1}[V_2] = ([Z_1] - [Z_2][Z_4]^{-1}[Z_3])[I_1] \tag{3.45}$$

Alternatively

$$[V'] = [Z'][I_1] \tag{3.46}$$

where

$$[V'] = [V_1] - [Z_2][Z_4]^{-1}[V_2] \tag{3.47}$$

and

$$[Z'] = [Z_1] - [Z_2][Z_4]^{-1}[Z_3] \qquad (3.48)$$

we have thus set up a new matrix equation referring to the wanted currents only. This is useful particularly if all the voltages represented by $[V_2]$ are zero when

$$[V'] = [V_1] \qquad (3.49)$$

Example 1:

$$
\begin{bmatrix} \begin{bmatrix} V_1 \\ V_2 \end{bmatrix} \\ \begin{bmatrix} 0 \\ 0 \\ 0 \end{bmatrix} \end{bmatrix}
=
\begin{bmatrix} \begin{bmatrix} 1 & 0 \\ 0 & 2 \end{bmatrix} \begin{bmatrix} 2 & 0 & 3 \\ 0 & 4 & 0 \end{bmatrix} \\ \begin{bmatrix} 2 & 0 \\ 0 & 4 \\ 3 & 0 \end{bmatrix} \begin{bmatrix} 3 & 0 & 0 \\ 0 & 4 & 6 \\ 0 & 6 & 5 \end{bmatrix} \end{bmatrix}
\cdot
\begin{bmatrix} \begin{bmatrix} I_1 \\ I_2 \end{bmatrix} \\ \begin{bmatrix} I_3 \\ I_4 \\ I_5 \end{bmatrix} \end{bmatrix}
$$

and we are concerned only with I_1 and I_2.

$$[V'] = \begin{bmatrix} V_1 \\ V_2 \end{bmatrix} = [Z'] \cdot \begin{bmatrix} I_1 \\ I_2 \end{bmatrix}$$

and

$$[Z'] = [Z_1] - [Z_2][Z_4]^{-1}[Z_3]$$

$$
= \begin{bmatrix} 1 & 0 \\ 0 & 2 \end{bmatrix} - \begin{bmatrix} 2 & 0 & 3 \\ 0 & 4 & 0 \end{bmatrix} \cdot \begin{bmatrix} 3 & 0 & 0 \\ 0 & 4 & 6 \\ 0 & 6 & 5 \end{bmatrix}^{-1} \cdot \begin{bmatrix} 2 & 0 \\ 0 & 4 \\ 3 & 0 \end{bmatrix}
$$

$$
= \begin{bmatrix} 1 & 0 \\ 0 & 2 \end{bmatrix} + \frac{1}{48} \begin{bmatrix} 2 & 0 & 3 \\ 0 & 4 & 0 \end{bmatrix} \cdot \begin{bmatrix} -16 & 0 & 0 \\ 0 & 15 & -18 \\ 0 & -18 & 12 \end{bmatrix} \cdot \begin{bmatrix} 2 & 0 \\ 0 & 4 \\ 3 & 0 \end{bmatrix}
$$

$$
= \begin{bmatrix} 1 & 0 \\ 0 & 2 \end{bmatrix} + \frac{1}{48} \begin{bmatrix} 2 & 0 & 3 \\ 0 & 4 & 0 \end{bmatrix} \cdot \begin{bmatrix} -32 & 0 \\ -54 & 60 \\ 36 & -72 \end{bmatrix}
$$

$$
= \begin{bmatrix} 1 & 0 \\ 0 & 2 \end{bmatrix} + \frac{1}{48} \begin{bmatrix} 44 & -216 \\ -216 & 240 \end{bmatrix}
$$

$$
= \begin{bmatrix} 1 & 0 \\ 0 & 2 \end{bmatrix} + \frac{1}{12} \begin{bmatrix} 11 & -54 \\ -54 & 60 \end{bmatrix} = \frac{1}{12} \left[\begin{bmatrix} 12 & 0 \\ 0 & 24 \end{bmatrix} + \begin{bmatrix} 11 & -54 \\ -54 & 60 \end{bmatrix} \right]
$$

$$
= \frac{1}{12} \begin{bmatrix} 23 & -54 \\ -54 & 84 \end{bmatrix}
$$

$$\begin{bmatrix} V_1 \\ V_2 \end{bmatrix} = \frac{1}{12} \begin{bmatrix} 23 & -54 \\ -54 & 84 \end{bmatrix} \cdot \begin{bmatrix} I_1 \\ I_2 \end{bmatrix}$$

from which I_1 and I_2 can be found if V_1 and V_2 are known.

Example 2:

If

$$\begin{bmatrix} V_1 \\ V_2 \\ V_3 \\ V_4 \\ 0 \end{bmatrix} = \begin{bmatrix} 1 & 0 & 2 & 0 & 9 \\ 0 & 3 & 0 & 4 & 8 \\ 5 & 0 & 6 & 0 & 7 \\ 0 & 7 & 0 & 8 & 6 \\ 1 & 2 & 3 & 4 & 5 \end{bmatrix} \cdot \begin{bmatrix} I_1 \\ I_2 \\ I_3 \\ I_4 \\ I_5 \end{bmatrix}$$

to eliminate the last row and column

$$\begin{bmatrix} V_1 \\ V_2 \\ V_3 \\ V_4 \end{bmatrix} = [Z'] \cdot \begin{bmatrix} I_1 \\ I_2 \\ I_3 \\ I_4 \end{bmatrix}$$

where

$$[Z'] = [Z_1] - [Z_2][Z_4]^{-1}[Z_3]$$

$$= \begin{bmatrix} 1 & 0 & 2 & 0 \\ 0 & 3 & 0 & 4 \\ 5 & 0 & 6 & 0 \\ 0 & 7 & 0 & 8 \end{bmatrix} - \begin{bmatrix} 9 \\ 8 \\ 7 \\ 6 \end{bmatrix} \cdot [5^{-1}] \cdot \begin{bmatrix} 1 & 2 & 3 & 4 \end{bmatrix}$$

$$= \begin{bmatrix} 1 & 0 & 2 & 0 \\ 0 & 3 & 0 & 4 \\ 5 & 0 & 6 & 0 \\ 0 & 7 & 0 & 8 \end{bmatrix} - \frac{1}{5} \begin{bmatrix} 9 & 18 & 27 & 36 \\ 8 & 16 & 24 & 32 \\ 7 & 14 & 21 & 28 \\ 6 & 12 & 18 & 24 \end{bmatrix}$$

$$= \begin{bmatrix} -0.8 & -3.6 & -3.4 & -7.2 \\ -1.6 & -0.2 & -4.8 & -2.4 \\ 3.6 & -2.8 & 1.8 & -5.6 \\ -1.2 & 4.6 & -3.6 & 3.2 \end{bmatrix}$$

3.8.1 Reduction of a matrix line by line

If only one line and column are to be eliminated, the reduced matrix can usually be written down by inspection in the following manner.

First, the order of the rows and columns are altered to bring in the unwanted ones at the bottom and on the right.

$$\begin{bmatrix} V_1 \\ V_2 \\ V_3 \\ V_4 \\ 0 \end{bmatrix} = \begin{bmatrix} p & & & & a \\ & & & & -b \\ & & q & & -c \\ & & & & d \\ e & f & g & h & k \end{bmatrix} \cdot \begin{bmatrix} I_1 \\ I_2 \\ I_3 \\ I_4 \\ I_5 \end{bmatrix} \qquad (3.50)$$

For example, in the above 5×5 matrix (in which only a few elements are shown for clarity), since V_5 is zero, the last line and column is to be eliminated. The resulting 4×4 matrix $[Z']$ is such that

$$[Z'] = [Z_1] - [Z_2][Z_4]^{-1}[Z_3] \qquad (3.48)$$

where

$$Z_2 = \begin{bmatrix} a \\ b \\ c \\ d \end{bmatrix}$$

$$Z_3 = [\, e \quad f \quad g \quad h \,] \qquad (3.51)$$

and

$$[Z_4]^{-1} = 1/k$$

Thus we see that each element of Z_1 has subtracted from it the fraction xy/k where x comes from the eliminated column and y from the eliminated row. The appropriate values of x and y are selected by projecting horizontally and vertically from each element in turn. Thus in Eq. (3.50), b and e correspond to p, c and g corresponds to q and two of the new elements in (Z') are given by

$$[Z'] = \begin{bmatrix} p - \dfrac{be}{k} & & & \\ & q - \dfrac{cg}{k} & & \\ & & & \\ & & & \end{bmatrix} \qquad (3.52)$$

The remaining elements are similarly obtained.

The procedure is the same whether or not the order of the rows and columns is altered but it is easier to follow the process if the unwanted rows and columns are placed at the ends.

Diagram 3.53 corresponds to 3.50 for the elimination of an intermediate row and column.

$$
\begin{bmatrix} V_1 \\ V_2 \\ V_3 \\ 0 \\ V_4 \end{bmatrix} = \begin{array}{c} I_1 \quad I_2 \quad I_3 \quad I_5 \quad I_4 \end{array} \begin{bmatrix} & & & a \\ p - - - - - & b \\ & & q - & c \\ - | - - - - - | - \\ e \quad f \quad g & k & h \\ & & d \end{bmatrix}
$$

—Row to be eliminated

Column to be
eliminated (3.53)

If more than one line has to be eliminated, this process can be repeated as many times as is required.

3.9 The Gauss-Seidel method

Another method of determining the inverse of a matrix arises from the Gauss–Seidel procedure for solving simultaneous equations. It is particularly suited to the impedance matrices arising in power networks where several elements are zero and where non-diagonal elements are usually small.

The method is best explained with reference to an example:

Let

$$
[Y] = \begin{bmatrix} Y_{11} & Y_{12} & Y_{13} \\ Y_{21} & Y_{22} & Y_{23} \\ Y_{31} & Y_{32} & Y_{33} \end{bmatrix}
$$

be the inverse of $[Z]$ where

$$
[Z] = \begin{bmatrix} 8 & 1 & 2 \\ 1 & 10 & 3 \\ 2 & 3 & 12 \end{bmatrix}
$$

The product of $[Y]$ and $[Z]$ is the unit matrix

$$
[1] = [Y][Z]
$$

or

$$
\begin{bmatrix} 1 & 0 & 0 \\ 0 & 1 & 0 \\ 0 & 0 & 1 \end{bmatrix} = \begin{bmatrix} 8 & 1 & 2 \\ 1 & 10 & 3 \\ 2 & 3 & 12 \end{bmatrix} \cdot \begin{bmatrix} Y_{11} & Y_{12} & Y_{13} \\ Y_{21} & Y_{22} & Y_{23} \\ Y_{31} & Y_{32} & Y_{33} \end{bmatrix} \quad (3.54)
$$

This matrix equation represents three sets of simultaneous equations

$$
\begin{bmatrix} 1 \\ 0 \\ 0 \end{bmatrix} = \begin{bmatrix} 8 & 1 & 2 \\ 1 & 10 & 3 \\ 2 & 3 & 12 \end{bmatrix} \cdot \begin{bmatrix} Y_{11} \\ Y_{21} \\ Y_{31} \end{bmatrix} \quad (3.55)
$$

$$\begin{bmatrix} 0 \\ 1 \\ 0 \end{bmatrix} = \begin{bmatrix} 8 & 1 & 2 \\ 1 & 10 & 3 \\ 2 & 3 & 12 \end{bmatrix} \cdot \begin{bmatrix} Y_{12} \\ Y_{22} \\ Y_{32} \end{bmatrix} \tag{3.56}$$

$$\begin{bmatrix} 0 \\ 0 \\ 1 \end{bmatrix} = \begin{bmatrix} 8 & 1 & 2 \\ 1 & 10 & 3 \\ 2 & 3 & 12 \end{bmatrix} \cdot \begin{bmatrix} Y_{13} \\ Y_{23} \\ Y_{33} \end{bmatrix} \tag{3.57}$$

and the three columns of $[Y]$ can be found by solving these sets of equations.

To solve Eqs. (3.55) by the Gauss–Seidel method, we make a series of approximations. Thus in the first row of (3.55),

$$1 = 8Y_{11} + Y_{21} + 2Y_{31} \tag{3.58}$$

putting $Y_{21} = Y_{31} = 0$ we see that Y_{11} is approximately equal to $\frac{1}{8} = 0 \cdot 125$. Proceeding to the second row,

$$0 = Y_{11} + 10Y_{21} + 3Y_{31} \tag{3.59}$$

we substitute this approximate value for Y_{11} and retaining $Y_{31} = 0$, we solve for Y_{21} thus

$$Y_{21} = (0 - 1 \times 0 \cdot 125 - 0)/10 = -0 \cdot 0125$$

In the final row

$$0 = 2Y_{11} + 3Y_{21} + 12Y_{31} \tag{3.60}$$

we substitute the above values of Y_{11} and Y_{21} and so obtain Y_{31}
Thus

$$Y_{31} = (-2 \times 0 \cdot 125 + 3 \times 0 \cdot 0125)/12$$
$$= (-0 \cdot 25 + 0 \cdot 0375)/12$$
$$= 0 \cdot 2125/12 = -0 \cdot 0177$$

We have now obtained the first approximation to the first column of the inverse

$$\begin{bmatrix} Y_{11} \\ Y_{21} \\ Y_{31} \end{bmatrix} = \begin{bmatrix} 0 \cdot 1250 \\ -0 \cdot 0125 \\ -0 \cdot 0177 \end{bmatrix} \tag{3.61}$$

Now we return to the first line of Eq. (3.55) and obtain an improved value for Y_{11} using the latest values obtained for Y_{21} and Y_{31}. The second approximation to Y_{11} is thus given by

$$1 = 8Y_{11} - 0 \cdot 0125 - 2 \times 0 \cdot 0177$$
$$8Y_{11} = 1 + 0 \cdot 0125 + 0 \cdot 0354$$
$$= 1 \cdot 0479$$
$$Y_{11} = 0 \cdot 131$$

Substituting in the second line of Eq. (3.55) to obtain the second approximation to Y_{21}

$$0 = 0.131 + 10\,Y_{21} - 3 \times 0.0177$$
$$= 0.131 + 10\,Y_{21} - 0.0531$$
$$10\,Y_{21} = -0.0779$$
$$Y_{21} = -0.00779$$

and from the third line of Eq. (3.55) the second time around

$$0 = 2 \times 0.131 - 3 \times 0.00779 + 12\,Y_{31}$$
$$12\,Y_{31} = -0.262 + 0.02337$$
$$= -0.23863$$
$$Y_{31} = -0.0199$$

The second approximation to Eq. (3.61) is thus

$$\begin{bmatrix} Y_{11} \\ Y_{21} \\ Y_{31} \end{bmatrix} = \begin{bmatrix} 0.1310 \\ -0.0078 \\ -0.0199 \end{bmatrix} \tag{3.62}$$

These steps are repeated in sequence each time obtaining an improvement in the value of an element by using the latest values obtained for the others. Three or four cycles are usually sufficient to obtain the desired accuracy as will be observed from (3.63).

	Y_{11}	Y_{21}	Y_{31}
1st approximation	0.125000	−0.012500	−0.017708
2nd approximation	0.130989	−0.007786	−0.019885
3rd approximation	0.130945	−0.007129	−0.020041
4th approximation	0.130901	−0.007078	−0.020047
5th approximation	0.130897	−0.007075	−0.020047
6th approximation	0.130896	−0.007075	−0.020047

$$(3.63)$$

The repeated calculations are easily carried out if a desk calculator is available.

3.10 Matrix terms

The following terms are also extensively used in matrix algebra. These are given with definitions.

3.10.1 A singular matrix

A matrix is said to be singular if its determinant is zero.

$$[A] \text{ is singular if } |A| = 0 \tag{3.64}$$
$$[A] \text{ is non-singular if } |A| \neq 0 \tag{3.65}$$

Since we have noted in Sec. 3.6.1 that it is necessary to find the determinant of a matrix as one of the steps in obtaining the inverse, it follows that a singular matrix cannot be inverted.

Only non-singular matrices have finite inverses.

3.10.2 A skew symmetric matrix

A symmetric matrix has already been defined in section 3.2.3 as one where all the non-diagonal terms consist of pairs symmetrical about the diagonal so that

$$a_{jk} = a_{kj} \tag{3.66}$$

If the pairs are equal in magnitude but opposite in sign so that

$$a_{jk} = -a_{kj} \tag{3.67}$$

the matrix is said to be skew symmetric.

Example:

$$\begin{bmatrix} 1 & 2 & -3 \\ -2 & 5 & 4 \\ 3 & -4 & 6 \end{bmatrix} \quad \text{is skew symmetric}$$

3.10.3 A hermitian matrix

A matrix of complex numbers which is equal to its conjugate transpose.

$$\text{If } [A] = [A_t{}^*] \tag{3.68}$$

$[A]$ is a hermitian matrix.

Example:

$$\begin{bmatrix} 1 & -j \\ j & 1 \end{bmatrix} \quad \text{is hermitian}$$

3.10.4 An orthogonal matrix

A matrix is said to be orthogonal if its inverse and transpose are equal.

$$[A] \text{ is orthogonal if } [A_t] = [A]^{-1} \tag{3.69}$$

$$\text{that is if } [A] \cdot [A_t] = 1 \tag{3.70}$$

3.11 Eigenvalues and eigenvectors

If $[X]$ is a column vector (not zero) of order n and $[A]$ is a $n \times n$ matrix the product $[A] \cdot [X]$ is a column matrix spoken of as $[X]$ transformed by $[A]$.

If an equation can be determined in which $[X]$ transformed by $[A]$ is equal to $[X]$ multiplied by a simple constant, we have

$$[A] \cdot [X] = \lambda[X] \tag{3.71}$$

where λ is a number simple or complex, λ is said to be an *eigenvalue* of matrix $[A]$ and $[X]$ is a corresponding *eigenvector*. Equation (3.71) is referred to as formulating the *eigenvalue problem*. For example, if

$$[A] = \begin{bmatrix} 4 & -2 \\ 1 & 1 \end{bmatrix} \tag{3.72}$$

it is true that

$$\begin{bmatrix} 4 & -2 \\ 1 & 1 \end{bmatrix} \cdot \begin{bmatrix} 1 \\ 1 \end{bmatrix} = 2 \begin{bmatrix} 1 \\ 1 \end{bmatrix} \tag{3.73}$$

and also

$$\begin{bmatrix} 4 & -2 \\ 1 & 1 \end{bmatrix} \cdot \begin{bmatrix} 2 \\ 1 \end{bmatrix} = 3 \begin{bmatrix} 2 \\ 1 \end{bmatrix} \tag{3.74}$$

consequently in Eq. (3.73) we have an eigenvalue of 2 with eigenvector $\begin{bmatrix} 1 \\ 1 \end{bmatrix}$

and in (3.74) the eigenvalue is 3 and corresponding eigenvector $\begin{bmatrix} 2 \\ 1 \end{bmatrix}$.

(It will be seen that although for each eigenvalue there is a value of $[X]$ satisfying the equation $[A][X] = \lambda[X]$ any multiple of $[X]$ will also satisfy the equation. The fact that an eigenvector may be divided by any factor k does not destroy its intrinsic pattern.) The importance of eigenvalues and eigenvectors will be seen in later sections.

3.11.1 To determine the eigenvalues of a matrix $[A]$.

If

$$[A] \cdot [X] = \lambda[X] \tag{3.71}$$

$$[A] \cdot [X] = [1] \cdot \lambda[X]$$

Therefore $[A - 1\lambda][X] = 0$ $\tag{3.75}$

It will be appreciated that $[X] = 0$ is a solution of Eq. (3.75) but this is known as a trivial solution since it gets us nowhere.

$$[A - 1\lambda] = \begin{bmatrix} a_{11} - \lambda & a_{12} & \cdot & a_{1n} \\ a_{21} & a_{22} - \lambda & \cdot & \cdot \\ \cdot & \cdot & \cdot & \cdot \\ a_{n1} & \cdot & \cdot & a_{nn} - \lambda \end{bmatrix}$$

and for the non trivial solution of Eq. (3.75) the determinant*

$$|A - 1\lambda| = 0 \tag{3.77}$$

* Let $[B].[X] = 0$ which is the product of a square matrix $[B]$ and the column vector $[X]$. If the determinant $|B| \neq 0$, the matrix is non-singular and $[B]^{-1}$ exists, so that

$$[B]^{-1} [B] [X] = [X] = 0$$

and only the trivial solution is possible. For non trivial solutions $[X] \neq 0$ hence the determinant $|B| = 0$.

$[A - 1\lambda]$ is an $n \times n$ matrix with λ occurring once only in each row. There are therefore n roots to Eq. (3.77) and consequently solving this equation for λ yields the n eigenvalues of matrix $[A]$

$$|A - 1\lambda| = (1 - \lambda_1)(1 - \lambda_2) \ldots (1 - \lambda_n) = 0 \qquad (3.78)$$

Several of the factors of the determinant may be obtained from manipulation of the determinant by recognised methods.

The corresponding eigenvector to a given eigenvalue λ_1 is obtained by substituting this in Eq. (3.75) to yield a set of equations each equal to zero.

$$[A - 1\lambda_1] \begin{bmatrix} x_1 \\ x_2 \\ x_n \end{bmatrix} = \begin{bmatrix} 0 \\ 0 \\ 0 \end{bmatrix} \qquad (3.79)$$

from which x_2 x_3 - - x_n can be determined in terms of x_1. The corresponding eigenvector is $[1 \, . \, x_2/x_1 \ldots]$ or any multiple.

When n independent eigenvectors have been determined corresponding to the n eigenvalues, these may be displayed in an $n \times n$ matrix $[Q]$ which is the *eigenvector matrix*.

3.11.2 Diagonalisation of a matrix

If a non singular matrix $[A]$ has distinct eigenvalues the product $[Q]^{-1}[A][Q]$ is a diagonal matrix. This can be shown in the following way.

For each eigenvalue

$$[A][Q_1] = \lambda_1[Q_1]$$
$$[A][Q_2] = \lambda_2[Q_2]$$
$$[A][Q_n] = \lambda_n[Q_n] \qquad (3.80)$$

Writing this in matrix form

$$[A][Q] = [Q] \begin{bmatrix} \lambda_1 & \cdot & \cdot & \cdot \\ \cdot & \lambda_2 & \cdot & \cdot \\ \cdot & \cdot & \cdot & \cdot \\ \cdot & \cdot & \cdot & \lambda_n \end{bmatrix} \qquad (3.81)$$

$$= [Q][\Lambda] \qquad (3.82)$$

where

$$[\Lambda] = \begin{bmatrix} \lambda_1 & \cdot & \cdot & \cdot \\ \cdot & \lambda_2 & \cdot & \cdot \\ \cdot & \cdot & \cdot & \cdot \\ \cdot & \cdot & \cdot & \lambda_n \end{bmatrix} \qquad (3.83)$$

Premultiplying by $[Q]^{-1}$

$$[Q]^{-1}[A][Q] = [Q]^{-1}[Q][\Lambda] = [\Lambda] \qquad (3.84)$$

and the original matrix has thus been transformed to a diagonal matrix.

Example:

Let
$$[A] = \begin{bmatrix} 1 & 1 & 1 \\ -1 & 3 & 4 \\ -3 & 3 & 2 \end{bmatrix}$$

$$[A - 1\lambda] = \begin{bmatrix} (1-\lambda) & 1 & 1 \\ -1 & (3-\lambda) & 4 \\ -3 & 3 & (2-\lambda) \end{bmatrix}$$

Equating the determinant of $[A - 1\lambda]$ to zero

$$0 = \begin{vmatrix} (1-\lambda) & 1 & 1 \\ -1 & (3-\lambda) & 4 \\ -3 & 3 & (2-\lambda) \end{vmatrix}$$

Replace column 1 by column 1 + column 2
Replace column 2 by column 2 − column 3

$$0 = \begin{vmatrix} (2-\lambda) & 0 & 1 \\ 2-\lambda & (-1-\lambda) & 4 \\ 0 & 1+\lambda & (2-\lambda) \end{vmatrix} = (2-\lambda)(1+\lambda) \begin{vmatrix} 1 & 0 & 1 \\ 1 & -1 & 4 \\ 0 & 1 & (2-\lambda) \end{vmatrix}$$

$$0 = (2-\lambda)(1+\lambda)(-2+\lambda+1-4)$$

and the eigenvalues are

$$\lambda_1 = 2 \quad \lambda_2 = -1 \quad \lambda_3 = 5$$

$$\Lambda = \begin{bmatrix} 2 & 0 & 0 \\ 0 & -1 & 0 \\ 0 & 0 & 5 \end{bmatrix}$$

To find the corresponding eigenvectors

for $\lambda_1 = 2$
$$\begin{bmatrix} -1 & 1 & 1 \\ -1 & 1 & 4 \\ -3 & 3 & 0 \end{bmatrix} \cdot \begin{bmatrix} x_1 \\ x_2 \\ x_3 \end{bmatrix}_1 = 0$$

hence $\begin{matrix} x_1 = x_2 \\ x_3 = 0 \end{matrix}$
$$[Q_1] = \begin{bmatrix} 1 \\ 1 \\ 0 \end{bmatrix}$$

For $\lambda_2 = -1$
$$\begin{bmatrix} 2 & 1 & 1 \\ -1 & 4 & 4 \\ -3 & 3 & 3 \end{bmatrix} \cdot \begin{bmatrix} x_1 \\ x_2 \\ x_3 \end{bmatrix}_2 = 0$$

$$x_2 = -x_3$$
$$x_1 = 0$$

$$[Q_2] = \begin{bmatrix} 0 \\ 1 \\ -1 \end{bmatrix}$$

For $\lambda_3 = 5$

$$\begin{bmatrix} -4 & 1 & 1 \\ -1 & -2 & 4 \\ -3 & 3 & -3 \end{bmatrix} \cdot \begin{bmatrix} x_1 \\ x_2 \\ x_3 \end{bmatrix}_3 = 0$$

$$4x_1 = x_2 + x_3$$
$$x_1 = -2x_2 + 4x_3$$
$$x_1 = x_2 - x_3 \qquad [Q_3] = \begin{bmatrix} 2 \\ 5 \\ 3 \end{bmatrix}$$
$$5x_1 = 2x_2$$
$$3x_1 = 2x_3$$

Hence $$[Q] = \begin{bmatrix} 1 & 0 & 2 \\ 1 & 1 & 5 \\ 0 & -1 & 3 \end{bmatrix}$$

Inverting $$[Q]^{-1} = \frac{1}{6} \begin{bmatrix} 8 & -2 & -2 \\ -3 & 3 & -3 \\ -1 & 1 & 1 \end{bmatrix}$$

Multiplying out will check that

$$[\Lambda] = [Q]^{-1}[A][Q]$$

$$\begin{bmatrix} 2 & 0 & 0 \\ 0 & -1 & 0 \\ 0 & 0 & 5 \end{bmatrix} = \frac{1}{6} \begin{bmatrix} 8 & -2 & -2 \\ -3 & 3 & -3 \\ -1 & 1 & 1 \end{bmatrix} \begin{bmatrix} 1 & 1 & 1 \\ -1 & 3 & 4 \\ -3 & 3 & 2 \end{bmatrix} \begin{bmatrix} 1 & 0 & 2 \\ 1 & 1 & 5 \\ 0 & -1 & 3 \end{bmatrix}$$

Converting a matrix to diagonal form is of importance in the solution of equations.

If $$[y] = [A][x] \qquad\qquad (3.85)$$

these are a set of simultaneous equations which requires the process of inversion to obtain the solution

$$[x] = [A]^{-1}[y] \qquad\qquad (3.86)$$

But if $$[\Lambda] = [Q]^{-1}[A][Q] \qquad\qquad (3.84)$$
$$[A] = [Q][\Lambda][Q]^{-1} \qquad\qquad (3.87)$$

and substituting for $[A]$ in Eq. (3.85)

$$[y] = [Q][\Lambda][Q]^{-1}[x]$$

or $$[y'] = [\Lambda][x'] \qquad\qquad (3.86)$$
where $$[x'] = [Q]^{-1}[x] \qquad\qquad (3.87)$$
and $$[y'] = [Q]^{-1}[y] \qquad\qquad (3.88)$$

We therefore note that $[x]$ and $[y]$ can be transformed by $[Q]^{-1}$ to alternative forms $[x']$ and $[y']$ and these are related by Eq. (3.86) a series of independent equations.

The involved calculations required for the solution of simultaneous equations have been eliminated. The symmetrical component transform used in Chap. 13 is an example of how these advantages can be applied.

3.12 Algorithms

An algorithm is the solution of a type of problem reduced to a statement of the step-by-step procedure to be used for solving a specific example of this type. It is important that the solution should be achieved in a finite number of steps and that the method should be applicable to all problems of the same type.

For example, a multiplication algorithm could be devised by showing the steps in a long multiplication sum. It would probably begin by stating how the multiplicand would first be written down and go on to show exactly what has to be done using the first digit of the multiplier. It would then state how to move to the next digit and deal with that. It might then continue by stating 'Go on to the next digit in the multiplier and treat this in the same way repeating the procedure until all the digits have been used'.

Algorithms are conveniently illustrated graphically by means of flow charts. These are diagrams showing (a) the operations to be undertaken (b) the decisions that have to be made and (c) the order in which these items are connected. Flow charts are particularly useful when planning any sequence of operations and especially in the early stages of writing computer programs.

A computer program in itself is a very detailed list of instructions assembled in such a form as to be acceptable to the computer. It may be written in machine code which the machine can accept directly or in a computer language which the machine can translate to its own order code by means of a language compiler. An algorithm is less detailed and not necessarily complete but it should show how the work is organised and particularly how repetitive calculations are to be handled. Before discussing a number of simple algorithms it will be useful to examine some very elementary principles applied to computer programming.

Fundamentally a computer consists of a number of stores in which numbers can be placed. The machine can be instructed to read input data, that is to transfer numbers from punched cards or punched paper tape and place them in particular stores.

The instruction [READ A] means read the next number on the input tape and place it in location A cancelling whatever number was previously in store A. The machine can then be programmed to read these numbers in the stores, perform the arithmetical functions of addition, subtraction, multiplication and division using these numbers and to store the results in the same or other locations.

The instruction $[B := C + D]$ means take the contents of stores C and D, add them together and place in store location B, without destroying C and D. Such an instruction is a simple algebraic expression.

The instruction $[A : = A + B]$ is not algebraic but is interpreted as 'read the contents of stores A and B, add together and place in store A.' The original number in store A is now lost.

When calculations are completed the result must be displayed either on punched card, tape, by teleprinter, line printer or by cathode ray tube display.

The instruction [OUTPUT A] or [PRINT A] means take the number from store A and display it by the chosen output device.

Numbers can also be compared and on the basis of A being greater or less than B simple yes/no decisions can be made. Such decisions can be used to instruct the computer what to do next either to continue with a given sequence or to transfer to another sequence. This property is extremely important, since it is also possible to return the computer to an earlier point in the program thereby creating a loop and causing the program to be repeated with modified input. A single program can thus generate a whole series of results.

Questions are usually displayed in a flowchart in a diamond frame.

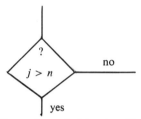

FIG. 3.1 Diamond frame for a decision in a flow chart

This instruction entered from the top asks the question 'Is j greater than n?' If the answer is 'yes' then we go on to the next instruction following on the route marked 'yes'. On the other hand if the answer is 'no' the alternative route marked 'no' must be taken.

In order to undertake repetitive work it is usually necessary to arrange a counting sequence. This is done by allocating a number to a store, say j and commencing with the instruction $[j : = 0]$ followed by $[j : = j + 1]$.

After a sequence of instructions is completed for the first time the program loop returns to the instruction $[j : = j + 1]$ thereby making $j = 2$.

The cycle is repeated only to return again and make $j = 3$ thus counting the sequences. Eventually when j is equal to the required number of counts a decision order must be included to exit from the loop.

It is also important to remember that a computer has not only to be told to start but also to stop.

The following examples of flowcharts are included to show how simple repetitive work can be controlled. They must not be regarded as complete programs nor are the instructions necessarily compatible with any specific computer or computer language.

3.12.1 To select the diagonal terms in a matrix $n \times n$

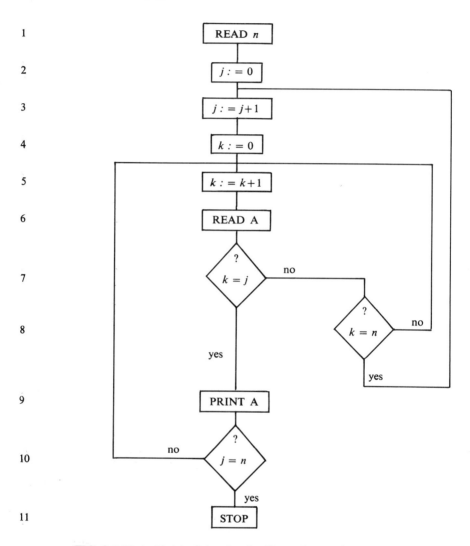

1

2

3

4

5

6

7

8

9

10

11

FIG. 3.2 Flow chart to determine the diagonal terms in a matrix

It is assumed we have an $n \times n$ matrix. The first number on the input tape must be n followed by the n^2 elements of the matrix in sequence line by line.

The instructions are interpreted as follows.

1. Read the first number on the input tape to establish n.
2. Put j equal to zero.
3. Add 1 making $j = 1$.

4. Put k equal to zero.
5. Add 1 making $k = 1$.
6. Read the next input which is Z_{11}.
7. Test. Is $k = j$? Answer yes—so continue.
9. Print A, which prints out the first diagonal term.
10. Test. Have we got to the end? No! so we return to 5 and make $k = 2$.
6. The next number is read in (Z_{12}).
7. Test. Is $k = j$? Answer no. Proceed to 8.
8. Test. Is $k = n$? (Are we at the end of the first line?) Answer no—so back again to 5.

This cycle proceeds until the answer to 8 is yes and we move back to 3.

3. This starts the second line of the matrix.
4. k is returned to zero. We are considering the first term in the second line.
   By this means we note that all the elements of the matrix will be read in line by line and only those numbers for which $k = j$ will be printed at instruction 9.
   The cycles will stop when the last term $Z_{nn}(j = k = n)$ has been printed.

3.12.2 Matrix multiplication

$$[V_j) = [Z_{jk}] . [I_k]$$

The purpose of this program is to obtain n terms in the vector $[V]$ given the n^2 terms in $[Z_{jk}]$ and the n terms of $[I_k]$.

To perform matrix multiplication we have seen that each element of the first row in $[Z_{jk}]$ must be multiplied by the corresponding element in $[I_k]$; they must be added and entered as the first element of $[V_j]$. The cycle must be repeated for the second row and so on. (See Fig. 3.3)

The instructions are interpreted as follows.

1. Read and store the first number on the input tape to establish n.
2. Read and store the next n numbers on the input tape to be called up as required by the identifier $I(k)$.
3. Put j equal to zero.
4. Add 1 making $j = 1$.
5. Put k equal to zero.
6. Add 1 making $k = 1$.
7. Read the next number on the input tape and store as Z.
8. Multiply Z by $I(k)$ and add to the present value in store A and replace in A.

This operation in the first cycle gives

$$A : = 0 + Z_{11}I(1).$$

9. Test. Is $k = n$? (Are we at the end of the first row?) Answer no—so return

to 6. k is increased to 2. The next value of Z is read in and at instruction 8 the value in store A is increased by $Z_{12}I(2)$. The cycles continue until

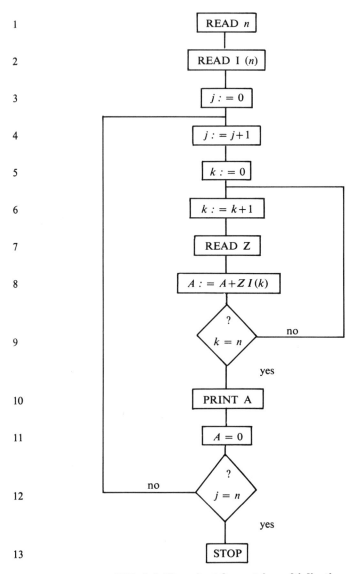

FIG. 3.3 Flow chart for matrix multiplication

$k = n$ when all the products of ZI for the first row will have been added in store A. We now answer yes to instruction 8 and proceed.

10. Print A which is now equal to the required $V(1)$.
11. Store A is returned to zero.
12. Test. Is $j = n$? Have we printed the last voltage $V(n)$? If not we are returned

to 4 where the value of j is increased by one and the cycles repeated for the next line.

When the answer to 12 is yes the computer advances to 13.

13. Stop!

3.12.3 Family of resonance curves

To determine graphs giving the value of the current in a series $L\,R\,C$ circuit with applied a.c. voltage V and for which we are given a series of different values for R and f.

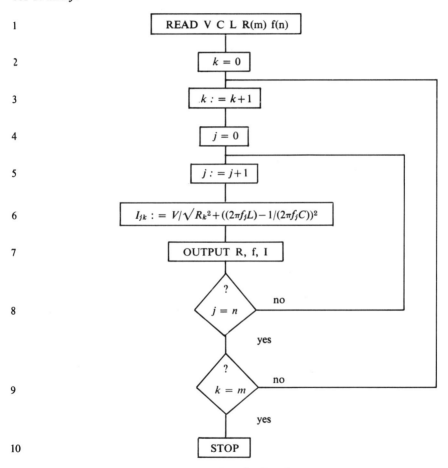

FIG. 3.4 Flow chart to obtain a family of resonance curves

The explanation of this algorithm is as follows.

1. The items to be read and stored in the computer from the input tape are in order V, C, L, m values of resistance R and n values of frequency f. It is

proposed to provide values so that a current/frequency graph can be drawn corresponding to each of the m values of R.

A separate program could have been used to generate these values of f and R if the limiting values and the number of steps had been given instead.

2. Put k equal to zero.
3. Add 1 to make $k = 1$.
4. Put j equal to zero.
5. Add 1 to make $j = 1$.
6. This calculates the value of current I_{jk} corresponding to the first values of R and f. Here the actual computer program may have several steps in the calculation.
7. Print those values of R, f and I.
8. Test. Is $j = n$? Is this the final frequency of the series? If the answer is no, the loop repeats at 5 adding 1 to j to go to the next frequency.
   When the answer to 8 is yes the program proceeds to
9. Test. Is $k = m$? Is this the last value of R? If no, the loop repeats at 3 adding 1 to k and going to the next value of resistance. When the answer to 8 is yes the program proceeds to
10. Stop.

The above algorithms are not intended to act as complete computer programs but to show how the sequence of the cycles or loops can be achieved.

CHAPTER 4

The primitive network

4.1 Introduction

We have seen in Chap. 1 that the network approach to an electrical problem requires the replacement of the actual system by a number of elementary two-terminal units which are then suitably inter-connected by impedanceless conductors.

Properties of the single elements have been discussed: attention must now be concentrated on the effect of interconnection. The reader is again reminded that the actual substitution of an electrical network, in place of a system in which parameters are distributed, involves a degree of approximation which may or may not be completely justified. Judgment on the part of the engineer must be exercised in this respect.

Figure 4.1 shows three coils or conductors

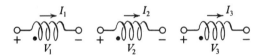

FIG. 4.1 Three independent or primitive coils

each with two terminals and which, for the moment, are not connected to each other. Voltages and currents have been labelled in accordance with the conventions of Chap. 2.

Mutual inductances are assumed to exist between each pair of coils and designated by subscripts. M_{12} is the voltage produced in coil **1** due to unit rate of change of current in coil **2**. M_{21} is the voltage produced in coil **2** due to unit rate of change of current in coil **1**.

M_{21} and M_{12} are identical in the case of passive elements, but they can be unequal when the coils are in relative motion, or when they include effects due to a third coil moving relatively to them. It will be necessary to distinguish carefully between the two cases.

The voltages and currents in the coils are, therefore, related by

$$[V] = [Z][I] \tag{4.1}$$

or

$$\begin{bmatrix} V_1 \\ V_2 \\ V_3 \end{bmatrix} = \begin{bmatrix} Z_{11} & Z_{12} & Z_{13} \\ Z_{21} & Z_{22} & Z_{23} \\ Z_{31} & Z_{32} & Z_{33} \end{bmatrix} \cdot \begin{bmatrix} I_1 \\ I_2 \\ I_3 \end{bmatrix} \tag{4.2}$$

These voltages and currents are known as the primitive voltages and currents respectively, and the matrix $[Z]$ is known as the primitive unconnected impedance matrix.

The elements such as Z_{11} have both resistance and inductance, hence

$$Z_{11} = R_1 + pL_1 \tag{4.3}$$

also

$$Z_{12} = pM_{12} \tag{4.4}$$

The impedance matrix can therefore be built up by superposition of three component matrices, resistance $[R]$ self inductance $[L]$ and mutual inductance $[M]$

$$[Z] = [R] + p[L] + p[M] \tag{4.5}$$

$$= \begin{bmatrix} R_1 & 0 & 0 \\ 0 & R_2 & 0 \\ 0 & 0 & R_3 \end{bmatrix} + p \begin{bmatrix} L_1 & 0 & 0 \\ 0 & L_2 & 0 \\ 0 & 0 & L_3 \end{bmatrix} + p \begin{bmatrix} 0 & M_{12} & M_{13} \\ M_{21} & 0 & M_{23} \\ M_{31} & M_{32} & 0 \end{bmatrix} \tag{4.6}$$

This is a very satisfactory way of writing down the values by inspection, ensuring that each term is placed in its proper place.

If the branches also contain capacitance in series, then a further capacitance impedance matrix must be added

$$(1/p)[1/C] = (1/p) \begin{bmatrix} 1/C_1 & 0 & 0 \\ 0 & 1/C_2 & 0 \\ 0 & 0 & 1/C_3 \end{bmatrix} \tag{4.7}$$

It is also noted that if there is no relative movement between coils then $M_{12} = M_{21}$, $M_{23} = M_{32}$, $M_{31} = M_{13}$ and the primitive impedance matrix is symmetrical.

4.2 Kron's connection matrix

4.2.1 Constraints

In the primitive network the currents in the unconnected coils are independent of each other. Connecting the coils together in an actual network produces constraints on the currents since Kirchhoff's first law must be obeyed at all nodes.

For example, in Fig. 4.2(a), five branches are shown with five currents $I_a I_b I_c I_d I_e$. These currents are obviously related since now the current I_c in the central branch is such that

$$I_a - I_b = I_c = I_e - I_d \qquad (4.8)$$

4.2.2 Loop currents

At any node in a network, the algebraic sum of the branch currents entering the node is zero. For example, at node **A**, Fig. 4.2(a) (repeated in Fig. 4.2(b)), we have

$$I_a - I_c - I_b = 0. \qquad (4.9)$$

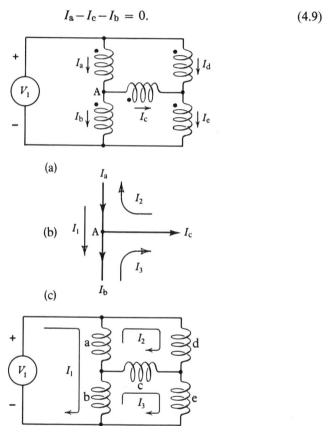

(a)

(b)

(c)

FIG. 4.2 The branch currents shown in (a) are related to the loop currents shown in (c). Equation 4.10 relates to the currents associated with node A as shown in (b)

On the other hand, each of these branch currents can be replaced by two currents. I_a is replaced by I_1 flowing into the node from branch **a** along branch **b**, together with I_2 flowing along branch **c** and out from the node along branch **a**. Thus

$$I_a = I_1 - I_2$$

also

$$I_b = I_1 - I_3 \tag{4.10}$$

and

$$I_c = I_3 - I_2$$

Thus the algebraic sum of the currents at the node **A**

$$I_a - I_b - I_c = (I_1 - I_2) - (I_1 - I_3) - (I_3 - I_2) \tag{4.11}$$

This is equal to zero whatever the values of I_1 I_2 and I_3.

When we show each component current flowing into and out of a node without gain or loss, we assume Kirchhoff's Law. It is not surprising, therefore, to find the sum (Eq. (4.11)) expresses this Law.

If the three-loop currents of Fig. 4.2(*b*) are applied to the circuits, as in Fig. 4.2(*c*), all of the branch currents may be expressed in terms of them. This is a convenient simplification. It is an example of the advantage to be gained by stating the physical restrictions or constraints in the most convenient mathematical terms at the start.

Inspection of the Figs. 4.2(*a*) and 4.2(*c*) shows that

$$
\begin{aligned}
I_a &= I_1 - I_2 \\
I_b &= I_1 - I_3 \\
I_c &= I_3 - I_2 \\
I_d &= I_2 \\
I_e &= I_3
\end{aligned} \tag{4.12}
$$

or in matrix form

$$
\begin{bmatrix} I_a \\ I_b \\ I_c \\ I_d \\ I_e \end{bmatrix} =
\begin{bmatrix} 1 & -1 & 0 \\ 1 & 0 & -1 \\ 0 & -1 & 1 \\ 0 & 1 & 0 \\ 0 & 0 & 1 \end{bmatrix} \cdot
\begin{bmatrix} I_1 \\ I_2 \\ I_3 \end{bmatrix} \tag{4.13}
$$

Thus

$$[I] = [c][I'] \tag{4.14}$$

where $[I]$ represents branch currents and $[I']$ the loop currents. The matrix $[c]$ is known as the connection matrix. It is usually simple in form as shown above, consisting of elements which are either 0, $+1$ or -1. Essentially it states a relationship which enables us to transform loop currents $[I']$ into branch currents $[I]$.

4.3 Transforms

The principle of transformation is an important concept. In the above problem, a set of currents in the branches is replaced by a related set in the loops. The latter are not additional currents, but are the effect of looking at the system in a different way. This alteration of viewpoint is termed transformation to a new frame of reference. The prime is commonly used to indicate the new sets of terms in the new frame or, alternatively, new subscripts are applied.

A similar relationship exists between loop voltages and branch voltages. Again referring to Fig. (4.2c), summing the voltage drops around each loop in turn, we have by Kirchhoff's second law

$$V_1 = V_a + V_b$$
$$0 = V_d - V_c - V_a \qquad (4.15)$$
$$0 = V_c + V_e - V_b$$

or

$$
\begin{bmatrix} V_1 \\ 0 \\ 0 \end{bmatrix} =
\begin{bmatrix} 1 & 1 & 0 & 0 & 0 \\ -1 & 0 & -1 & 1 & 0 \\ 0 & -1 & 1 & 0 & 1 \end{bmatrix} \cdot
\begin{bmatrix} V_a \\ V_b \\ V_c \\ V_d \\ V_e \end{bmatrix}
\qquad (4.16)
$$

or

$$[V'] = [d][V] \qquad (4.17)$$

Matrix $[d]$ is the transform matrix transforming branch voltages into loop voltages. By inspection it will be seen that the matrix $[d]$ is the transpose of $[c]$. The reason for this is given in the next section.

4.3.1 Invariance of power

In terms of the primitive network, if $[V]$ and $[I]$ are the branch voltages and currents, the power P is such that

$$P = V_a I_a + V_b I_b + V_c I_c + \cdots \qquad (4.18)$$

In matrix notation

since

$$[V] = \begin{bmatrix} V_a \\ V_b \\ V_c \end{bmatrix} \qquad [V_t] = [V_a \quad V_b \quad V_c] \qquad (4.19)$$

$$P = [V_a \quad V_b \quad V_c] \cdot \begin{bmatrix} I_a \\ I_b \\ I_c \end{bmatrix} = [V_t][I] \qquad (4.20)$$

Note that P is a single number not a matrix.

Alternatively, the power could be computed from loop currents and loop voltages which would give

$$P = [V'_t][I'] \qquad (4.21)$$

It is logical to assume that power is the same whichever currents and voltages have been used for the computation. It is the same system operating under the same conditions. Power is thus said to be *invariant*, that is to say it is unaffected by the transforms from the branch to the loop frames of reference. Hence

$$[V'_t][I'] = [V_t][I] \tag{4.22}$$

from this it follows, substituting from Eq. (4.14)

$$[V'_t][I'] = [V_t][c][I'] \tag{4.23}$$

or

$$[V'_t] = [V_t][c] \tag{4.24}$$

To invert the order of multiplication, all matrices must be transposed (Eq. (3.12), and thus

$$[V'] = [c_t][V] \tag{4.25}$$

and so comparing with Eq. (4.17)

$$[d] = [c_t] \tag{4.26}$$

4.3.2 Relationship between loop voltages and loop currents

The relationship between branch voltages and branch currents was given by Eq. (4.1).

$$[V] = [Z][I] \tag{4.1}$$

We are led to expect that a similar relationship exists between the loop voltages and loop currents.

$$[V'] = [Z'][I'] \tag{4.27}$$

This we observe since

$$[I] = [c][I'] \tag{4.14}$$

and

$$[V'] = [c_t][V] \tag{4.25}$$

from (4.1) and (4.14)

$$[V] = [Z][I] = [Z][c][I']$$

and from (4.25)

$$[V'] = [c_t][Z][c][I'] \tag{4.28}$$
$$= [Z'][I'] \tag{4.27}$$

where

$$[Z'] = [c_t][Z][c] \tag{4.29}$$

$[Z']$ is referred to as the impedance matrix of the connected network and it shows the relationship between the loop currents and the loop voltages.

Equation (4.29) shows how the unconnected impedance matrix $[Z]$ is modified by $[c]$ and $[c_t]$ to form $[Z']$ in other words it shows precisely the way in which the constraints introduced by the connections modify the relationships between voltage and current.

The matrix multiplication involved in computing $[c_t][Z][c]$ effects this transformation from branch to loop. The individual elements comprising $[Z']$ are usually more complicated than those of the $[Z]$ matrix but the number of elements is usually much smaller.

Equation (4.28) now refers to the actual network and the solution of this equation provides the analysis of the network. This will be attempted in the next chapter but at this point we must pause to consider further the principle involved in transformation.

It might be that the loop voltage and currents of this network are related to some other system of voltages and currents where

$$[I'] = (a)[I''] \tag{4.14a}$$

and

$$[V''] = [a_t][V'] \tag{4.25a}$$

and from this we observe

$$[V''] = [a_t][Z'][a][I''] \tag{4.28a}$$
$$= [Z''][I''] \tag{4.27a}$$

where

$$[Z''] = [a_t][Z'][a] = [a_t][c_t][Z][c][a] \tag{4.29a}$$

Continued transformation in this manner is possible if further constraints are introduced or reference to other frames of reference is required to simplify the problem. Such transforms are only valid if power remains invariant, justifying the use of a transform such as $[a]$ together with its transpose $[a_t]$.

4.3.3 Special case for steady state alternating current

When the impedance matrix is used to represent steady state a.c. conditions, the operator p is replaced by $j\omega$, and the matrix becomes an array of complex numbers operating on the phasors of voltage and current. The invariant now is the 'complex power'* which, according to I.E.C. definition is given by,

$$S = P - jQ = V^*I \tag{4.30}$$

where V^* is the conjugate of the phasor V.

Complex power

If the alternating voltage drop across a load and the current in it are represented by the phasors V and I respectively
where

$$V = V_d + jV_q \tag{4.34}$$

and

$$I = I_d + jI_q \tag{4.35}$$

The component voltages and currents can be shown (Fig. 4.3) by representing the source components as two generators in series and the load as two paths in parallel.

Considering the volt–ampere products of the components taken separately, we note:

1. V_d and I_d are in phase, hence the product $V_d I_d$ represents power;
2. jV_q and jI_q are in phase, hence the product $V_q I_q$ also represents power;
3. V_d and jI_q are in quadrature, (jI_q leading), hence the product $V_d I_q$ represents leading volt–amperes reactive;
4. jV_q and I_d are in quadrature, (I_d lagging), hence the product $V_q I_d$ represents lagging volt–amperes reactive.

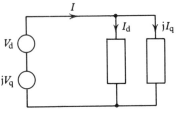

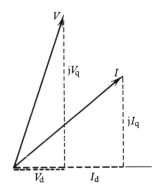

FIG. 4.3 The components of the phasors of current and voltage $I_d\, I_q\, V_d$ and V_q illustrating complex power

The total power is, therefore

$$P = V_d I_d + V_q I_q \tag{4.36}$$

and if volt–amperes lagging is taken to be positive (this is the convention adopted by the I.E.C.), the total volt–amperes reactive is given by

$$Q = V_q I_d - V_d I_q \tag{4.37}$$

Now if V^* is the conjugate of V
i.e.

$$V^* = V_d - jV_q \tag{4.38}$$

the product of the phasors V^*I is given by

$$
\begin{aligned}
V^*I &= (V_d - jV_q)(I_d + jI_q) \\
&= (V_d I_d + V_q I_q) - j(V_q I_d - V_d I_q) \\
&= P - jQ = S
\end{aligned}
\tag{4.39}
$$

The complex number $S = P - jQ$ is a very convenient expression since its two components are identified as the power and volt–amperes reactive respectively. Its magnitude

$$S = \sqrt{(P^2 + Q^2)} \tag{4.40}$$

is the total volt–amperes and P/S is the power factor.

The complex number S is termed 'complex power' and from equation (4.39) is obtained by multiplying the current phasor by the *conjugate* of the voltage phasor.

Further information relating to complex power is given in section 22.4.

Equation (4.22) now becomes

$$[V^{*'}_t][I'] = [V^*_t][I] \tag{4.31}$$

and therefore

$$[V^{*'}_t][I'] = [V^*_t][c][I']$$

hence

$$[V^{*'}_t] = [V^*_t][c]$$
$$[V^{*'}] = [c_t][V^*]$$

Finally

$$[V'] = [c^*_t][V] \tag{4.32}$$

$[c^*_t]$ being the conjugate transpose of $[c]$.
and therefore

$$[Z'] = [c_t^*][Z][c] \tag{4.33}$$

Equations (4.32) and (4.33) revert to (4.25) and (4.29), if the elements of c contain only real terms.

If the connection matrix is complex such as may be the case if phase transformation is involved, then Eqs. (4.32) and (4.33) must be used.

4.4 The 'dot' convention for mutual inductance

Some ambiguity may arise with respect to the positive or negative sign attributed to the mutual inductance between two coils. When the two coils are depicted in the manner of Fig. 4.4, it is clearly observed that the two coils are both wound

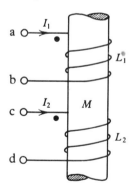

FIG. 4.4 The dot convention as used to ensure that there is no ambiguity in specifying the mutual inductance between two coils

in the same direction about the same former, and that the terminal **c** of the second coil corresponds to terminal **a** of the first coil. Terminals **d** and **b** also correspond.

Moreover, if the current entering terminal **a** is I_1 the voltage drops due to

inductance will be

$$V_{ab} = L_1 \, dI_1/dt \qquad (4.41)$$

$$V_{cd} = M \, dI_1/dt \qquad (4.42)$$

which is equivalent to

$$V_{dc} = -M \, dI_1/dt \qquad (4.43)$$

Unless the terminals of the secondary coil **cd** have been carefully marked it is impossible to decide which equation (4.42) or (4.43) is the correct one to be used to determine the induced voltage due to mutual induction. In other words, if it is uncertain whether V_{cd} or V_{dc} is required, ambiguity exists as to whether M is positive or negative.

FIG. 4.5 The dot convention as used to ensure that there is no ambiguity in specifying the mutual inductance between two coils

For example, in Fig. 4.5, it is not at all clear whether terminal **e** corresponds to terminal **g** or whether **eh** and **fg** are the corresponding pairs.

The 'dot' convention is used to ensure that this ambiguity does not arise. Dots are used to identify one pair of corresponding terminals. This is shown in Fig. 4.6, which is an alternative to Fig. 4.4. By this convention we infer that

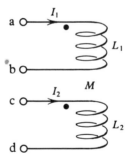

FIG. 4.6 The dot convention as used to ensure that there is no ambiguity in specifying the mutual inductance between two coils

terminals **a** and **c** in the figure correspond to each other, as do the undotted pair **b** and **d**.

If I_1 and I_2 are the currents entering the two coils at the dotted terminals we have from Eq. (4.6)

$$\begin{bmatrix} V_{ab} \\ V_{cd} \end{bmatrix} = \begin{bmatrix} R_1 + pL_1 & pM_{12} \\ pM_{21} & R_2 + pL_2 \end{bmatrix} \cdot \begin{bmatrix} I_1 \\ I_2 \end{bmatrix} \qquad (4.44)$$

But if the currents I_1 and I_2 correspond to Fig. 4.7 where **e** and **h** are the dotted terminals, then

$$\begin{bmatrix} V_{ef} \\ V_{gh} \end{bmatrix} = \begin{bmatrix} R_1 + pL_1 & -pM_{12} \\ -pM_{21} & R_2 + pL_2 \end{bmatrix} \cdot \begin{bmatrix} I_1 \\ I_2 \end{bmatrix} \tag{4.45}$$

To avoid any confusion, the *primitive currents* in coils are consistently defined as the currents entering the dotted terminals (this has been done in Sec. 4.1 and

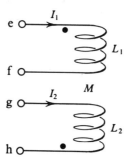

FIG. 4.7 A similar pair of coils to those of Fig. 4.6 but with currents associated with different terminals

Fig. 4.1). The mutual inductance terms in Eq. (4.6) are then always positive and Eq. (4.45) is never needed.

CHAPTER 5

Mesh analysis

5.1 Symbols

The symbols to be used in this chapter and also Chap. 6 are as follows.

$$V_b = \text{branch voltages.}$$
$$I_b = \text{branch currents.}$$
$$E_1 = \text{loop e.m.f's.}$$
$$I_1 = \text{loop currents.}$$
$$V_N = \text{nodal voltages.}$$
$$I_s = \text{source currents.}$$
$$Z_b = \text{branch impedances.}$$
$$Y_b = \text{branch admittances.}$$
$$Z_g = \text{generator impedances.}$$
$$Y_g = \text{generator admittances.}$$
$$E_s = \text{source e.m.f's.}$$
$$I_g = \text{generator currents.}$$

Lower case letters **a b c** will be used for branches.
Upper case letters **A B C** will be used to designate nodes.
Numbers **1, 2, 3** will enumerate loops.

5.2 Introduction

The basic principles of mesh analysis have already been enumerated in Chaps. 2 and 4. In Chap. 2 we saw how to prepare the graph of a network and to recognise the number of independent loops. In Chap. 4 the matrix transformation from branch currents to loop currents was explained.

The method of mesh analysis to be developed in this chapter is to concentrate upon these loop currents and to establish in an orderly manner a sufficient number of simultaneous equations, the solution of which will enable the loop currents to be determined.

We must not forget, going back to general principles, that all the excitation voltages in practical systems will be functions of time, and all the equations will be linear integro-differential equations. Each value of Z is a differential operator and the operator p is not an algebraic quantity.

For exponential excitation, where voltages are of the form

$$v(t) = \mathrm{Re}\, V \exp (st),$$

where s is a complex number, we shall see in Chap. 7 that the operator p can be replaced in the equations by s, and the differential equations become algebraic equations in complex numbers. This enables many steady-state problems to be solved.

A similar method will be used later (Chap. 7) to deal with many transient problems, but at this stage we will confine ourselves to developing the equilibrium equations for a network and obtaining the steady-state solution for exponential excitation.

Since much of the arithmetical work can be done by computer, it is important to set out logically the sequence of steps in the analysis so that the computer programming can be compiled at a later stage.

5.3 The procedures of mesh analysis

We must assume that the initial data is such that the configuration of the network is known, the sources and sinks have been identified and the necessary lumping of inductances and capacitances has been carried out.

For a network with B branches, mesh analysis proceeds in the following steps.

1. The graph of the network is constructed identifying the nodes.
2. Directions are assigned to branch currents.
3. The unconnected branch impedance matrix $[Z_b]$ is compiled (Eqs. (4.5) and (4.6).
4. The nodes are connected by a tree and the tree links chosen (Sec. 2.2.7). This establishes the M loops.
   If it is not necessary to determine the value of all branch currents but only a limited number are required, these branches should be chosen as tree-links if possible.
5. Having identified the loops the directions of the loop currents are next decided.
6. The connection matrix $[c]$ is now determined

$$[\text{branch currents}] = [c][\text{loop currents}] \tag{5.1}$$

$$[I_b] = [c][I_1] \tag{5.2}$$

7. The transpose $[c_t]$ is now formulated.

8. The column matrix $[E_1]$ of the loop e.m.f's or driving voltages is now assembled by inspection. Each loop is inspected in turn following the assigned direction of the loop current, the values and direction of the *source voltages* in the loops are observed and added algebraically. It usually happens that many loops are passive and the matrix $[E_1]$ will contain several zeros.

9. The equilibrium equations for the network are given by

$$[E_1] = [c_t][Z_b][c][I_1] \tag{5.3}$$

This step is equivalent to applying Kirchhoff's Second Law to each of the loops.

10. To proceed to a solution for a given excitation the matrix $[Z_b]$ is made algebraic by replacing p by s, $j\omega$ or 0 according to whether the excitation is exponential, steady-state a.c., or steady-state d.c. (Sec. 7.1).

11. The loop impedance matrix is obtained from

$$[Z_1] = [c_t][Z_b][c] \tag{5.4}$$

12. The loop admittance matrix is obtained by inversion

$$[Y_1] = [Z_1]^{-1} \tag{5.5}$$

13. The loop currents are determined from

$$[I_1] = [Y_1][E_1] \tag{5.6}$$

14. If the branch currents are required these are calculated from

$$[I_b] = [c][I_1] \tag{5.7}$$

If the currents in the tree-links only are required this step is unnecessary as each tree-link carries only one loop current.

The procedure of mesh analysis will be carried out step by step in the following examples before further simplification is introduced.

Example:

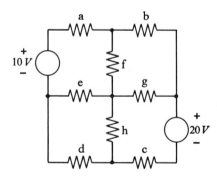

FIG. 5.1 A four-loop network

Figure 5.1 represents a d.c. network with eight branches, two of which are active.

Branches	a	b	c	d	e	f	g	h
R (ohms)	10	15	20	25	30	35	40	45

Steps 1 and 2

Graph of the network and assigned direction of branch currents. There are five nodes (Fig. 5.2).

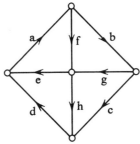

FIG. 5.2 The network graph corresponding to Fig. 5.1

Step 3

$$[Z] = \begin{bmatrix} 10 & 0 & 0 & 0 & 0 & 0 & 0 & 0 \\ 0 & 15 & 0 & 0 & 0 & 0 & 0 & 0 \\ 0 & 0 & 20 & 0 & 0 & 0 & 0 & 0 \\ 0 & 0 & 0 & 25 & 0 & 0 & 0 & 0 \\ 0 & 0 & 0 & 0 & 30 & 0 & 0 & 0 \\ 0 & 0 & 0 & 0 & 0 & 35 & 0 & 0 \\ 0 & 0 & 0 & 0 & 0 & 0 & 40 & 0 \\ 0 & 0 & 0 & 0 & 0 & 0 & 0 & 45 \end{bmatrix}$$

Steps 4 and 5

There are four loops. Branches **a, b, c** and **d** chosen as tree links. (Fig. 5.3).

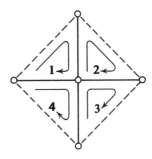

FIG. 5.3 The trees and links of Fig. 5.2

Steps 6 and 7

$$[I] = [c][I'] \qquad [c] = \overset{\text{Loop}}{\underset{\text{Branch}}{\begin{bmatrix} 1 & 0 & 0 & 0 \\ 0 & 1 & 0 & 0 \\ 0 & 0 & 1 & 0 \\ 0 & 0 & 0 & 1 \\ 1 & 0 & 0 & -1 \\ 1 & -1 & 0 & 0 \\ 0 & 1 & -1 & 0 \\ 0 & 0 & -1 & 1 \end{bmatrix}}}$$

$$[c_t] = \overset{\text{Branch}}{\underset{\text{Loop}}{\begin{bmatrix} 1 & 0 & 0 & 0 & 1 & 1 & 0 & 0 \\ 0 & 1 & 0 & 0 & 0 & -1 & 1 & 0 \\ 0 & 0 & 1 & 0 & 0 & 0 & -1 & -1 \\ 0 & 0 & 0 & 1 & -1 & 0 & 0 & 1 \end{bmatrix}}}$$

Each tree branch carries the difference between two loop currents and each tree link current is a loop current.

Step 8

$$[E_1] = \begin{bmatrix} 10 \\ 0 \\ -20 \\ 0 \end{bmatrix}$$

Note that the applied voltage is negative in the third loop.

Step 9

The equation to the network

$$\begin{bmatrix} 10 \\ 0 \\ -20 \\ 0 \end{bmatrix} = \begin{bmatrix} 1 & 0 & 0 & 0 & 1 & 1 & 0 & 0 \\ 0 & 1 & 0 & 0 & 0 & -1 & 1 & 0 \\ 0 & 0 & 1 & 0 & 0 & 0 & -1 & -1 \\ 0 & 0 & 0 & 1 & -1 & 0 & 0 & 1 \end{bmatrix} \cdot$$

$$\begin{bmatrix} 10 & 0 & 0 & 0 & 0 & 0 & 0 & 0 \\ 0 & 15 & 0 & 0 & 0 & 0 & 0 & 0 \\ 0 & 0 & 20 & 0 & 0 & 0 & 0 & 0 \\ 0 & 0 & 0 & 25 & 0 & 0 & 0 & 0 \\ 0 & 0 & 0 & 0 & 30 & 0 & 0 & 0 \\ 0 & 0 & 0 & 0 & 0 & 35 & 0 & 0 \\ 0 & 0 & 0 & 0 & 0 & 0 & 40 & 0 \\ 0 & 0 & 0 & 0 & 0 & 0 & 0 & 45 \end{bmatrix} \cdot \begin{bmatrix} 1 & 0 & 0 & 0 \\ 0 & 1 & 0 & 0 \\ 0 & 0 & 1 & 0 \\ 0 & 0 & 0 & 1 \\ 1 & 0 & 0 & -1 \\ 1 & -1 & 0 & 0 \\ 0 & 1 & -1 & 0 \\ 0 & 0 & -1 & 1 \end{bmatrix} \cdot \begin{bmatrix} I'_1 \\ I'_2 \\ I'_3 \\ I'_4 \end{bmatrix}$$

which reduces to

$$
\begin{bmatrix} 10 \\ 0 \\ -20 \\ 0 \end{bmatrix} = \begin{bmatrix} 75 & -35 & 0 & -30 \\ -35 & 90 & -40 & 0 \\ 0 & -40 & 105 & -45 \\ -30 & 0 & -45 & 100 \end{bmatrix} \cdot \begin{bmatrix} I'_1 \\ I'_2 \\ I'_3 \\ I'_4 \end{bmatrix}
$$

and after inversion

$$
\begin{bmatrix} I'_1 \\ I'_2 \\ I'_3 \\ I'_4 \end{bmatrix} = \frac{1}{1000} \begin{bmatrix} 25 \cdot 136 & 14 \cdot 622 & 10 \cdot 905 & 12 \cdot 448 \\ 14 \cdot 622 & 22 \cdot 566 & 12 \cdot 980 & 10 \cdot 228 \\ 10 \cdot 905 & 12 \cdot 980 & 19 \cdot 663 & 12 \cdot 120 \\ 12 \cdot 448 & 10 \cdot 228 & 12 \cdot 120 & 19 \cdot 188 \end{bmatrix} \cdot \begin{bmatrix} 10 \\ 0 \\ -20 \\ 0 \end{bmatrix}
$$

$$
\begin{bmatrix} I'_1 \\ I'_2 \\ I'_3 \\ I'_4 \end{bmatrix} = \begin{bmatrix} 0 \cdot 25136 - 0 \cdot 21810 \\ 0 \cdot 14622 - 0 \cdot 25960 \\ 0 \cdot 10905 - 0 \cdot 39326 \\ 0 \cdot 12448 - 0 \cdot 24240 \end{bmatrix} = \begin{bmatrix} 0 \cdot 03326 \\ -0 \cdot 11338 \\ -0.28421 \\ -0 \cdot 11792 \end{bmatrix}
$$

Step 14. Branch currents

$$
\begin{bmatrix} I_a \\ I_b \\ I_c \\ I_d \\ I_e \\ I_f \\ I_g \\ I_h \end{bmatrix} = \begin{bmatrix} 0 \cdot 03326 \\ -0 \cdot 11338 \\ -0 \cdot 28421 \\ -0.11792 \\ 0 \cdot 03326 + 0 \cdot 11792 \\ 0 \cdot 03326 + 0 \cdot 11338 \\ -0 \cdot 11338 + 0 \cdot 28421 \\ 0 \cdot 28421 - 0 \cdot 11792 \end{bmatrix} = \begin{bmatrix} 0 \cdot 03326 \\ -0 \cdot 11338 \\ -0 \cdot 28421 \\ -0 \cdot 11792 \\ 0 \cdot 15118 \\ 0 \cdot 14664 \\ 0 \cdot 17083 \\ 0 \cdot 16629 \end{bmatrix}
$$

5.4 Stigant's rule

Although the method used in Sec. 5.3 for finding $[Z'] = [c_t][z][c]$ has the great virtue of proceeding one step at a time with exceptional clarity, a method is available which enables the matrix $[Z']$ to be written down by inspection if the loops of the network are easily defined

$$
[Z'] = \begin{bmatrix} Z'_{11} & Z'_{12} & Z'_{13} & \cdot & Z'_{1n} \\ Z'_{21} & Z'_{22} & Z'_{23} & \cdot & Z'_{2n} \\ Z'_{31} & Z'_{32} & Z'_{33} & \cdot & Z'_{3n} \\ \cdot & \cdot & \cdot & \cdot & \cdot \\ Z'_{n1} & Z'_{n2} & Z'_{n3} & \cdot & Z'_{nn} \end{bmatrix} \tag{5.8}
$$

This impedance matrix refers directly to the loops and is the relationship between loop driving voltages and loop currents. The diagonal terms Z'_{11},

$Z'_{22} \dots Z'_{nn}$ are known as the *contour impedances* and are obtained by summing the impedance around the individual loops.

In the previous example, Z'_{11} is the sum of three branch impedances, branch **a**, branch **f**, branch **e** forming the first loop hence

$$Z'_{11} = 10+30+35 = 75.$$

Similarly Z'_{22}, the contour impedance of the second loop is the sum of branch **b**, branch **g** and branch **f** hence

$$Z'_{22} = 15+40+35 = 90.$$

Obviously each contour impedance is that value of Z' which has to be multiplied by the loop current in order to obtain the component of voltage drop, due to that current.

When the contour impedances have all been determined and written in the diagonal positions, it will next be observed that the matrix is symmetrical and the other terms are in pairs

$$Z'_{mn} = Z'_{nm}.$$

This is always the case for a bilateral passive network.

Z'_{mn} is the impedance which causes a volt drop in loop **m** due to loop current I'_n; Z'_{nm} is the impedance which causes a volt drop in loop **n** due to loop current I'_m. They are the same, namely, the impedance which is common to both loops. These are known as *mutual impedances*. They are positive or negative according to whether the currents in the loops that they connect flow through them in the same or opposite direction respectively.

Again referring to the previous example, branch **f** $(R = 35)$ is common to loops **1** and **2** and the two loop currents flow in opposite directions through this branch
hence

$$Z'_{12} = Z'_{21} = -35.$$

Similarly

$$Z'_{34} = Z'_{43} = -45 \text{ (branch h)}.$$

Also there is no common branch between loops **1** and **3** or between **2** and **4**, and so

$$Z'_{13}, Z'_{31}, Z'_{24}, Z'_{42} \text{ are all zero.}$$

Stigant's rule can be summarised as follows:

To determine the loop impedance matrix of a network
(a) Sum the impedance of the elements included in the individual loop. For loop **m** this is the contour impedance Z_{mm} and should be entered in its appropriate place in the matrix.
(b) Determine the impedance of all branches common to two loops. For loops **m** and **n**, this is the mutual impedance Z_{mn} and also Z_{nm} the suffixes indicating the position in the loop impedance matrix. Z_{mn} and Z_{nm} are positive

or negative according to whether the loop currents I_m and I_n flow in the same direction or the opposite direction in this impedance.
If an element is common to three loops, say m n and p, its value is part of the mutual impedances

$$Z_{mn} \ Z_{nm}, \ Z_{np} \ Z_{pn}, \ Z_{mp} \ Z_{pm}.$$

5.4.1 Example:

On the use of Stigant's rule.

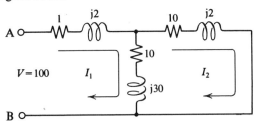

FIG. 5.4 A two-loop network illustrating Stigant's rule

The loop equations are written down by inspection

$$\begin{bmatrix} 100 \\ 0 \end{bmatrix} = \begin{bmatrix} 11+j32 & -10-j30 \\ -10-j30 & 20+j32 \end{bmatrix} \cdot \begin{bmatrix} I_1 \\ I_2 \end{bmatrix}$$

This can be inverted by the method of the previous example or alternatively using

$$\begin{bmatrix} V_d \\ V_q \end{bmatrix} = \begin{bmatrix} R & -X \\ X & R \end{bmatrix} \cdot \begin{bmatrix} I_d \\ I_q \end{bmatrix} \tag{3.36}$$

$$\begin{bmatrix} 100 \\ 0 \\ 0 \\ 0 \end{bmatrix} = \begin{bmatrix} 11 & -32 & -10 & 30 \\ 32 & 11 & -30 & -10 \\ -10 & 30 & 20 & -32 \\ -30 & -10 & 32 & 20 \end{bmatrix} \cdot \begin{bmatrix} I_{d1} \\ I_{q1} \\ I_{d2} \\ I_{q2} \end{bmatrix}$$

Only the first column of $[Z']^{-1}$ is needed giving

$$\begin{bmatrix} I_{d1} \\ I_{q1} \\ I_{d2} \\ I_{q2} \end{bmatrix} = \frac{1}{100} \begin{bmatrix} 8 \cdot 110 \\ -5 \cdot 185 \\ 7 \cdot 626 \\ -2 \cdot 629 \end{bmatrix} 100$$

$$I_1 = 8 \cdot 110 - j5.185$$
$$I_2 = 7 \cdot 626 - j2 \cdot 629$$

5.4.2 Example:

This example illustrates the effect of mutual inductance. It is a steady-state a.c. problem and is seen to be a two-loop network with mutual inductance between branches **a** and **c**. In this simple example many steps in the procedure are self-evident.

$$Z_a = 2+j6$$
$$Z_b = 3-j2$$
$$Z_c = 1+j5$$
$$Z_{ac} = j4$$
$$V_{AB} = 10$$

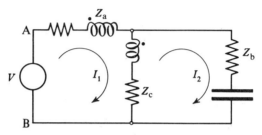

FIG. 5.5 A two-loop circuit with mutual inductance

Step 3. The unconnected branch impedance matrix is given by

$$[Z_b] = \begin{bmatrix} 2+j6 & 0 & j4 \\ 0 & 3-j2 & 0 \\ j4 & 0 & 1+j5 \end{bmatrix}$$

The mutual induction terms are seen in the non-diagonal elements.

Step 6. The connection matrix

$$\begin{bmatrix} I_a \\ I_b \\ I_c \end{bmatrix} = \begin{bmatrix} 1 & 0 \\ 0 & 1 \\ 1 & -1 \end{bmatrix} \begin{bmatrix} I_1 \\ I_2 \end{bmatrix}$$

Step 11. The loop impedance matrix

$$[Z] = \begin{bmatrix} 1 & 0 & 1 \\ 0 & 1 & -1 \end{bmatrix} \begin{bmatrix} 2+j6 & 0 & j4 \\ 0 & 3-j2 & 0 \\ j4 & 0 & 1+j5 \end{bmatrix} \begin{bmatrix} 1 & 0 \\ 0 & 1 \\ 1 & -1 \end{bmatrix}$$

$$= \begin{bmatrix} 1 & 0 & 1 \\ 0 & 1 & -1 \end{bmatrix} \begin{bmatrix} 2+j10 & 0 \\ 0 & 3-j2 \\ 1+j9 & -1-j5 \end{bmatrix}$$

$$= \begin{bmatrix} 3+j19 & -1-j9 \\ -1-j9 & 4+j3 \end{bmatrix}$$

Step 12. Inverting the loop impedance matrix

If
$$[Z] = \begin{bmatrix} Z_1 & Z_{12} \\ Z_{21} & Z_2 \end{bmatrix} \qquad \text{where} \quad Z_{21} = Z_{12}$$

$$[Y] = \frac{1}{\Delta} \begin{bmatrix} Z_2 & -Z_{12} \\ -Z_{12} & Z_1 \end{bmatrix}$$

where
$$\Delta = Z_1 Z_2 - Z_{12}{}^2$$

Step 13

$$\begin{bmatrix} I_1 \\ I_2 \end{bmatrix} = \frac{1}{\Delta} \begin{bmatrix} Z_2 & -Z_{12} \\ -Z_{12} & Z_1 \end{bmatrix} \cdot \begin{bmatrix} V \\ 0 \end{bmatrix}$$

$$I_1 = VZ_2/(Z_1 Z_2 - Z_{12}{}^2)$$

$$I_2 = V(-Z_{12})/(Z_1 Z_2 - Z_{12}{}^2)$$

where
$$Z_1 = 3 + j19$$
$$Z_2 = 4 + j3$$
$$Z_{12} = -1 - j9$$
$$V = 10$$

CHAPTER 6

Nodal analysis

6.1 Comparison with mesh analysis

The method of mesh analysis depends on two fundamental principles, firstly the recognition of independent loops in the network and the assumption that the branch currents can be transformed to loop currents. This step assumes Kirchhoff's First Law.

Secondly the equilibrium equations for the loops

$$[V'] = (Z')[I'] \tag{4.27}$$

are set up by applying Kirchhoff's Second Law. The loop currents and eventually the branch currents are then determined from the solution of these equations.

An alternative approach is the method of nodal analysis when attention is focused on the differences in potential between the nodes of the system. The p.d. between pairs of nodes (known as nodal pair voltages) are treated as the unknowns, and the branch currents are expressed in terms of these voltages and the branch admittances. The step involves Kirchhoff's Second Law.

A set of stability equations is now set up by equating to zero the currents entering each individual node and from the solution of these equations, the nodal pair voltages can be determined and from these the branch currents.

The two methods mesh and nodal analysis have many points of similarity and the steps are analogues of each other. The reason for this we shall see later in chapter 8, but first we enumerate the steps taken in carrying out the procedure of nodal analysis.

As we have done previously we must assume that the initial data is known in the form required for mesh analysis. There are however two differences.

(a) Admittances are required in place of impedances for the passive elements.

(b) The method refers to current sources rather than voltage sources. In power system work, for which nodal analysis is particularly applicable, practical sources are usually considered as constant voltage sources E with series impedance Z. It is always possible to replace such sources by equivalent constant current sources I_s with parallel admittance Y (Fig. 6.1) having the following values.

$$Y = 1/Z \qquad (6.1)$$
$$I_s = EY \qquad (6.2)$$

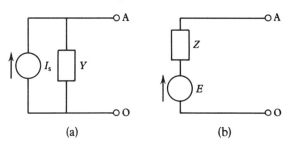

(a)   (b)

FIG. 6.1 (a) A current source with parallel admittance. (b) The corresponding voltage source with series impedance

6.2 The method of nodal analysis

1. The graph of the network is constructed and the nodes are identified. One node is marked as the reference node. For convenience this is usually one of the nodes normally at earth potential.
   The graph must show the sources as current sources with parallel admittances.
2. Directions are assigned to all branch currents.
3. The unconnected branch admittance matrix (Y_b] is compiled. If there is no mutual inductance, this can be done by arranging the individual branch admittances along the diagonal.
   If mutual inductance exists, the impedance matrix $[Z_b]$ must first be obtained as for mesh analysis and this must then be inverted.
   The matrix $[Y_b]$ determines the relationship between branch voltages and branch currents.
   Hence

$$[I_b] = [Y_b][V_b] \qquad (6.3)$$

4. The *incidence matrix* is now constructed. This is a rectangular matrix in which the number of columns is equal to the number of nodes (omitting the reference node) and the number of rows is equal to the number of branches. The elements of the matrix are either 1, 0 or -1. $+1$ is entered in the matrix to show a particular branch connected to a particular node if the assigned direction of the branch current is away from the node. -1 is entered for a branch where the assigned current is toward the node. The incidence matrix contains all passive branches including those corresponding to the equivalent parallel admittances of the generators but the current generators themselves are omitted.

The incidence matrix in nodal analysis has a similar function to the connection matrix of mesh analysis and has a similar appearance.

Since the incidence matrix shows the connections of the branches to the nodes we observe that

$$[\text{branch voltages}] = [A] \, [\text{nodal voltages}]$$
$$[V_b] = [A] \, [V_N] \tag{6.4}$$

The matrix $[V_b]$ gives the branch voltages (the potential differences across the branches) and $[V_N]$ the nodal voltages which are the potential differences between each node and the reference node.

The validity of Eq. (6.4) will be recognised if we take note of the line corresponding to the branch which is connected between nodes **P** and **Q**. This line shows that the branch voltage is equal to the difference between the nodal voltages V_P and V_Q; a statement which is obviously true.

The transpose of the incidence matrix $[A_t]$ is next determined.

Since $[A]$ transforms nodal voltages to branch voltages, we may infer from he principle of power invariance that $[A_t]$ transforms branch currents to odal currents.

$$\begin{bmatrix} \text{Total current} \\ \text{leaving individual} \\ \text{nodes} \end{bmatrix} = [A_t][I_b] \tag{6.5}$$

Inspection of any line of these equations will show the truth of the statement.

6. The column matrix of the source currents $[I_s]$ is next compiled. This is done by listing the nodes in order and observing the magnitude of the current *entering* the node from the source. Where the direction of the source current is away from the node, the value of I_s is negative. For non-generator nodes, zero is entered in the column.

7. The equilibrium equations for the system are obtained by applying Kirchhoff's First Law to the nodes by observing this law in the following form.

$$\begin{bmatrix} \text{Total} \\ \text{source current} \\ \text{entering a node} \end{bmatrix} = \begin{bmatrix} \text{sum of branch} \\ \text{currents leaving} \\ \text{the node} \end{bmatrix} \tag{6.6}$$

Hence
$$[I_s] = [A_t][I_b]$$
$$= [A_t][Y_b][V_b]$$
$$= [A_t][Y_b][A][V_N] \tag{6.8}$$

or
$$[I_s] = [Y_N][V_N] \tag{6.9}$$

where
$$[Y_N] = [A_t][Y_b][A] \tag{6.10}$$

8. To proceed to a solution, the matrix $[Y_b]$ must be available with numerical elements which may be simple or complex if the problem is steady-state d.c. or steady-state a.c. respectively.

9. The matrix $[Y_N]$ is evaluated from

$$[Y_N] = [A_t][Y_b][A] \qquad (6.10)$$

10. The nodal impedance matrix is obtained by inversion

$$[Z_N] = [Y_N]^{-1} \qquad (6.11)$$

11. The nodal voltages are obtained from

$$[V_N] = [Z_N][I_s] \qquad (6.12)$$

This result may be a sufficient solution to the problem in hand.

12. If the branch currents are required

$$[I_b] = [Y_b][V_b] \qquad (6.13)$$
$$= [Y_b][A][V_N] \qquad (6.14)$$

13. Actual generator currents will be obtained by selecting the currents in the generator admittances from the total matrix of branch currents. These values must then be subtracted from the source currents. In other words

$$[I_g] = [I_s] - [I_b]_g \qquad (6.15)$$

6.3 Examples of nodal analysis

6.3.1 Example

Nodal analysis applied to the problem solved by mesh analysis on page 73.

Step 1. The voltage generators are converted to current generators.

$$E = 10 \text{ and } Z = 1 \quad \text{becomes} \quad I_s = 1 \quad Y = 0.1$$
$$E = 20 \quad Z = 20 \quad \text{become} \quad I_s = 1 \quad Y = 0.05$$

Step 2. The graph of the network is given in Fig. 6.2.

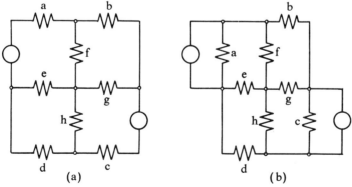

(a)   (b)

FIG. 6.2 (a) The network of Fig. 5.1. (b) The same network with current generators replacing the voltage generators

Branch	a	b	c	d	e	f	g	h
Admittance $Y =$	$\dfrac{1}{10}$	$\dfrac{1}{15}$	$\dfrac{1}{20}$	$\dfrac{1}{25}$	$\dfrac{1}{30}$	$\dfrac{1}{35}$	$\dfrac{1}{40}$	$\dfrac{1}{45}$

Step 3. Assigned directions of branch currents are given in Fig. 6.3.

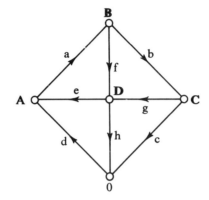

FIG. 6.3 The graph of the network of Fig. 6.2

$$[Y] = \begin{bmatrix} 1/10 & 0 & 0 & 0 & 0 & 0 & 0 & 0 \\ 0 & 1/15 & 0 & 0 & 0 & 0 & 0 & 0 \\ 0 & 0 & 1/20 & 0 & 0 & 0 & 0 & 0 \\ 0 & 0 & 0 & 1/25 & 0 & 0 & 0 & 0 \\ 0 & 0 & 0 & 0 & 1/30 & 0 & 0 & 0 \\ 0 & 0 & 0 & 0 & 0 & 1/35 & 0 & 0 \\ 0 & 0 & 0 & 0 & 0 & 0 & 1/40 & 0 \\ 0 & 0 & 0 & 0 & 0 & 0 & 0 & 1/45 \end{bmatrix} \quad (6.16)$$

Step 4

$$[A] = \begin{bmatrix} 1 & -1 & 0 & 0 \\ 0 & 1 & -1 & 0 \\ 0 & 0 & 1 & 0 \\ -1 & 0 & 0 & 0 \\ -1 & 0 & 0 & 1 \\ 0 & 1 & 0 & -1 \\ 0 & 0 & 1 & -1 \\ 0 & 0 & 0 & 1 \end{bmatrix} \quad (6.17)$$

Note that for each branch 1 and −1 can be entered once only.

Step 5

$$[A_t] = \begin{bmatrix} 1 & 0 & 0 & -1 & -1 & 0 & 0 & 0 \\ -1 & 1 & 0 & 0 & 0 & 1 & 0 & 0 \\ 0 & -1 & 1 & 0 & 0 & 0 & 1 & 0 \\ 0 & 0 & 0 & 0 & 1 & -1 & -1 & 1 \end{bmatrix} \quad (6.18)$$

Step 6

$$[I_s] = \begin{bmatrix} -1 \\ 1 \\ 1 \\ 0 \end{bmatrix}$$

Step 7

$$[Y_N] = [A_t][Y_b][A] \tag{6.10}$$

$$= \begin{bmatrix} 1 & 0 & 0 & -1 & -1 & 0 & 0 & 0 \\ -1 & 1 & 0 & 0 & 0 & 1 & 0 & 0 \\ 0 & -1 & 1 & 0 & 0 & 0 & 1 & 0 \\ 0 & 0 & 0 & 0 & 1 & -1 & -1 & 1 \end{bmatrix}$$

$$\begin{bmatrix} 1/10 & -1/10 & 0 & 0 \\ 0 & 1/15 & -1/15 & 0 \\ 0 & 0 & 1/20 & 0 \\ -1/25 & 0 & 0 & 0 \\ -1/30 & 0 & 0 & 1/30 \\ 0 & 1/35 & 0 & -1/35 \\ 0 & 0 & 1/40 & -1/40 \\ 0 & 0 & 0 & 1/45 \end{bmatrix}$$

$$[Y_N] =$$

$$\begin{bmatrix} 1/10+1/25+1/30 & -1/10 & 0 & -1/30 \\ -1/10 & 1/10+1/15+1/35 & -1/15 & -1/35 \\ 0 & -1/15 & 1/15+1/20+1/40 & -1/40 \\ -1/30 & -1/35 & -1/40 & 1/30+1/35+1/40 \\ & & & +1/45 \end{bmatrix}$$

Step 10. By inversion

$$[Z_N] = \begin{bmatrix} 13 \cdot 0070 & 9 \cdot 8952 & 6 \cdot 0599 & 7 \cdot 9521 \\ 9 \cdot 8952 & 14 \cdot 2700 & 8 \cdot 2409 & 8 \cdot 6465 \\ 6 \cdot 0599 & 8 \cdot 2409 & 12 \cdot 1350 & 6 \cdot 7886 \\ 7 \cdot 9521 & 8.6465 & 6 \cdot 7886 & 15 \cdot 4120 \end{bmatrix} \tag{6.19}$$

Step 11

$$[V_N] = [Z_N][I_s] = \begin{bmatrix} -13 \cdot 0070 + 9 \cdot 8952 + 6 \cdot 0599 \\ -9 \cdot 8952 + 14 \cdot 2700 + 8 \cdot 2409 \\ -6 \cdot 0599 + 8 \cdot 2409 + 12 \cdot 1350 \\ -7 \cdot 9521 + 8 \cdot 6465 + 6 \cdot 7886 \end{bmatrix} = \begin{bmatrix} 2.9481 \\ 12 \cdot 6157 \\ 14.3160 \\ 7 \cdot 4830 \end{bmatrix}$$

Step 12

$$[V_b] = [A][V_N] = \begin{bmatrix} 2.9481 - 12.6157 \\ 12.6157 - 14.3160 \\ 14.3160 \\ -2.9481 \\ -2.9481 + 7.4830 \\ 12.6157 - 7.4830 \\ 14.3160 - 7.4830 \\ 7.4830 \end{bmatrix} = \begin{bmatrix} -9.6676 \\ -1.7003 \\ 14.3160 \\ -2.9481 \\ 4.5349 \\ 5.1327 \\ 6.8330 \\ 7.4830 \end{bmatrix}$$

$$[I_b] = [Y][V_b] = \begin{bmatrix} -0.96676 \\ -0.11335 \\ 0.71580 \\ -0.11792 \\ 0.15116 \\ 0.14665 \\ 0.17082 \\ 0.16629 \end{bmatrix} \tag{6.20}$$

$$[I_g] = [I_b]_g - [I_s] = \begin{bmatrix} -0.96676 + 1 \\ 0.71580 - 1 \end{bmatrix} = \begin{bmatrix} 0.03324 \\ -0.28420 \end{bmatrix} \tag{6.21}$$

6.3.2 Example

Nodal analysis applied to load sharing in a power system

Figure 6.4 is the line diagram of a power network showing two power stations and a sub-station. The generator branches are labelled **a** and **b** and loads **c**, **d** and

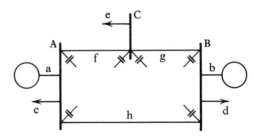

FIG. 6.4 A typical power-system network with two generating stations and a sub-station

e are connected at the busbars which are shown as nodes **A**, **B**, and **C**. Transmission lines are shown as branches **f**, **g** and **h**.

It is assumed that we are considering balanced three-phase conditions using line-to-neutral values and that the admittance of all branches including the generators is known. The admittance of a line is taken to be the reciprocal of its series impedance and the capacitance of the line is allowed for by increasing the admittance of the load at each end by half the shunt admittance of the line.

If the generator e.m.f.s. are known (in magnitude and phase) the busbar voltages can be obtained in the following manner.

The graph of the network is given in Fig. 6.5.

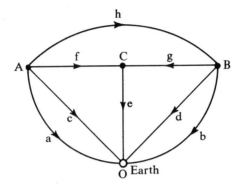

FIG. 6.5 The graph of the network of Fig. 6.4

Arrow heads show the directions considered to be positive and from this graph the connection or incidence matrix is drawn.

For this network

$$[A] = \begin{bmatrix} 1 & \cdot & \cdot \\ \cdot & 1 & \cdot \\ 1 & \cdot & \cdot \\ \cdot & 1 & \cdot \\ \cdot & \cdot & 1 \\ 1 & \cdot & -1 \\ \cdot & 1 & -1 \\ 1 & -1 & \cdot \end{bmatrix} \quad [Y_b] = \begin{bmatrix} Y_a & \cdot & \cdot & \cdot & \cdot & \cdot & \cdot & \cdot \\ \cdot & Y_b & \cdot & \cdot & \cdot & \cdot & \cdot & \cdot \\ \cdot & \cdot & Y_c & \cdot & \cdot & \cdot & \cdot & \cdot \\ \cdot & \cdot & \cdot & Y_d & \cdot & \cdot & \cdot & \cdot \\ \cdot & \cdot & \cdot & \cdot & Y_e & \cdot & \cdot & \cdot \\ \cdot & \cdot & \cdot & \cdot & \cdot & Y_f & \cdot & \cdot \\ \cdot & \cdot & \cdot & \cdot & \cdot & \cdot & Y_g & \cdot \\ \cdot & \cdot & \cdot & \cdot & \cdot & \cdot & \cdot & Y_h \end{bmatrix}$$

$$[Y_N] = [A_t][Y_b][A] = \begin{bmatrix} Y_a + Y_c + Y_f + Y_h & -Y_h & -Y_f \\ -Y_h & Y_b + Y_d + Y_g + Y_h & -Y_g \\ -Y_f & -Y_g & Y_e + Y_f + Y_g \end{bmatrix}$$

Numerical values must be inserted at this stage and the matrix inverted to give $[Z_N]$. This can be done as a 3×3 matrix in complex numbers or as a 6×6 matrix in simple numbers.

The source currents connected at **A** and **B** are

$$[I_s] = \begin{bmatrix} E_1 Y_1 \\ E_2 Y_2 \\ 0 \end{bmatrix} \tag{6.22}$$

and the nodal voltages (busbar voltages) can be found from

$$[V_N] = [Z_N][I_s] \tag{6.12}$$

Branch currents are given by

$$[I_b] = [Y_b][A][V_N] = \begin{bmatrix} Y_a V_A \\ Y_b V_B \\ Y_c V_A \\ Y_d V_B \\ Y_e V_C \\ Y_f(V_A - V_C) \\ Y_g(V_B - V_C) \\ Y_h(V_A - V_C) \end{bmatrix} \tag{6.23}$$

Generator currents are given by

$$[I_g] = [I_s] - [I_b]_g = \begin{bmatrix} E_a Y_a - I_a \\ E_b Y_b - I_b \end{bmatrix} \tag{6.24}$$

6.4 Alternative method of obtaining $[Y_N]$

We have seen that it is possible, with mesh analysis, to obtain the loop impedance matrix directly using Stigant's rule (Sec. 5.3). A similar method is available for compiling $[Y_N]$.

The previous example shows clearly that the matrix $[Y_N]$ consists of
(a) diagonal terms known as *nodal admittances*;
(b) non diagonal terms known as *nodal-pair admittances*.

The non-diagonal terms are in equal pairs symmetrical about the diagonal.

The nodal admittance for a particular node is the sum of the admittances of all branches (including generators) connected to that node.

The nodal-pair admittances between two nodes is the total admittance of all branches directly connected between these particular nodes *with a negative sign attached.*

6.5 Advantages and disadvantages of the nodal method

Mesh analysis is usually the simpler method to be applied in a general case but with power system networks nodal analysis is a powerful tool for the following reasons.

1. The number of nodes in power system network is often considerably less than the number of loops. The matrix equation is thus easier to invert.
2. Nodes are easily identified, whereas loops may be complicated especially if the map cannot be drawn without cross-overs.
3. Where lines or transformers are in parallel, admittances can be added without complication.
4. The method lends itself to network analyser techniques where potentiometer

measurements of nodal voltages are made on actual model networks representing the system.

The method has the disadvantages that:

1. Currents are not evaluated directly.
2. Computation often involves the multiplication of the difference between two quantities which are nearly alike, with consequent magnification of any errors.
3. The effect of mutual inductance between branches cannot be introduced without considerable complication.

6.6 Load flow studies in power networks

The previous example assumes that the generator e.m.f.s. are known. In the analysis of much more complicated networks, it is likely that the voltages of the generator busbars are known (or assumed) and the problem is to determine the voltages of the non-generator bars and the currents in all branches.

The incidence matrix of the network and its transpose $[A_t]$ are first determined. If the $[A_t]$ matrix is carefully assembled by first arranging the generator branches in order followed by the non-generator branches, and with the nodes arranged in the same sequence (this has been done in the previous example), it will be observed that the matrix can be partitioned in the following way.

$$[A_t] = \begin{array}{c} \text{Generator Nodes} \\ \\ \text{Non-Generator Nodes} \end{array} \begin{array}{cc} \overset{\text{Generator}}{\underset{\text{Branches}}{}} & \overset{\text{Non-Generator}}{\underset{\text{Branches}}{}} \\ \begin{bmatrix} [1] & [a_{12}] \\ [0] & [a_{22}] \end{bmatrix} \end{array} \qquad (6.25)$$

$[A_t]$ is seen to consist of four sub-matrices. The unit matrix refers to the connection of the generator branches to their respective nodes, and the zero shows that the generator branches are not connected to the non-generator nodes.

The sub-matrix $[a_{12}]$ shows the non-generator branch connections to the generator nodes and $[a_{22}]$ the connection of non-generator branches to non-generator nodes.

The transpose of this matrix is

$$[A] = \begin{bmatrix} [1] & [0] \\ [a_{12t}] & [a_{22t}] \end{bmatrix} \qquad (6.26)$$

If $[I_g]$ and $[I_b]$ are the generator and the non-generator branch currents, then we have

$$0 = \begin{bmatrix} [1] & [a_{12}] \\ [0] & [a_{22}] \end{bmatrix} \cdot \begin{bmatrix} [I_g] \\ [I_b] \end{bmatrix} \qquad (6.27)$$

in accordance with Kirchhoff's current law at the nodes.

From the second row of equation (6.27)

$$[a_{22}][I_b] = 0 \qquad (6.28)$$

The incidence matrix $[A]$ is also the transform from branch to nodal voltages, hence

$$\begin{bmatrix} [V_{gn}] \\ [V_b] \end{bmatrix} = \begin{bmatrix} [1] & [0] \\ [a_{12t}] & [a_{22t}] \end{bmatrix} \cdot \begin{bmatrix} [V_{gn}] \\ [V_{ngn}] \end{bmatrix} \qquad (6.29)$$

where $[V_{gn}]$ is the voltage matrix of the generator nodes

$[V_{ngn}]$ is the voltage matrix of the non-generator nodes

$[V_b]$ is the voltage matrix of the non-generator branches.

From the bottom row of equation (6.29)

$$[V_b] = [a_{12t}][V_{gn}] + [a_{22t}][V_{ngn}] \qquad (6.30)$$

The relationship between branch currents and branch voltages is

$$[I_b] = [Y_b][V_b] \qquad (6.31)$$

where $[Y_b]$ is the admittance matrix of the non-generator branches. This matrix is usually diagonal but could include mutual coupling between lines.

Substituting Eq. (6.31) and (6.30) in (6.28)

$$0 = [a_{22}][Y_b][V_b]$$
$$= [a_{22}][Y_b][a_{12t}][V_{gn}] + [a_{22}][Y_b][a_{22t}][V_{ngn}]$$

Thus

$$[V_{ngn}] = -[[a_{22}][Y_b][a_{22t}]]^{-1}[a_{22}][Y_b][a_{12t}][V_{gn}] \qquad (6.32)$$
$$= -[\beta][V_{gn}] \qquad (6.33)$$

where $\qquad (6.34)$

$$[\beta] = [[a_{22}][Y_b][a_{22t}]]^{-1}[a_{22}][Y_b][a_{12t}]$$

Equation (6.33) thus enables the voltages of the non-generator bus-bars to be found from the matrix of generator busbar voltages without any further inversion.

Although matrix $[\beta]$ is complicated, its evaluation need be done once only for a particular system and is then available for use with any combination of generator voltages.

Computer programmes are available so that expressions such as the right-hand side of Eq. (6.34) can be evaluated by routine.

To obtain the branch voltages, we see from the second row of equation (6.29)

$$[V_b] = [a_{12t}][V_{gn}] + [a_{22t}][V_{ngn}] \qquad (6.35)$$

and by substitution of equation (6.33)

$$[V_b] = [[a_{12t}] - [a_{22t}][\beta]][V_{gn}] \qquad (6.36)$$

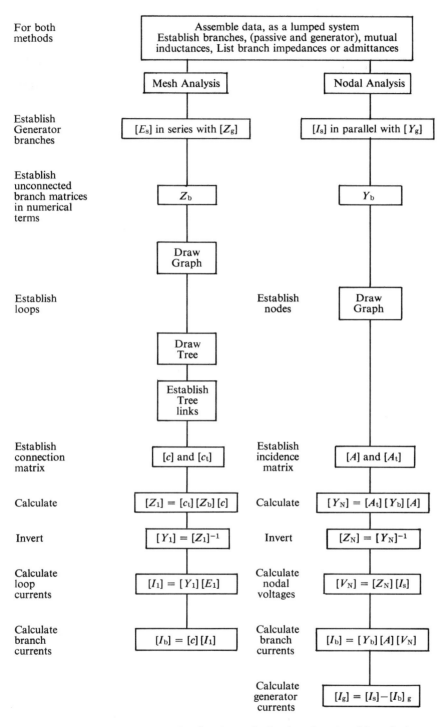

FIG. 6.6 The flow chart showing the methods of mesh and nodal analysis

and the non-generator branch currents are given by

$$[I_b] = [Y_b][[a_{12t}] - [a_{22t}][\beta]][V_{gn}] \tag{6.37}$$

The generator currents can then be obtained by substitution in the first row of Eq. (6.27)

$$[I_g] = -[a_{12}][I_b] \tag{6.38}$$

6.7 Mesh and nodal analysis compared

The methods of mesh and nodal analysis are compared in Fig. (6.6) (see p. 91) giving flow charts showing the steps of both methods. These charts are of assistance in memorising the steps and the comparison is useful when the principle of duality is studied in Chap. 8.

CHAPTER 7

The Laplace
transform

7.1 Exponential stimulus

It is to be expected that exponential functions will play a vital part in the solution of problems involving differential equations since they are one of the few functions which are substantially unchanged after repeated differentiation.

Only this type of function can be added to its derivatives to build up a differential equation that is capable of analytic solution.

We begin, therefore, by discussing the application of an exponential stimulus to the familiar *L.R.C.* circuit (Fig. 7.1).

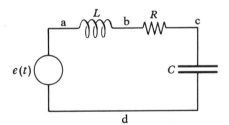

FIG. 7.1 A series *L.R.C.* circuit

Let the current in this circuit be represented by

* $$i(t) = \mathrm{Re}\, I \exp(st) \tag{7.1}$$

* According to Euler's formula
$$\exp(jx) = \cos x + j \sin x$$
$$= a + jb$$

where a is known as the real part of the exponential and b its imaginary part, the symbols Re and Im stand for 'the real part of' and 'the imaginary part of' respectively. Thus

$$\cos x = \mathrm{Re}\, \exp(jx)$$
$$\sin x = \mathrm{Im}\, \exp(jx)$$

where s is a complex number with components

$$s = \alpha + j\omega \tag{7.2}$$

Expanding Eq. (7.1)

$$i(t) = I \exp(\alpha t) \operatorname{Re} \exp(j\omega t) \tag{7.3}$$

$$= I \exp(\alpha t) \cos \omega t \tag{7.4}$$

The interpretation of this expression is that $i(t)$ is an alternating quantity of angular frequency ω, but whose amplitude increases with $\exp(\alpha t)$. If α is negative (which is the more common condition) then the amplitude decreases exponentially. Such a function is sketched in Fig. 7.2.

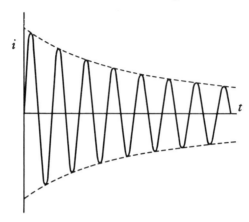

FIG. 7.2 An alternating quantity whose amplitude is decreasing exponentially

Now if this current flows in the circuit of Fig. 7.1

$$v_{\text{bc}} = Ri = R \operatorname{Re} I \exp(st) \tag{7.5}$$

$$v_{\text{ab}} = L \, di/dt = L \, d(\operatorname{Re} I \exp(st))/dt \tag{7.6}$$

$$= L \operatorname{Re} Is \exp(st) \tag{7.7}$$

and

$$v_{\text{cd}} = (1/C) \int i \, dt = (1/C) \operatorname{Re}(I/s) \exp(st) \tag{7.8}$$

The applied voltage v_{ad} is given by

$$v_{\text{ad}} = v_{\text{ab}} + v_{\text{bc}} + v_{\text{cd}}$$

$$= \operatorname{Re}(Ls + R + 1/Cs) I \exp(st) \tag{7.9}$$

This equation should be compared with the original differential equation (Eq. (1.47)) repeated below

$$v(t) = (Lp + R + 1/Cp)i(t) \tag{1.47}$$

The similarity in form is obvious. Equation (1.47) is a differential equation, but

Eq. (7.9) is a much simpler one; it is only algebraic although it involves complex numbers.

It is, therefore, possible to suggest that provided the current $i(t)$ is a function described by

$$i(t) = \text{Re } I \exp (st) \tag{7.1}$$

it is permissible to replace p in the original differential equation by s thus converting equation (1.47) to equation (7.9) from which the applied voltage can be found without recourse to differentiation and integration.

The same procedure is permissible in the case where the applied voltage is exponential and it is required to determine the current.

Now we have

$$v(t) = \text{Re } V \exp (st) \tag{7.10}$$

and since

$$i(t) = \frac{1}{Lp+R+(1/Cp)} v(t) \tag{1.48}$$

and replacing p by s

$$i(t) = \text{Re } \frac{V}{Ls+R+(1/Cs)} \exp (st) \tag{7.11}$$

enabling $i(t)$ to be determined by complex algebra.

It must be emphasised that this method of solution of the differential equation applies only when the stimulus is exponential.

Example:

A voltage $v = \text{Re } 100 \exp (-2+j100)t$ is applied to a coil of resistance 50 ohms and inductance of 0·5 henries. Find an expression for the current:

$$i(t) = \text{Re} \frac{100 \exp (-2+j100)t}{50+(-2+j100)0·5}$$

$$= \text{Re} \frac{100 \exp (-2+j100)t}{49+j50}$$

$$= \text{Re} \frac{100(49-j50)}{49^2 + 50^2} \exp (j100t) \exp (-2t)$$

$$= \text{Re } (0·9998 - j1·0202) (\cos 100t + j \sin 100t) \exp (-2t)$$

$$= (0·9998 \cos 100t + 1·0202 \sin 100t) \exp (-2t)$$

$$= \sqrt{(0·9998^2 + 1·0202^2)} \cos \left(100t - \tan^{-1} \frac{1·0202}{0·9998}\right) \exp (-2t)$$

In this simple example it is assumed that the voltage and current have been in existence for some considerable time before $t = 0$. It must not be assumed that this voltage is suddenly applied at $t = 0$ since the value of the current obtained by putting $t = 0$ in the solution gives

$$i(0) = 0·9998$$

If the voltage were suddenly applied at $t = 0$ this current could not be created instantaneously and a transient condition would intervene. The case of suddenly applied alternating voltage is considered in Sec. 7.8.7.

7.1.1 Steady-state alternating current

Sinusoidal excitation is a special case of exponential excitation where the decrement α is zero. Thus for sinusoidal excitation

$$s = j\omega \tag{7.12}$$

and Eq. (7.10) becomes

$$v(t) = \mathrm{Re}\, V \exp j\omega t \tag{7.13}$$

The geometrical interpretation of Eq. (7.13) is that $V \exp j\omega t$ is represented by a line of length V drawn from the origin and rotating with angular velocity ω.

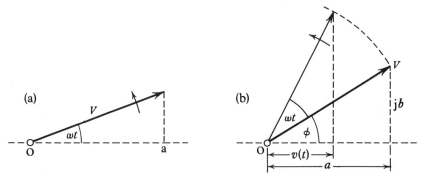

FIG. 7.3 (a) The geometrical interpretation of $v(t) = \mathrm{Re}\, V \exp j\omega t$. (b) The geometrical interpretation of $v(t) = \mathrm{Re}\, V \exp j\phi \exp j\omega t$

At the instant t it is inclined to the horizontal by an angle θ where $\theta = \omega t$ as shown in Fig. 7.3.

The projection of this line on the horizontal* is known as its real part (**oa** in Fig. 7.3). Thus

$$v(t) = \mathrm{Re}\, V \exp j\omega t$$
$$= V \cos \omega t \tag{7.14}$$

7.1.2 Phasors

A more general case arises when

$$v(t) = \mathrm{Re}\, V \exp j(\omega t + \phi)$$
$$= \mathrm{Re}\, V \exp j\phi \exp j\omega t$$
$$= \mathrm{Re}\, V \exp j\omega t \tag{7.15}$$

* The reader will probably be familiar with the use of rotating lines (vectors) for representing alternating qualities. In an elementary treatment, the instantaneous value is usually obtained by projecting on the vertical axis. With complex number representation, it is conventional to project on the horizontal axis. The term *phasor* is used instead of *rotating vector* since a vector quantity has a different meaning in field theory.

where
$$V = V \exp j\phi$$
$$= V(\cos \phi + j \sin \phi) = a + jb \qquad (7.16)$$

The complex number V is known as a *phasor* and this number is sufficient to represent the sinusoidal alternating voltage since the instantaneous value can always be obtained by multiplying by $\exp j\omega t$ and taking the real part (Eq. 7.15)).

The geometrical interpretation of Eqs. (7.15) and (7.16) is shown in Fig. 7.3(*b*).

7.1.3 Phasor diagrams

If the equilibrium equations of a system are
$$[v(t)] = [Z(\mathrm{p})][i(t)] \qquad (7.17)$$

then for steady-state response to sinusoidal excitation, p is replaced by $j\omega$ in the operational impedance matrix $[Z(\mathrm{p})]$ which then becomes a matrix of complex numbers $[Z]$

Equation (7.17) becomes
$$[v(t)] = \mathrm{Re}\,[V] \exp j\omega t = \mathrm{Re}\,[Z]\,[I] \exp j\omega t \qquad (7.18)$$

we have thus transformed the differential equations (7.17) into the phasor equations
$$[V] = [Z][I] \qquad (7.19)$$

Each equation can be given geometrical interpretation by familiar phasor or 'vector' diagrams. The impedance matrix $[Z]$ being composed of complex numbers is now capable of inversion, giving
$$[I] = [Z]^{-1}[V] \qquad (7.20)$$

thus determining the unknown currents.

It is usually sufficient to retain the solution in phasor notation but if instantaneous values of current are required
$$[i(t)] = \mathrm{Re}\,[I] \exp j\omega t \qquad (7.21)$$

The transform to phasors is suitable for the majority of problems involving the determination of steady-state responses to single frequency sinusoidal stimuli but for other types of stimuli, particularly step and ramp functions, another type of transform will be investigated in the next section.

7.2 The transform concept

7.2.1 Comparison with logarithms

The use of a logarithmic transform to facilitate arithmetic when multiplying numbers together is well known. Instead of operating directly on the numbers, they are first converted to logarithms. Addition and subtraction of these logarithms then follows and the resulting logarithm is re-converted to an ordinary number.

The additional time involved in this double transformation is justified since the process of adding and subtracting logarithms is much simpler than multiplication and division of the original numbers. The scheme of this process is shown in the diagram Fig. 7.4.

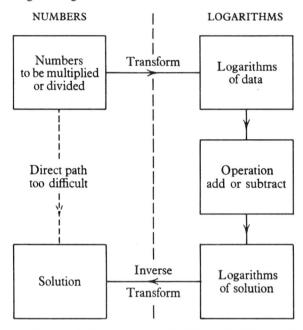

FIG. 7.4 The transform method used when multiplying or dividing numbers with the aid of logarithms

A similar device can be applied to reduce the labour in solving differential equations. Figure 7.5 shows the logical steps we shall try to take to solve linear differential equations by means of Laplace transforms.

The problem begins with a given circuit and a stimulus which is a function of time. If by a mathematical device this particular function of time can be transformed to a special function of the complex number s, then the circuit equations are reduced to algebraic equations. These must then be solved and the response so obtained will be a function of (s). Inverse transformation then yields the response as a function of time.

7.3 The Laplace transform

If $f(t)$ is a function of time its Laplace transform $L[f(t)]$ is defined by

$$\int_0^\infty f(t) \exp(-st)\, dt \tag{7.22}$$

and is written as $F(s)$, where s is a complex number as in Sec. 7.1. This means that the original time function is multiplied by the exponential decay factor $\exp(-st)$ and the resulting function integrated over the range from 0 to infinity, it being inferred that the integral is convergent. This integration eliminates the time variable, and, therefore, the resulting transform is solely a function of (s). We will now investigate a few simple functions and their transforms.

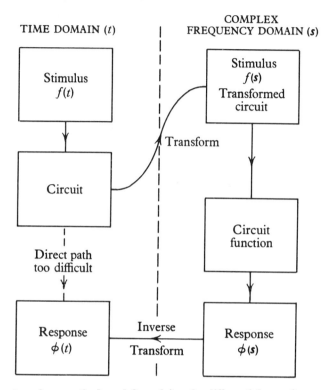

FIG. 7.5 The transform method used for solving the differential equations encountered in electrical circuits

7.3.1 The exponential function

Let

$$f(t) = \exp(-at) \tag{7.23}$$

then

$$F(s) = \int_0^\infty \exp(-at)\exp(-st)\,dt$$

$$= \int_0^\infty \exp(-(s+a)t)\,dt$$

$$= (-1/(s+a))[\exp(-(s+a)t)]_0^\infty$$

$$= 1/(s+a) \tag{7.24}$$

Note that the transform is a simpler function than the original.

7.3.2 The singularity function or step function

Let

$$f(t) = 0]^0_{-\infty} + V]^\infty_0 \qquad (7.25)$$

This is a very important function, sometimes known as the 'step function' and shown in Fig. 7.6. It is the effect of the connection of a constant potential source at the time t is equal to zero.

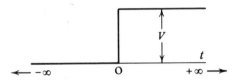

FIG. 7.6 A step-function voltage

Thus

$$\begin{aligned}
F(s) &= \int_0^\infty V \exp(-st)\, dt \\
&= (-1/s)\, V\, [\exp(-st)]^\infty_0 \\
&= V/s \qquad (7.26)
\end{aligned}$$

again a very simple function of s.

7.3.3 Trigonometrical functions

Let

$$\begin{aligned}
f(t) &= \sin at \qquad (7.27) \\
&= (1/j2)\, (\exp(jat) - \exp(-jat))
\end{aligned}$$

Hence

$$\begin{aligned}
F(s) &= (1/j2)\, (1/(s-ja) - 1/(s+ja)) \\
&= a/(s^2+a^2) \qquad (7.28)
\end{aligned}$$

Similarly if

$$f(t) = \cos at \qquad (7.29)$$

$$F(s) = s/(s^2+a^2) \qquad (7.30)$$

7.3.4 The ramp function

Let

$$f(t) = 0]^0_{-\infty} + at]^\infty_0 \qquad (7.31)$$

which is known as a 'ramp function' and shown in Fig. 7.7.

$$F(s) = \int_0^\infty at \exp(-st)\, dt$$

$$= [(-at/s) \exp(-st)]_0^\infty - \int_0^\infty (-a/s) \exp(-st) \, dt$$

$$= 0 - [(a/s^2) \exp(-st)]_0^\infty$$

$$= a/s^2. \tag{7.32}$$

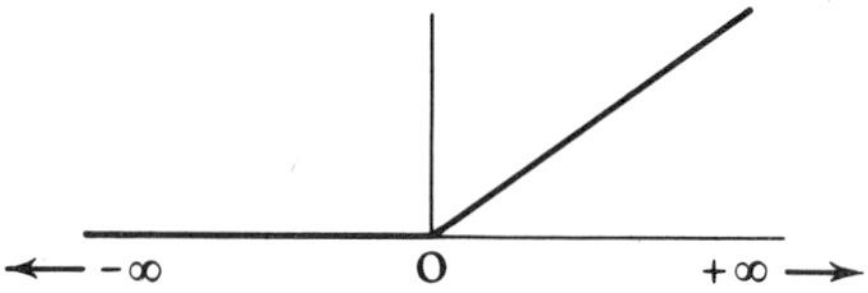

FIG. 7.7 A ramp function

7.4 Transform pairs

The Laplace transforms of other functions in common use are given in the following table* (see pp. 102 and 103). In each case it will be observed that the transform is a simpler function than the original.

Much of the algebraic work associated with Laplace transforms consists of reducing expressions into such forms that the inverse transform is immediately recognisable.

7.5 Differentiation and integration

We now investigate the relationship between Laplace transforms of (a) a function (b) its differential coefficient (c) its integral.

7.5.1 Differentiation

Let

$$y = f(t) \tag{7.33}$$

whose transform

$$F(s) = \int_0^\infty y \exp(-st) \, dt \tag{7.34}$$

The Laplace transform of the differential coefficient

$$L[dy/dt] = \int_0^\infty (dy/dt) \exp(-st) \, dt$$

$$= [y \exp(-st)]_0^\infty - \int_0^\infty y(-s) \exp(-st) \, dt$$

$$= [0 - y_0] + sF(s) \tag{7.35}$$

[continued on p. 104]

* This table is issued to candidates by the Institution of Electrical Engineers for use in their Part III Examination.

The author is indebted to the Examinations Committee of the Institution for permission to reproduce the table in this precise form.

TABLE OF LAPLACE TRANSFORMS

$\mathrm{L}[f(t)]$ is defined by $\int_0^\infty f(t)\exp(-st)\,dt$ and is written as $F(s)$.

	$f(t)$ from $t = 0$	$F(s) = \mathrm{L}[f(t)]$
1	$\dfrac{d}{dt}f(t)$	$sF(s)-f(0)$
2	$\dfrac{d^n}{dt^n}f(t)$	$s^nF(s)-s^{n-1}f(0)$ $-s^{n-2}f'(0)\ldots-f^{n-1}(0)$
3	$\displaystyle\int_0^t f(t)\,dt$	$\dfrac{1}{s}F(s)$
4	$\exp(-\alpha t)\,f(t)$	$F(s+\alpha)$
5	Unit impulse δ	1
6	Unit function or Step function	$\dfrac{1}{s}$
7	Delayed unit function	$\dfrac{\exp(-sT)}{s}$
8	Rectangular pulse	$\dfrac{1-\exp(-sT)}{s}$
9	Ramp function t	$\dfrac{1}{s^2}$
10	$t^{n-1}/(n-1)!$	$\dfrac{1}{s^n}$
11	$\dfrac{t^{n-1}}{(n-1)!}\exp(-\alpha t)$	$\dfrac{1}{(s+\alpha)^n}$
12	$\exp(-\alpha t)$	$\dfrac{1}{(s+\alpha)}$
13	$1-\exp(-\alpha t)$	$\dfrac{\alpha}{s(s+\alpha)}$
14	$t\exp(-\alpha t)$	$\dfrac{1}{(s+\alpha)^2}$
15	$\exp(-\alpha t)-\exp(-\beta t)$	$\dfrac{\beta-\alpha}{(s+\alpha)(s+\beta)}$
16	$\sin\omega t$	$\dfrac{\omega}{s^2+\omega^2}$

TABLE OF LAPLACE TRANSFORMS—contd.

$L[f(t)]$ is defined by $\int_0^\infty f(t)\exp(-st)$, d$t$ and is written as $F(s)$

	$f(t)$ from $t = 0$	$F(s) = L[f(t)]$
17	$\cos \omega t$	$\dfrac{s}{s^2+\omega^2}$
18	$1-\cos \omega t$	$\dfrac{\omega^2}{s(s^2+\omega^2)}$
19	$\omega t \sin \omega t$	$\dfrac{2\omega^2 s}{(s^2+\omega^2)^2}$
20	$\sin \omega t - \omega t \cos \omega t$	$\dfrac{2\omega^2}{(s^2+\omega^2)^2}$
21	$\exp(-\alpha t)\sin \omega t$	$\dfrac{\omega}{(s+\alpha)^2+\omega^2}$
22	$\exp(-\alpha t)\cos \omega t$	$\dfrac{s+\alpha}{(s+\alpha)^2+\omega^2}$
23	$\exp(-\alpha t)\left(\cos \omega t - \dfrac{\alpha}{\omega}\sin \omega t\right)$	$\dfrac{s}{(s+\alpha)^2+\omega^2}$
24	$\sin(\omega t + \phi)$	$\dfrac{s \sin \phi + \omega \cos \phi}{s^2+\omega^2}$
25	$\exp(-\alpha t)+(\alpha/\omega)\sin \omega t - \cos \omega t$	$\dfrac{\alpha^2+\omega^2}{(s+\alpha)(s^2+\omega^2)}$
26	$\sinh \beta t$	$\dfrac{\beta}{s^2-\beta^2}$
27	$\cosh \beta t$	$\dfrac{s}{s^2-\beta^2}$

Rectangular periodic wave

28	$f(t) = 1, 0 < t < \tfrac{1}{2}a; f(t) = 0, \tfrac{1}{2}a < t < a$	$\dfrac{1}{s}\tanh as$

Half-wave rectified sine

29	$\begin{aligned} f(t) &= \sin \omega t, & 0 < t < \pi/\omega \\ f(t) &= 0, & \pi/\omega < t < 2\pi/\omega \end{aligned}$	$\dfrac{\omega}{s^2+\omega^2}\tfrac{1}{2}\exp(\pi s/2\omega)\operatorname{cosech}(\pi s/2\omega)$

Full-wave rectified sine

30	$f(t) = |\sin \omega t|$	$\dfrac{\omega}{s^2+\omega^2}\coth(\pi s/2\omega)$

That is to say, the Laplace transform of the differential coefficient of a function is obtained from the transform of the function itself by multiplying by s and subtracting the zero value of the function.

7.5.2 Integration

Let

$$y = f(t) \tag{7.36}$$

whose transform $= F(s)$

$$L\left[\int_{-\infty}^{t} f(t)\,dt\right] = \int_{0}^{\infty}\left[\int_{-\infty}^{t} f(t)\,dt\right]\exp\left(-st\right)dt \tag{7.37}$$

$$= \int_{0}^{\infty} u\,dv$$

$$= uv - \int_{0}^{\infty} v\,du \tag{7.38}$$

where

$$u = \int_{-\infty}^{t} f(t)\,dt$$

$$du = f(t)\,dt$$

$$dv = \exp\left(-st\right)dt$$

$$v = (-1/s)\exp\left(-st\right)$$

Substituting in Eq. (7.38)

$$L\left[\int_{-\infty}^{t} f(t)\,dt\right] = \left[\left(\int_{-\infty}^{t} f(t)\,dt\right)(-1/s)\exp\left(-st\right)\right]_{0}^{\infty}$$

$$- (-1/s)\int_{0}^{\infty}\exp\left(-st\right)f(t)\,dt$$

$$= 0 + (1/s)\int_{-\infty}^{0} f(t)\,dt + (1/s)\,F(s) \tag{7.39}$$

In other words, if we integrate a function of time from $-\infty$ to t, the Laplace transform of the integral is obtained by adding the Laplace transform of the function to the value of the integral from $-\infty$ to 0 and dividing the total by s.

The value of the integral from minus infinity to zero is an initial condition for the function which is usually readily obtained before a particular problem is expressed. Obviously these conditions at time zero must be known in order to determine subsequent behaviour.

It should be noted that the relationships between a function and its differential coefficient or its integral are usually much more complicated than the corresponding relationships between the Laplace transforms of these items.

Before attempting the solution of differential equations in this way, one further theorem is introduced.

7.5.3 The shifting theorem

Let

$$y = f(t) \tag{7.40}$$

whose transform

$$L[f(t)] = F(s) \tag{7.41}$$

Then

$$L\left[\exp\left(-at\right)f(t)\right] = \int_0^\infty \exp\left(-at\right)f(t)\exp\left(-st\right)\mathrm{d}t$$

$$= \int_0^\infty f(t)\exp\left(-(s+a)t\right)\mathrm{d}t$$

$$= F(s+a) \tag{7.42}$$

This means that if the transform of a function is known, then to find the transform of the product of that function and $\exp\left(-at\right)$, s in the original transform must be replaced by $(s+a)$.

7.6 The transform circuit

We are now in a position to attempt the solution of the circuit discussed in Sec. 7.1, with any applied stimulus $e(t)$, provided that the function $e(t)$ has a Laplace transform. This includes step, ramp and impulse function inputs, together with suddenly applied sinusoidal functions, all of which are of practical importance.

Still considering the LRC circuit by way of example and commencing with the equation (1.32)

$$e(t) = L\left(\mathrm{d}i/\mathrm{d}t\right) + Ri + (1/C)\int_{-\infty}^t i\,\mathrm{d}t \tag{1.32}$$

The Laplace transform of each term is first obtained using, where possible, the table of transform pairs

$$e(t) \quad \text{becomes} \quad L[e(t)] \quad \text{or} \quad e(s)$$
$$L\,\mathrm{d}i/\mathrm{d}t \quad \text{becomes} \quad Lsi(s) - LI_0 \quad \text{from Eq. (7.35)}$$
$$Ri(t) \quad \text{becomes} \quad Ri(s)$$

$$\text{and } (1/C)\int_{-\infty}^t i(t)\,\mathrm{d}t \quad \text{becomes} \quad (1/Cs)\int_{-\infty}^0 i(t)\,\mathrm{d}t + (1/Cs)i(s)$$

$$\text{or} \quad Q_0/Cs + (1/Cs)i(s)$$

$$\text{or} \quad V_0/s + (1/Cs)i(s)$$

where $Q_0 = \int_\infty^0 i(t)\,\mathrm{d}t$ the charge on the capacitor at time zero and $V_0 = Q_0/C$

the capacitor voltage at time zero. Equation (1.32) is thus transformed to

$$e(s) + LI_0 - Q_0/Cs = (Ls + R + 1/Cs)i(s) \qquad (7.43)$$

which should be compared with the original differential equation in operational nomenclature

$$e(t) = (Lp + R + 1/Cp)i(t) \qquad (1.47)$$

The transformed Eq. (7.43) can be written down directly from the original four steps

1. replace p by s
2. replace the function $e(t)$ by its transform $e(s)$ using the table of transform pairs
3. add the terms LI_0 and $-Q_0/Cs$ on the left hand side of the equation
4. write $i(s)$ in place of $i(t)$.

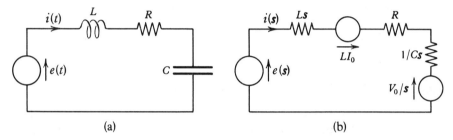

(a)   (b)

FIG. 7.8 (a) A simple $L.R.C.$ circuit. (b) The corresponding transform circuit

Although Eq. (7.43) is purely algebraic, it is not capable of solution in the normal way owing to the presence of the complex number s. We can think of it as arising from an equivalent circuit shown in Fig. 7.8(b). This circuit is very similar to the original circuit (7.8(a)) and can be termed its *transform circuit*. The transform circuit can be derived from the original in the following routine manner:

1. Source $e(t)$ is replaced by its transform $e(s)$
2. R remains R
3. L becomes a 'resistor' Ls
4. C becomes a 'resistor' $1/Cs$
5. Z becomes $(Ls + R + 1/Cs)$
6. additional sources $+LI_0$ and $-V_0/s$ are added if the stores have internal energy at time zero.

The expression for current in this circuit is

$$i(s) = \frac{v(s)}{Z(s)} = \frac{e(s) + LI_0 - V_0/s}{Ls + R + 1/Cs} \qquad (7.44)$$

which is seen to be derived from Eq. (7.43).

Although this expression can not be evaluated directly the right hand side can usually be converted* by straightforward algebraic manipulation to a recognisable form consisting of a series of terms taken from the second column in the table of transform pairs.

When $i(s)$ is expressed in this way it is a simple matter to replace each term by its *inverse transform* taken from the first column of the table of pairs and so the corresponding expression for $i(t)$ is obtained.

Subsequent examples using the table of transform pairs will illustrate this procedure.

7.6.1 Mutual inductance

We have seen that when mutual inductance exists between loops 1 and 2 in a network the voltage drop in loop 2 due to current in loop 1 is given by Eq. 4.43.

$$v_2 = M_{21} \cdot \mathrm{d}i_1/\mathrm{d}t \tag{4.42}$$

or
$$v_2(t) = M_{21}\mathrm{p} \cdot i_1(t) \tag{7.45}$$

This equation in the time domain relates to an original circuit and consequently by analogy with self inductance we can set up a circuit transformed to the s domain. The transformed voltage in the second loop is given by

$$v_2(s) = M_{21}s \cdot i_1(s) - M_{21}(I_1)_0 \tag{7.46}$$

The transform circuit is thus obtained by replacing mutual impedance $M\mathrm{p}$ in the original by Ms and inserting an equivalent source $M_{21}(I_1)_0$ if initial current exists.

7.6.2 Application to networks

The rules developed in the previous section for single circuits can also be applied to complete networks. Transform networks can be set by replacing p by s in the original network and also in the impedance matrix. Appropriate sources $[MI_0]$, $[LI_0]$ and $[-V_0/s]$ must be added to the transformed loop voltages $[V(s)]$ and the loop equations

$$[V_1(t)] = [Z(t)] \cdot [I_1(t)] \tag{5.3}$$

become
$$[V_1(s)] + [MI_0] + [LI_0] + [-V_0/s] = [Z(s)] \cdot [I_1(s)] \tag{7.47}$$

These can now be solved for $[I_1(s)]$ and subsequently retransformed to establish the loop currents $[I_1(t)]$ from which the branch currents can be obtained if required.

Nodal analysis can be applied in a similar manner.

---

* This step may present some difficulty and a generalised method for attacking the problem is given later in the chapter.

7.7 Partial fractions

Expressions such as Eq. (7.44) arising from manipulation of transformed circuits are generally of the form

$$i(s) = \frac{M(s)}{N(s)} \tag{7.48}$$

where M and N are both polynomials in s, N usually being of a higher order than M. If these can be reduced to a series of partial fractions, for example, if

$$i(s) = \frac{M(s)}{N(s)} = \frac{A}{(s+a_1)} + \frac{B}{(s+a_2)} + \frac{C}{(s+a_3)} \tag{7.49}$$

then the inverse transforms of the terms on the right-hand side are immediately recognised from line 12 of the transform table and

$$i(t) = A \exp(-a_1 t) + B \exp(-a_2 t) + C \exp(-a_3 t) \tag{7.50}$$

The problem of reducing $M(s)/N(s)$ to partial fractions consists of two parts, first finding the factors of the denominator $N(s)$ and secondly determining the coefficients A, B, C and so on.

Happily, in many of the simpler problems arising from electrical circuits $N(s)$ is already built up as the product of a number of factors. Some of these may be binomial expressions such as $as^2 + bs + c$ which factorises to $a(s+a_1)$ $(s+a_2)$ where a_1 and a_2 are conjugate complex numbers.

If the factors of $N(s)$ are not obvious, they must be obtained by determining the roots of the equation

$$N(s) = 0 \tag{7.51}$$

and for appropriate methods the reader is referred to mathematical textbooks.

To determine the coefficients A, B, C, etc. three methods are available.

7.7.1 Equating coefficients

This method is suitable if $N(s)$ has only a small number of factors. For example Let

$$\frac{M(s)}{N(s)} = \frac{s+3}{(s+1)(s+2)} = \frac{A}{(s+1)} + \frac{B}{s+2}$$

The coefficients A and B can be found by combining the factors on the right-hand side and establishing the identity of the numerators.
Thus

$$s+3 \equiv A(s+2) + B(s+1)$$

The coefficients of s and s^0 can, therefore, be equated, giving

$$\text{for } s^1 \quad 1 = A + B$$
$$\text{for } s^0 \quad 3 = 2A + B$$

Thus

$$A = 2 \quad \text{and} \quad B = -1$$

If

$$i(s) = \frac{M(s)}{N(s)} = \frac{2}{s+1} - \frac{1}{s+2}$$

the inverse transform comes from line 12 of the table

$$i(t) = 2 \exp(-t) - \exp(-2t)$$

A more difficult problem is one where one of the factors is a binomial expression, as in the next example
Let

$$\frac{M(s)}{N(s)} = \frac{s^2+2}{(s+1)(s^2+s+2)} = \frac{A}{s+1} + \frac{Bs+C}{s^2+s+2}$$

The coefficients, A, B and C, can be found by combining the fractions on the right-hand side and establishing the identity

$$s^2+2 \equiv A(s^2+s+2) + (Bs+C)(s+1)$$

The coefficients of s^0, s and s^2 can therefore be equated, giving

$$\begin{array}{lc} \text{for } s^2 & 1 = A+B \\ \text{for } s^1 & 0 = A+B+C \\ \text{for } s^0 & 2 = 2A+C \end{array}$$

giving

$$C = -1$$
$$A = 3/2$$
$$B = -1/2$$

$$i(s) = \frac{M(s)}{N(s)} = \frac{1\cdot5}{s+1} - \frac{0\cdot5s+1\cdot0}{s^2+s+2} = \frac{1\cdot5}{s+1} - \frac{0\cdot5(s+0\cdot5)+0\cdot75}{(s+0\cdot5)^2+1\cdot75}$$

and the inverse transforms can be found from lines 12, 21 and 22.

$$i(t) = 1\cdot5 \exp(-t) - 0\cdot5 \exp(-0\cdot5t) \cos \sqrt{1\cdot75}t$$
$$-\frac{0\cdot75}{\sqrt{1\cdot75}} \exp(-0\cdot5t) \sin \sqrt{1\cdot75}t$$

7.7.2 The 'cover-up' rule

If the expression $N(s)$ has several factors, this method has many advantages. Suppose

$$i(s) = \frac{M(s)}{N(s)} = \frac{A}{(s+a_1)} + \frac{B}{(s+a_2)} + \frac{C}{(s+a_3)} + \cdots \qquad (7.52)$$

then

$$M(s) = A(s+a_2)(s+a_3)(s+a_4)$$
$$+ B(s+a_1)(s+a_3)(s+a_4)$$
$$+ C(s+a_1)(s+a_2)(s+a_4)$$
$$+ \cdots$$

as before. Now if we put $s = -a_1$ in equation (7.52)

$$M(-a_1) = A(-a_1+a_2)(-a_1+a_3)(-a_1+a_4)$$

all other terms vanishing.

Hence

$$A = M(-a_1)/(-a_1+a_2)(-a_1+a_3)(-a_1+a_4)$$

and similarly putting $s = -a_2$

$$B = M(-a_2)/(-a_2+a_1)(-a_2+a_3)(-a_2+a_4)$$

and so on.

The rule is, therefore, 'to find the coefficients of the various fractions due to the factors of $N(s)$, "cover up" each factor in turn and evaluate the rest of the expression putting for s that value of s which makes the covered up factor zero'.

Example

If

$$i(s) = \frac{s^2+2s+3}{(s+1)(s+2)(s+3)}$$

Then

$$i(s) = \frac{(1-2+3)/(-1+2)(-1+3)}{(s+1)}$$
$$+ \frac{(4-4+3)/(-2+1)(-2+3)}{(s+2)}$$
$$+ \frac{(9-6+3)/(-3+1)(-3+2)}{(s+3)}$$

$$i(s) = \frac{1}{s+1} - \frac{3}{s+2} + \frac{3}{s+3}$$

and

$$i(t) = \exp(-t) - 3\exp(-2t) + 3\exp(-3t)$$

7.7.3 Heaviside's expansion formula

An extension of the last method is Heaviside's expansion formula.
If the denominator $N(s)$ is differentiated by the usual product law

$$N'(s) = (s+a_2)(s+a_3)(s+a_4)$$
$$+(s+a_1)(s+a_3)(s+a_4)$$
$$+(s+a_1)(s+a_2)(s+a_4)$$

and again putting $s = -a_1$ all terms vanish but the first.
Hence

$$A = \frac{M(-a_1)}{N'(-a_1)} \qquad B = \frac{M(-a_2)}{N'(-a_2)} \text{ and so on.}$$

The inverse transform is therefore given by

$$i(t) = \sum_{k=1}^{k=n} \frac{M(-a_k)}{N'(-a_k)} \exp(-a_k t) \qquad (7.53)$$

7.8 Examples

7.8.1 *R.C.* circuit

In the circuit shown in Fig. 7.9(a) the capacitor is uncharged and the switch closed at time $t = 0$. Derive an expression for the capacitor voltage v.

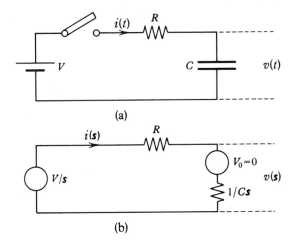

(a)

(b)

FIG. 7.9 (a) A simple *R.C.* circuit with applied step-function voltage. (b) The corresponding transform circuit

The transform circuit is given in Fig. 7.9(b), where $V(s) = V/s$ since the switching of the battery causes a step function voltage to be applied to the circuit. Since the capacitor is initially uncharged, $V_0 = 0$. Thus

$$i(s) = \frac{V}{s} \cdot \frac{1}{R+1/Cs}$$

and

$$v_0(s) = i(s)(1/Cs)$$

therefore

$$v_0(s) = \frac{V}{s} \cdot \frac{1/Cs}{R + 1/Cs}$$

$$= \frac{V}{s} \cdot \frac{1}{1 + RCs}$$

$$= \frac{V}{s} \cdot \frac{a}{s + a} \qquad \text{where} \qquad a = 1/RC$$

Thus from the table of transforms line 13

$$v_0(t) = V[1 - \exp(-t/RC)]$$

The procedure in this example should be compared with Sec. 1.8.1 where the same problem is solved by classical methods.

It should be noted how all the information is used including initial conditions and both the transient and steady-state responses appear in the answer.

7.8.2 *L.R.C.* circuit

Determine an expression for the current in the *L.R.C.* circuit shown in Fig. 7.10(*a*), if the switch is closed at time $t = 0$ the capacitor being uncharged.

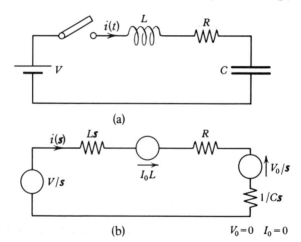

(a)

(b) $V_0 = 0 \quad I_0 = 0$

FIG. 7.10 (a) A simple *L.R.C.* circuit with applied step-function voltage. (b) The corresponding transform circuit

The initial conditions are such that $V_0 = 0$ and $I_0 = 0$. Thus

$$i(s) = \frac{V}{s} \cdot \frac{1}{Ls + R + 1/Cs}$$

$$= \frac{V}{L} \cdot \frac{1}{s^2 + (R/L)s + (1/LC)}$$

$$= \frac{V}{L} \cdot \frac{1}{(s+a)^2+\omega^2}$$

where

$$a = R/2L \quad \text{and} \quad \omega^2 = (1/LC) - (R/2L)^2$$

The following types of expression arise depending on the value of ω.

(a) If ω^2 is positive it is convenient to write the expression for $i(s)$ as

$$i(s) = \frac{V}{\omega L} \cdot \frac{\omega}{(s+a)^2+\omega^2}$$

and so from the transform table line 21

$$i(t) = \frac{V}{\omega L} \exp(-at) \sin \omega t$$

(b) If $\omega^2 = 0$ the expression for $i(s)$ becomes

$$i(s) = \frac{V}{L} \cdot \frac{1}{(s+a)^2}$$

and so from line 14

$$i(t) = \frac{V}{L} t \exp(-at)$$

(c) If ω^2 is negative $(-\beta^2)$ then

$$i(s) = \frac{V}{\beta L} \cdot \frac{\beta}{(s+a)^2-\beta^2}$$

and therefore from lines 4 and 26

$$i(t) = \frac{V}{\beta L} \exp(-at) \sinh \beta t$$

7.8.3 Ignition circuit

Find an expression for the secondary current i_2 in the circuit of Fig. 7.11(a) when the switch in the transformer primary circuit is opened.

Initial conditions

$$(I_1)_0 = V/R_1$$
$$(I_2)_0 = 0$$

When the switch is opened the transform circuit is given by Fig. 7.11(b).
Thus

$$i_1(s) = 0$$

and

$$L_2(I_2)_0 + M(I_1)_0 = (R_2+L_2s)i_2(s) + Msi_1(s)$$

$$i_2(s) = \frac{M(I_1)_0}{R_2+L_2s}$$

$$= \frac{MV}{R_1L_2} \cdot \frac{1}{s+R_2/L_2}$$

From line 12 of the transform table

$$i_2(t) = \frac{MV}{R_1L_2} \exp\left(-t(R_2/L_2)\right)$$

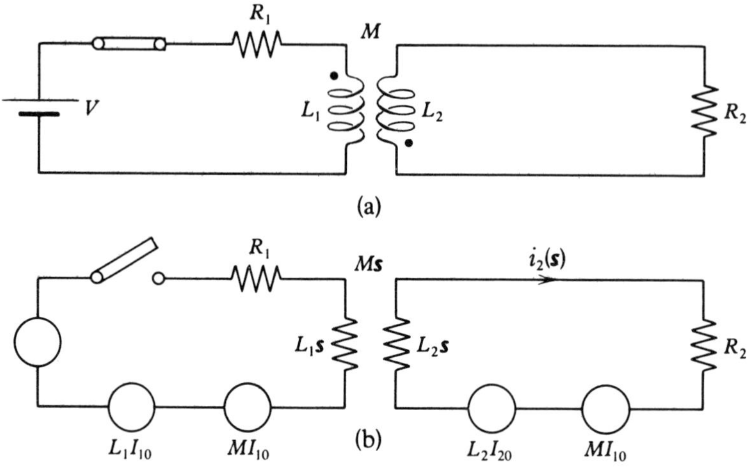

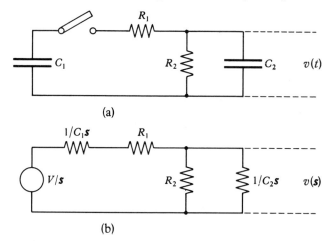

(a)

(b)

FIG. 7.11 (a) A simplified circuit consisting of an induction coil and a contact breaker. (b) The corresponding transform circuit

7.8.4 Impulse generator circuit

Derive an expression for the voltage across the capacitor C_2 (Fig. 7.12), when the switch is closed connecting the capacitor C_1 charged to potential V.

(a)

(b)

FIG. 7.12 (a) The circuit of a single-stage impulse generator. (b) The corresponding transform circuit

This is the impulse generator circuit where C_1 is the generator C_2 the load, R_1 the wave front resistor and R_2 the tail resistance.

Analysis shows how the shape of the impulse wave applied to C_2 depends on R_1 and R_2.

The transform circuit is given by Fig. 7.12(b).

Thus

$$v(s) = \frac{V}{s} \cdot \frac{Z_2}{Z_1 + Z_2}$$

where

$$Z_1 = 1/C_1 s + R_1$$

$$Z_2 = \frac{R_2/C_2 s}{R_2 + 1/C_2 s}$$

Expanding

$$v(s) = \frac{V}{s} \cdot \frac{R_2/(R_2 C_2 s + 1)}{R_1 + 1/C_1 s + R_2/(R_2 C_2 s + 1)}$$

$$= \frac{V}{s} \cdot \frac{R_2}{(R_1 + 1/C_1 s)(R_2 C_2 s + 1) + R_2}$$

$$= \frac{V}{s} \cdot \frac{R_2}{R_1 R_2 C_2 s + 1/C_1 s + R_2 C_2/C_1 + R_1 + R_2}$$

$$= \frac{V}{R_1 C_1} \cdot \frac{1}{s^2 + (1/R_1 C_1 + 1/R_2 C_2 + 1/R_1 C_2)s + (1/R_1 R_2 C_1 C_2)}$$

or

$$v(s) = \frac{V}{R_1 C_2} \cdot \frac{1}{s^2 + as + b}$$

where $a = (1/R_1 C_1 + 1/R_2 C_2 + 1/R_1 C_2)$ and $b = (1/R_1 R_2 C_1 C_2)$

$$v(s) = \frac{V}{R_1 C_2} \cdot \frac{1}{s_1 - s_2} \cdot \left[\frac{1}{s - s_1} - \frac{1}{s - s_2} \right]$$

where s_1 and s_2 are the roots of the equation

$$s^2 + as + b = 0$$

and both will be negative.

Thus from the transform table, line 12

$$v(t) = \frac{V}{R_1 C_2 (s_1 - s_2)} [\exp (s_1 t) - \exp (s_2 t)]$$

An approximate solution for a practical case where R_2 is much greater than R_1 and C_1 much greater than C_2, is obtained by re-examining the auxiliary equation

$$s^2 + (1/R_1 C_1 + 1/R_2 C_2 + 1/R_1 C_2)s + (1/R_1 R_2 C_1 C_2) = 0$$

where the value of $(1/R_1 C_1 + 1/R_2 C_2)$ is much smaller than $1/R_1 C_2$.

Thus the equation becomes approximately

$$s^2 + (1/R_1 C_2)s + 1/R_1 R_2 C_1 C_2 = 0$$

and the roots are seen to be

$$s_1 \simeq -1/R_1C_2$$
$$s_2 \simeq -1/R_2C_1$$
$$\text{and} \quad s_1 \gg s_2$$

The equation for $v(t)$ thus becomes

$$v(t) = V\left[\exp\left(-t/R_2C_1\right) - \exp\left(-t/R_1C_2\right)\right]$$

The graph of this expression is shown in Fig. 7.13.

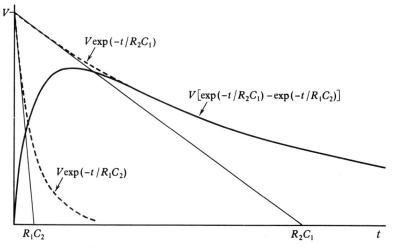

FIG. 7.13 The voltage waveforms of an impulse generator

7.8.5 Circuit-breaker restriking transient

The alternator shown in Fig. 7.14 is suddenly short-circuited and the circuit-breaker opens, causing an arc at the breaker contacts.

Determine an expression for the restriking voltage transient at the instant of zero current when successful interruption occurs. The alternator inductance is L and the system capacitance to neutral is represented by C.

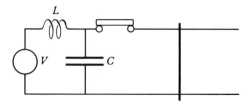

FIG. 7.14 A simplified circuit of an alternator and circuit-breaker on a short-circuit test

The short circuit current will be given by $I = V/\omega L$ (r.m.s. value) and the rate of change of this current when the instantaneous value is zero

$$[di/dt]_{i=0} = \sqrt{2}I\omega = \sqrt{2}V/L$$

Successful current interruption is the equivalent of injecting a ramp function current at the instant of current pause which has equal and opposite slope to the short circuit current (Fig. 7.15). The equation to the ramp function current is thus $i = \sqrt{2}(V/L)t$.

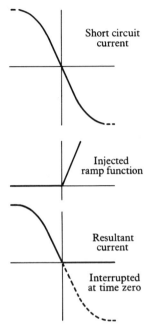

Short circuit current

Injected ramp function

Resultant current

Interrupted at time zero

FIG. 7.15 Current waveforms corresponding to successful interruption at a current zero

The transform circuit thus shows a ramp function current generator connected at the breaker terminals and the required voltage transient is the voltage developed across it (Fig. 7.16).

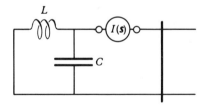

FIG. 7.16 The transform circuit corresponding to Fig. 7.14

From line 9 in the transform pair tables, if the ramp function is given by

$$i(t) = (\sqrt{2}V/L)t$$

its transform is

$$i(s) = (\sqrt{2}V/L)/s^2$$

Thus

$$v(s) = i(s)Z(s)$$

$$= \frac{(\sqrt{2}V/L)}{s^2} \cdot \frac{L/C}{Ls + 1/Cs}$$

$$= \sqrt{2}V \frac{\omega_0^2}{s(s^2 + \omega_0^2)}$$

where

$$\omega_0^2 = 1/LC$$

Expanding

$$v(s) = \sqrt{2}V \left[\frac{A}{s} + \frac{Bs + D}{s^2 + \omega_0^2} \right]$$

where

$$A(s^2 + \omega_0^2) + s(Bs + D) = \omega_0^2$$

so

$$D = 0 \qquad A = 1 \qquad B = -1$$

Thus

$$v(s) = \sqrt{2}V \left[\frac{1}{s} - \frac{s}{s^2 + \omega_0^2} \right]$$

$$v(t) = \sqrt{2}V[1 - \cos \sqrt{(1/LC)}t]$$

If the short circuit arises further along the system the impedance $Z(s)$ will be more complicated as may be seen from Fig. 7.17 where a Π section representing a transmission line has been included.

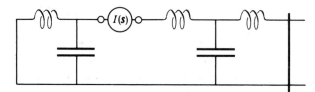

FIG. 7.17 The transform circuit of an alternator, circuit breaker and a transmission line

The restriking transient will then include a series of oscillatory terms which considerably affect the circuit breaker performance.

7.8.6 Coupled circuits

In the circuit of Fig. 7.18(a) the switch is closed for sufficient time to establish steady-state conditions. Determine an expression for the current in L_1 subsequent to the switch being opened.

Transform circuit (Fig. 7.18(b))

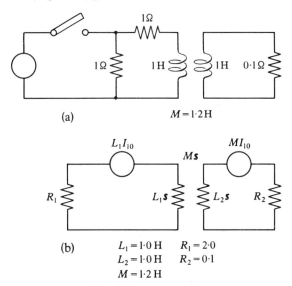

(a) $M = 1·2\,H$

(b) $L_1 = 1·0\,H$ $R_1 = 2·0$
            $L_2 = 1·0\,H$ $R_2 = 0·1$
            $M = 1·2\,H$

FIG. 7.18 (a) A circuit involving mutual inductance. (b) The corresponding transform circuit

By mesh analysis the loop equations of the transform circuit are given by

$$\begin{bmatrix} L_1I_0 \\ MI_0 \end{bmatrix} = \begin{bmatrix} R_1+L_1s & Ms \\ Ms & R_2+L_2s \end{bmatrix} \cdot \begin{bmatrix} i_1(s) \\ i_2(s) \end{bmatrix}$$

Inverting

$$\begin{bmatrix} i_1(s) \\ i_2(s) \end{bmatrix} = \frac{1}{(R_1+L_1s)(R_2+L_2s)-M^2s^2} \begin{bmatrix} R_2+L_2s & -Ms \\ -Ms & R_1+L_1s \end{bmatrix} \cdot \begin{bmatrix} L_1 \\ M \end{bmatrix} I_0$$

$$i_1(s) = I_0 \frac{(R_2+L_2s)L_1-M^2s}{R_1R_2+(L_1R_2+L_2R_1)s+(L_1L_2-M^2)s^2}$$

Putting

$$T_1 = L_1/R_1$$

$$T_2 = L_2/R_2$$

$$k^2 = M^2/L_1L_2$$

$$i_1(s) = I_0 \frac{(1/T_2)+(1-k^2)s}{(1/T_1T_2)+(1/T_1+1/T_2)s+(1-k^2)s^2}$$

$$= I_0 \frac{1/T_2(1-k^2)+s}{s^2+\dfrac{(1/T_1+1/T_2)s}{(1-k^2)}+\dfrac{1}{T_1T_2(1-k^2)}}$$

If $1/T_2 \ll 1/T_1$*

$$i_1(s) \simeq I_0 \frac{1/T_2(1-k^2)+s}{(s+a_1)(s+a_2)}$$

where

$$a_1 = \frac{1/T_1+1/T_2}{(1-k^2)} \quad \text{and} \quad a_2 = \frac{1}{T_1+T_2}$$

In the problem

$$T_1 = 0{\cdot}5 \quad T_2 = 20$$

$$k^2 = 1{\cdot}44/2 = 0{\cdot}72$$

$$(1-k^2) = 0{\cdot}28$$

$$a_1 = (2+0{\cdot}05)/0{\cdot}28 = 7{\cdot}34$$

$$a_2 = 1/20{\cdot}5 \qquad = 0{\cdot}0488$$

$$i_1(s) = I_0\frac{(1/20+0{\cdot}28)+s}{(s+7{\cdot}34)(s+0{\cdot}0488)} = I_0\frac{0{\cdot}18+s}{(s+7{\cdot}34)(s+0{\cdot}0488)}$$

$$= I_0\left(\frac{(-7{\cdot}16)/(-7{\cdot}29)}{s+7{\cdot}34}+\frac{0{\cdot}131/7{\cdot}29}{s+0{\cdot}0488}\right)$$

$$= I_0\left(\frac{0{\cdot}98}{s+7{\cdot}34}+\frac{0{\cdot}02}{s+0{\cdot}0488}\right)$$

$$i_1(t) = 0{\cdot}98\,I_0\exp\left(-7{\cdot}3t\right)+0{\cdot}02\,I_0\exp\left(-0{\cdot}0488t\right)$$

Similarly,

$$i_2(s) = I_0\frac{(R_1+L_1s)M-L_1Ms}{R_1R_2+(L_1R_2+L_2R_1)s+(L_1L_2-M^2)s^2}$$

$$= \frac{I_0}{(1-k^2)}\frac{M/T_1L_2}{(s+a_1)(s+a_2)}$$

Numerically

$$= I_0\frac{1{\cdot}2/0{\cdot}5\times0{\cdot}28\times2}{(s+a_1)(s+a_2)} = I_0\frac{4{\cdot}3}{(s+7{\cdot}34)(s+0{\cdot}0488)}$$

$$= I_0\left(\frac{-4{\cdot}3/7{\cdot}3}{s+7{\cdot}34}+\frac{4{\cdot}3/7{\cdot}3}{s+0{\cdot}0488}\right)$$

and so

$$I_2(t) = 0{\cdot}59\,I_0\left[\exp\left(-0{\cdot}0488t\right)-\exp\left(-7{\cdot}34t\right)\right]$$

7.8.7 Suddenly applied alternating voltage

A source of alternating voltage $100\sin(\omega t+\phi)$ where $\phi = 60$ degrees is suddenly switched at time zero to a circuit consisting of a resistance $10\ \Omega$ in series with an

---

* If one of the coefficients in a binomial expression is much greater than the other, approximate values of the factors can be found by inspection.

If $b \gg c$

$$s^2+bs+c \simeq (s+b)(s+c/b)$$

inductance of 1 H. Find an expression for the current if $\omega = 100$. Transform circuit (Fig. 7.19).

$$v(t) = 100 \sin (\omega t + \phi) = 100 \cos \phi \sin \omega t$$
$$+ 100 \sin \phi \cos \omega t$$

$$v(s) = 100 \frac{\omega \cos \phi + s \sin \phi}{(s^2 + \omega^2)} = \frac{5000 + 86 \cdot 6s}{s^2 + 100^2}$$

$$i(s) = \frac{5000 + 86 \cdot 6s}{(s^2 + 100^2)(s + 10)}$$

$$= \frac{As + B}{s^2 + 100^2} + \frac{C}{s + 10}$$

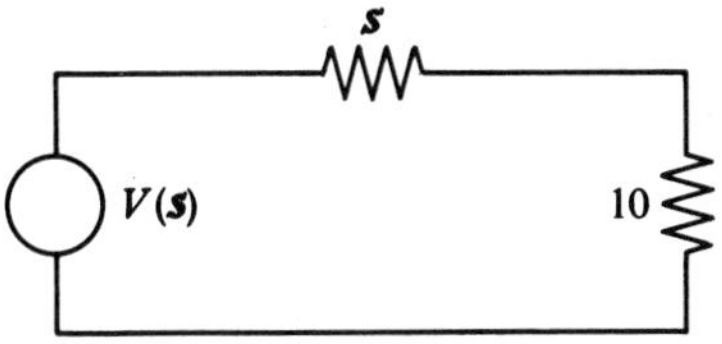

FIG. 7.19 The transform circuit for an *L.R.* circuit suddenly connected to an alternating voltage source

By the 'cover up' rule
put

$$s = -10 \qquad C = \frac{5000 - 866}{10^2 + 100^2} = \frac{41 \cdot 33}{101} = 0 \cdot 4092$$

put

$$s = j100$$

$$j100 A + B = \frac{100(50 + j86 \cdot 6)}{10 + j100} = \frac{10(50 + j86 \cdot 6)(1 - j10)}{1 + 100}$$

$$= \frac{10}{101} (50 + j86 \cdot 6 - j500 + 866)$$

$$= \frac{10}{101} (916 - j413 \cdot 3)$$

$$A = -\frac{10}{101} \frac{j413 \cdot 3}{j100} = -0 \cdot 4092$$

$$B = \frac{10}{101} 916 = 90 \cdot 69$$

$$i(s) = \frac{-0 \cdot 4092s + 0 \cdot 9069 \times 100}{(s^2 + 100^2)} + \frac{0 \cdot 4092}{s + 10}$$

$$i(t) = -0 \cdot 4092 \cos 100t + 0 \cdot 9069 \sin 100t + 0 \cdot 4092 \exp (-10t)$$

CHAPTER 8

Duality

8.1 Introduction

All students studying circuit theory as part of their professional training are quick to realise that many fundamental ideas seem to come in pairs. Kirchhoff's two Laws are one example, and the rules 'add impedances in series circuits, add admittances in parallel circuits' are another. We now attempt to investigate reasons, if any, for this 'twinning'.

Fundamentally, voltage and current are a very intimate pair. Sometimes it is convenient to regard one as the cause and the other the effect, and sometimes vice versa. In linear circuits these quantities are related by two equations

$$V = ZI \tag{8.1}$$

or
$$I = YV \tag{8.2}$$

It is just a matter of convenience which equation is used. Equation (8.1) is useful if the voltage across an impedance is required when the current is known and Eq. (8.2) is probably better for the determination of current if voltage and admittances are known.

The arithmetical processes used in manipulating these equations are the same. The equations are said to be dual equations. The individual terms occurring in pairs are said to be duals of each other (viz. voltage and current, also impedance and admittance), with the result that starting with one equation, the other can be written down merely by replacing each term in the first equation by its dual.

$$
\begin{array}{c}
V = I\ Z \\
\downarrow \quad \downarrow \downarrow \\
I\ = V\ Y
\end{array}
\tag{8.3}
$$

8.2 Dual pairs

We are thus encouraged to prepare a table showing these pairs, side by side, in order to find if other quantities have similar relationships. Table 8.1 is developed in this way. Beginning with the statement that voltage and current are dual terms, admittance is seen to be the dual of impedance and, consequently, the components of these, resistance and inductance on the one hand and conductance and capacitance on the other, are two more dual pairs.

Table 8.1

ORIGINAL		DUAL	
Voltage	V	Current	I
Current	I	Voltage	V
Impedance	Z	Admittance	Y
$Z = R + pL$		$Y = G + pC$	
Resistance	R	Conductance	G
Inductance	L	Capacitance	C
Energy of an inductance		Energy of a capacitance	
$\frac{1}{2}LI^2$		$\frac{1}{2}CV^2$	

The dual for 'energy of an inductance' become 'energy of a capacitance' by writing the dual term inductance in place of the term capacitance.

Thus we see that $\frac{1}{2}LI^2$ and $\frac{1}{2}CV^2$ are dual expressions, the latter being derived from the former by substituting C for L and V for I.

Capacitance is a property of an electric field and inductance is a similar property of a magnetic field. Consequently we are led to expect many dual pairs to arise, if we consider corresponding quantities in the two fields. These pairs are shown in Table 8.2.

It has been noted elsewhere* that flow lines and equipotentials are orthogonal for both electric and magnetic fields in uniform media, but the pattern is inverted in the case of the magnetic field. This also can be attributed to the duality principle.

Table 8.2 Field Duals

QUANTITY	MAGNETIC		ELECTRIC	
Potential gradient	H	(A/m)	E	(V/m)
Flux	ϕ	$\int V \, dt$	Q	$\int I \, dt$
Flux density	B	$(d\phi/da)$	D	(dQ/da)
Conductivity	Permeability $\mu\mu_0$		Permittivity $\epsilon\epsilon_0$	
	Ampere's Law		Maxwell's Law	
	$\oint H \, dl = I$		$\oint E \, dl = d\phi/dt$	

Electrical Machines, p. 328.

Apart from being an aid to the memory, the duality between electric and magnetic field quantities helps with an understanding of corresponding problems occurring in the two types of field.

So far the list of dual pairs has been derived solely from the duality between voltage and current. The principle can now be extended to circuits and network topology.

8.3 Dual circuits

We commence with two circuits (Fig. 8.1):

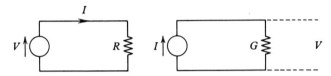

FIG. 8.1 A simple circuit and its dual

In the first circuit a potential source V is applied to a resistance R with the result that a current flows in the circuit. In the second circuit a current source I is applied to a conductance G and the voltage V across G is the system response.

If we apply Eq. (8.1) to the first circuit and (8.2) to the second and if the numerical values are the same in the two equations (say $10 = 2 \times 5$ in both cases) and the first circuit is referred to as the original circuit, the second circuit is known as its dual. Note that the second circuit is not the equivalent of the first, but the arithmetic is the same to determine the circuit response. A duality table for circuits can thus be drawn up:

Table 8.3

ORIGINAL	DUAL
Voltage source	Current source
Resistance	Conductance
Current response	Voltage response
Voltages are added when elements are in series	Currents are added when elements are in parallel
Series	Parallel
Open circuit (infinite R)	Short circuit (infinite G)
Switch open	Switch closed
Kirchhoff's 1st Law	Kirchhoff's 2nd Law
Node	Loop

Most of these dual statements will be obvious and the last two will be clearly

understood if we compare Kirchhoff's First Law—'the sum of the currents
entering a node is zero' with the Second Law 'the sum of the volt drops around
a loop is zero'.

It should now be observed that the methods of mesh analysis described in
Chap. 5 and those of nodal analysis of Chap. 6 are dual procedures.

In Fig. 8.2 are shown a number of dual circuits. In each case the dual has been

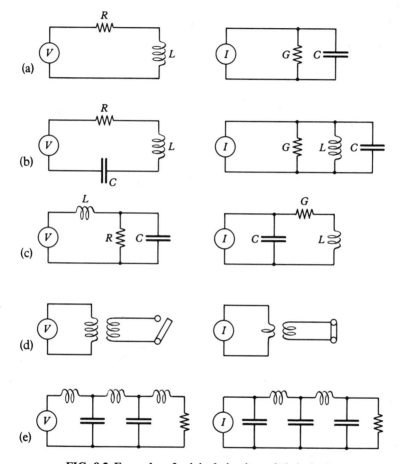

FIG. 8.2 Examples of original circuits and their duals

constructed from the original with the aid of the tables of dual pairs, series
circuits become parallel circuits, inductances become capacitances and so on.
Similar numerical values have been maintained. In each case a voltage source is
replaced by a current source of the same numerical value with the result that the
currents in the elements of the original will be equal to the voltages across the
corresponding elements in the dual.

Figure 8.2(a) and Fig. 8.2(b) are simple series circuits with their dual parallel
circuits.

In Fig. 8.2(c) the parallel R.C. circuit in the original is replaced by G and L in series in the dual and then C is added in parallel in place of L in series.

The original circuit of Fig. 8.2(d) represents a transformer on open circuit connected to a voltage source. The disastrous results of short-circuiting the secondary should be compared with the equally bad effects of open-circuiting the current transformer in the dual circuit.

Figure 8.3 shows a method of drawing a dual network corresponding to a complicated original.

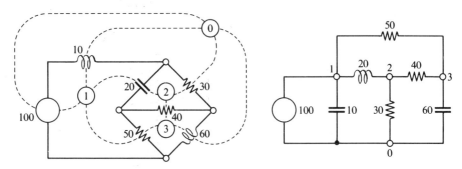

FIG. 8.3 (a) An original circuit shown by the continuous lines. (b) The corresponding dual circuit obtained from the dotted lines added to the circuit in (a)

The following procedure is adopted:

1. The independent loops in the original network are identified by placing a number in a small circle in the centre of each loop.
2. The number 0 is also placed in a small circle at a convenient position outside the network. These circles will become the nodes of the dual network with 0 as the reference.
3. The numbered circles are joined in pairs by dotted lines, one line for each element of the network and passing through that element. These dotted lines form the graph of the dual network.
4. The graph is now re-drawn in a more convenient shape, starting with an orderly arrangement of the nodes and connecting them according to the dotted lines.
5. Finally each dual element is drawn in each branch, being the dual of the original element crossed by the corresponding dotted line.

The Fig. 8.3. also helps to show that the principle of duality is really geometrical in character, arising out of fundamental relationships between lines and surfaces.

Dual circuits must not be confused with 'equivalent circuits' and it is seen that they are not necessarily simpler in configuration.

One of the uses of dual circuits will be shown in Chap. 14 where mechanical circuits are discussed.

Another use arises when a model of a system has to be constructed to enable laboratory tests to be conducted. If the original network contains constant-current sources or inductances with negligible resistance, these items are difficult, if not impossible, to obtain for the model.

The model of the dual circuit which will contain constant voltage sources and loss free capacitances is a more practical arrangement.

The principle of duality can also be applied to field phenomena. Indeed it can be argued that duality arises from the geometrical properties inherent in electric and magnetic fields.

These matters will be discussed in Chap. 21.

CHAPTER **9**

Two-port networks

9.1 Introduction

A two-port network is a passive network which, although it may have many
meshes, has only two places of entry which are known as ports. Alternatively,

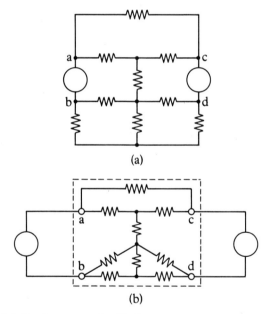

(a)

(b)

FIG. 9.1 (a) A five-loop network. (b) The same, drawn as a two-port network

this may be known as a four-terminal network, although two-port is more
explicit and more precise.

Figure 9.1(*a*) shows a five-loop network with active elements in loops 1 and
2 only. This has been re-drawn in Fig. 9.1(*b*) with the passive elements enclosed

by the dotted lines and the sources withdrawn. Access to the network is thus available only at the two ports **ab** and **cd**. It is assumed that we do not wish to solve the network completely but only to determine the external currents.

9.2 Matrix partitioning

The impedance matrix for the above network is given by

$$\begin{bmatrix} \begin{bmatrix} V_1 \\ V_2 \end{bmatrix} \\ \begin{bmatrix} 0 \\ 0 \\ 0 \end{bmatrix} \end{bmatrix} = \begin{bmatrix} \begin{bmatrix} Z_{11} & Z_{12} \\ Z_{21} & Z_{22} \end{bmatrix} & \begin{bmatrix} Z_{13} & Z_{14} & Z_{15} \\ Z_{23} & Z_{24} & Z_{25} \end{bmatrix} \\ \begin{bmatrix} Z_{31} & Z_{32} \\ Z_{41} & Z_{42} \\ Z_{51} & Z_{52} \end{bmatrix} & \begin{bmatrix} Z_{33} & Z_{34} & Z_{35} \\ Z_{43} & Z_{44} & Z_{45} \\ Z_{53} & Z_{54} & Z_{55} \end{bmatrix} \end{bmatrix} \cdot \begin{bmatrix} \begin{bmatrix} I_1 \\ I_2 \end{bmatrix} \\ \begin{bmatrix} I_3 \\ I_4 \\ I_5 \end{bmatrix} \end{bmatrix} \qquad (9.1)$$

If we are only interested in the currents entering the two ports, that is I_1 and I_2, the matrix can be partitioned (Sec. 3.18) and a 2×2 matrix $[Z']$ obtained such that

$$\begin{bmatrix} V_1 \\ V_2 \end{bmatrix} = [Z'] \cdot \begin{bmatrix} I_1 \\ I_2 \end{bmatrix} \qquad \begin{array}{c} (9.2) \\ \text{from } (3.46) \end{array}$$

This reduction is possible, no matter how many meshes there are in the original network.

9.3 Two-port network conventions

The two-port network may thus be regarded as a simple two-loop network which can be defined by four impedance parameters.

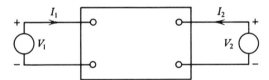

FIG. 9.2 The convention used for voltage and current when both ports are input ports

Such a two-loop network is shown in Fig. 9.2, where both ports are input ports, and it is noted that the directions of the currents I_1 and I_2 are drawn in accordance with the conventions of Chap. 4, applicable to generalised theory.

If the currents and voltages are all positive, energy is entering the network at both ports to be dissipated at sinks within the network.

But many items of electrical equipment, although they have two ports and are passive, have been specifically designed for the transmission of energy. Normally one of the ports is an input port where energy enters the device and the other is an output port where energy leaves and is passed on to other systems. Transformers, transmission lines, cables and regulators are all examples of this type of equipment. For these devices, diagrams such as Fig. 9.3 are appropriate and appear in many textbooks.

When this diagram is used and all currents and voltages are positive, the power entering the network at the left-hand port is the input power and the power leaving at the right-hand port is the output power to be dissipated in the load

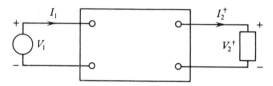

FIG. 9.3 The special convention for a two-port network when one port is used as an *input* **port and the other is an** *output* **port**

on the extreme right. The difference between these two powers is the power loss in the network.

Fig. 9.3 has been drawn in accordance with what is to be known as two-port network conventions, and it will be noted that the difference between Figs. 9.3 and 9.2 is in the direction of the current I_2. This is merely a question of convention. It does not matter which convention is used, so long as we state it clearly and stick to it.

Since both conventions have their uses, one in general network theory and the other in transmission theory, it is highly important to avoid ambiguity. This can be done by ensuring that we refer every time to the appropriate diagram, but to make doubly sure we shall use the symbols $V\dagger$ and $I\dagger$ when referring to output voltages and currents and when the conventional directions of Fig. 9.3 are required.*

Thus

$$V_2\dagger \equiv V_2$$

$$I_2\dagger \equiv -I_2 \qquad\qquad (9.3)$$

In diagrams the input port is usually drawn on the left and the output port on the right.

9.4 The transmission or chain matrix

The terms of the impedance matrix Eq. (9.2) can be rearranged as follows to show more clearly the relationship between input and output quantities. Rewriting Eq. (9.2)

$$\begin{vmatrix} V_1 \\ V_2 \end{vmatrix} = \begin{bmatrix} Z'_{11} & Z'_{12} \\ Z'_{21} & Z'_{22} \end{bmatrix} \cdot \begin{bmatrix} I_1 \\ I_2 \end{bmatrix} \qquad\qquad (9.4)$$

* If a reader feels that this convention is superfluous and does not find it helpful, it is relatively simple to ignore the $\dagger$ symbol attached to V_2 and to write $(-I_2)$ wherever $I_2\dagger$ occurs in the remainder of this chapter.

we have, omitting the primes from this point onwards for clarity,

$$V_1 = Z_{11}I_1 + Z_{12}I_2 \tag{9.5}$$

$$V_2 = Z_{21}I_1 + Z_{22}I_2 \tag{9.6}$$

Rearranging (9.6)

$$I_1 = (1/Z_{21})V_2 - (Z_{22}/Z_{21})I_2 \tag{9.7}$$

Substituting in (9.5)

$$V_1 = (Z_{11}/Z_{21})V_2 + (Z_{12} - Z_{11}Z_{22}/Z_{21})I_2 \tag{9.8}$$

and converting to V_1, I_1, $V_2\dagger$ and $I_2\dagger$ using Eq. (9.3)

$$\begin{bmatrix} V_1 \\ I_1 \end{bmatrix} = \begin{bmatrix} Z_{11}/Z_{21} & Z_{11}Z_{22}/Z_{21} - Z_{12} \\ 1/Z_{21} & Z_{22}/Z_{21} \end{bmatrix} \cdot \begin{bmatrix} V_2\dagger \\ I_2\dagger \end{bmatrix} \tag{9.9}$$

or

$$\begin{bmatrix} V_1 \\ I_1 \end{bmatrix} = [A] \begin{bmatrix} V_2\dagger \\ I_2\dagger \end{bmatrix} \tag{9.10}$$

where

$$[A] = \frac{1}{Z_{21}} \begin{bmatrix} Z_{11} & |Z| \\ 1 & Z_{22} \end{bmatrix} \tag{9.11}$$

since

$$|Z| = Z_{11}Z_{22} - Z_{21}Z_{12} \tag{9.12}$$

This presentation separates input and output quantities. Strictly

$$\text{neither} \quad \begin{bmatrix} V_1 \\ I_1 \end{bmatrix} \quad \text{nor} \quad \begin{bmatrix} V_2\dagger \\ I_2\dagger \end{bmatrix}$$

are true matrices since they are not sets of the same quantities. However, the first refers to the input to the two-port network and the second to the output, hence matrix $[A]$ converts output to input and is known, therefore, as the transmission matrix. The elements of $[A]$ are termed the transmission constants.

$$[A] = \begin{bmatrix} a_{11} & a_{12} \\ a_{21} & a_{22} \end{bmatrix} \tag{9.13}$$

An older nomenclature, more familiar with power engineers, is given by

$$[A] = \begin{bmatrix} A & B \\ C & D \end{bmatrix} \tag{9.14}$$

Inverting Eq. (9.10), we have

$$\begin{bmatrix} V_2\dagger \\ I_2\dagger \end{bmatrix} = [A]^{-1} \begin{bmatrix} V_1 \\ I_1 \end{bmatrix} \tag{9.15}$$

where

$$[A]^{-1} = \frac{1}{|A|} \begin{bmatrix} D & -B \\ -C & A \end{bmatrix} \qquad (9.16)$$

which enables output quantities to be determined in terms of the input.

9.4.1 Bilateral and symmetrical networks

The determinant of the $[A]$ matrix (from Eq. 9.11))

$$|A| = (1/Z_{21}{}^2)(Z_{11}Z_{22} - |Z|)$$

$$= Z_{12}/Z_{21} \qquad (9.17)$$

If the network is bilateral, that is to say if the $[Z]$ matrix is symmetrical, then $Z_{12} = Z_{21}$ and hence from (9.14) and (9.17)

$$|A| = AD - BC = 1 \qquad (9.18)$$

This is universally true for power system equipment, excluding rectifiers and rotating machines, and indicates that the four constants A, B, C and D are not independent. They have only three degrees of freedom: given three of them, the fourth can be calculated from Eq. 9.18

$$\begin{aligned} A &= (BC + 1)/D \\ B &= (AD - 1)/C \\ C &= (AD - 1)/B \\ D &= (BC + 1)/A \end{aligned} \qquad (9.18a)$$

It is also important to note that because Eq. 9.18 is applicable to a bilateral network, Eq. 9.16 is simplified to become

$$[A]^{-1} = \begin{bmatrix} D & -B \\ -C & A \end{bmatrix} \qquad (9.16a)$$

An additional simplification arises if a two-port network is symmetrical, that is to say, if the two ports can be interchanged without altering the relationship between input and output quantities. This occurs with a uniform transmission line. Under these conditions the two loops must have the same contour impedance, hence Z_{11} is equal to Z_{22} and so from Eq. (9.11)

$$A = D \qquad (9.19)$$

9.5 Determination of $[A]$ from open-circuit and short-circuit tests

Readings of voltage and current taken at one port when the other port is (a) open-circuited, (b) short-circuited, enable the elements of matrix $[A]$ to be determined.

In this section we shall assume networks to be bilateral and that Eq. (9.18) is applicable.

(*a*) Open-circuit test (Fig. 9.4). The input voltage and current is V_{10} and I_{10} respectively. The output voltage is $V_2\dagger$ but $I_2\dagger$ is zero.

FIG. 9.4 A two-port network with the output port open-circuited

Thus

$$\begin{bmatrix} V_{10} \\ I_{10} \end{bmatrix} = \begin{bmatrix} A & B \\ C & D \end{bmatrix} \cdot \begin{bmatrix} V_2\dagger \\ 0 \end{bmatrix} \tag{9.20}$$

$$V_{10} = A V_2\dagger \tag{9.21}$$

$$I_{10} = C V_2\dagger \tag{9.22}$$

and the open circuit input impedance

$$Z_{10} = V_{10}/I_{10} = A/C \tag{9.23}$$

(*b*) Short-circuit test (Fig. 9.5), with the output terminals short-circuited and a voltage V_{1x} applied to the input port sufficient to circulate $I_2\dagger$ in the short circuit.

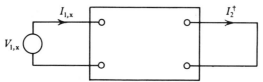

FIG. 9.5 A two-port network with the output port short-circuited

We have

$$\begin{bmatrix} V_{1x} \\ I_{1x} \end{bmatrix} = \begin{bmatrix} A & B \\ C & D \end{bmatrix} \cdot \begin{bmatrix} 0 \\ I_2\dagger \end{bmatrix} \tag{9.24}$$

and so

$$V_{1x} = B I_2\dagger \tag{9.25}$$

$$I_{1x} = D I_2 \tag{9.26}$$

and the short circuit input impedance

$$Z_{1x} = V_{1x}/I_{1x} = B/D \tag{9.27}$$

Equations (9.21), (9.22), (9.25) and (9.26) enable precise definitions to be given to the elements of matrix [*A*]

$A(= a_{11})$ is the open-circuit voltage ratio $V_{10}/V_2\dagger$
$B(= a_{12})$ is the short-circuit transfer impedance $V_{1x}/I_2\dagger$
$C(= a_{21})$ is the open-circuit transfer admittance $I_{10}/V_2\dagger$
$D(= a_{22})$ is the short-circuit current ratio $I_{1x}/I_2\dagger$

If measured values of $V_{10}/V_2\dagger$, Z_{10} and Z_{1x} are obtained from open- and short-circuit tests, A, B, C and D can be derived from (9.18), (9.21), (9.23) and (9.27).

$$[A] = \begin{bmatrix} A & Z_{10}Z_{1x}/A(Z_{10}-Z_{1x}) \\ A/Z_{10} & Z_{10}/Z(Z_{10}-Z_{1x}) \end{bmatrix} \qquad (9.28)$$

The voltage ratio A may be difficult to measure in some systems as, for example, in a transmission line, the ends of the line being a considerable distance apart.

This difficulty may be overcome by making a short-circuit test at the second port, with the first port short-circuited, and the impedance Z_{2x} determined. In this test the applied voltage will be $V_{2x}\dagger$ and the current $-I_{2x}\dagger$.

Hence from (9.15) and (9.16)

$$Z_{2x} = V_{2x}\dagger/(-I_{2x}\dagger) = B/A \qquad (9.29)$$

and so from (9.23), (9.27) and (9.29) the $[A]$ matrix can be determined in terms of Z_{10}, Z_{1x} and Z_{2x}

$$[A] = \sqrt{\left(\frac{Z_{1x}Z_{2x}Z_{10}}{Z_{10}-Z_{1x}}\right)} \begin{bmatrix} 1/Z_{2x} & 1 \\ 1/Z_{10}Z_{2x} & 1/Z_{1x} \end{bmatrix} \qquad (9.30)$$

In a symmetrical network $A = D$; $Z_{2x} = Z_{1x}$ and Eq. (9.30) reduces to

$$[A] = \sqrt{\left(\frac{Z_{10}}{Z_{10}-Z_{1x}}\right)} \begin{bmatrix} 1 & Z_{1x} \\ 1/Z_{10} & 1 \end{bmatrix} \qquad (9.31)$$

9.6 Two-port networks in cascade

Networks are connected in cascade when the output of the first becomes the input to the second and so on. A familiar example of three two-port networks in cascade is a transmission line with line transformers at both ends.

Figure 9.6 shows two networks A_a and A_b connected in cascade.

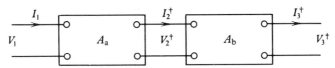

FIG. 9.6 Two two-port networks in cascade

For network A_a

$$\begin{bmatrix} V_1 \\ I_1 \end{bmatrix} = [A_a] \begin{bmatrix} V_2\dagger \\ I_2\dagger \end{bmatrix} \qquad (9.32)$$

and for A_b

$$\begin{bmatrix} V_2\dagger \\ I_2\dagger \end{bmatrix} = [A_b] \begin{bmatrix} V_3\dagger \\ I_3\dagger \end{bmatrix}$$

Thus

$$\begin{bmatrix} V_1 \\ I_1 \end{bmatrix} = [A_a][A_b]\begin{bmatrix} V_3\dagger \\ I_3\dagger \end{bmatrix}$$

$$= [A]\begin{bmatrix} V_3\dagger \\ I_3\dagger \end{bmatrix} \tag{9.33}$$

where

$$[A] = [A_a][A_b] \tag{9.34}$$

and $[A]$ is the equivalent transmission matrix for the cascade system. The process can be repeated if more sections are connected.

The overall transmission matrix corresponding to a series of sections connected in cascade is thus obtained by multiplying together the $[A]$ matrices of the sections in the same order in which they form the chain.

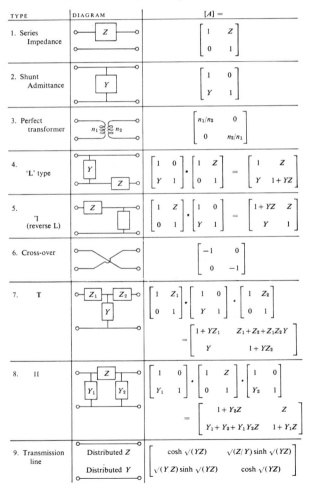

Table 9.1. Examples of two-port networks

Since many practical circuits are compounded of sections connected in this way, this method of synthesis can be very valuable. The table on p. 135 gives a number of common examples:

9.7 Two-port networks in series and parallel

Two-port networks may be connected in series or parallel as shown in Figs. 9.7 and 9.8, a familiar example of parallel connection being a duplicate feeder.

It is required to determine the equivalent $[A]$ matrices corresponding to these combinations. Direct manipulation of the individual $[A]$ matrices is laborious and it is preferable to convert to the corresponding $[Z]$ or $[Z]^{-1}$.

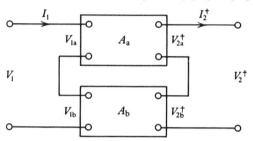

FIG. 9.7 Two two-port networks in series

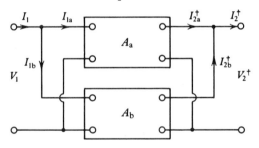

FIG. 9.8 Two two-port networks in parallel

9.7.1 Conversion from [A] to [Z]

The following conversions applicable to bilateral networks are derived from Eq. (9.9):

If

$$[A] = \begin{bmatrix} A & B \\ C & D \end{bmatrix} \qquad [A]^{-1} = \begin{bmatrix} D & -B \\ -C & A \end{bmatrix}$$

$$[Z] = \frac{1}{C} \begin{bmatrix} A & 1 \\ 1 & D \end{bmatrix}$$

$$[Z]^{-1} = \frac{1}{B} \begin{bmatrix} D & -1 \\ -1 & A \end{bmatrix} \tag{9.35}$$

If

$$[Z] = \begin{bmatrix} Z_{11} & Z_{12} \\ Z_{21} & Z_{22} \end{bmatrix} \quad [Z]^{-1} = \frac{1}{|Z|} \begin{bmatrix} Z_{22} & -Z_{12} \\ -Z_{21} & Z_{11} \end{bmatrix}$$

$$= \begin{bmatrix} Y_{11} & Y_{12} \\ Y_{21} & Y_{22} \end{bmatrix}$$

$$[A] = \frac{1}{Z_{21}} \begin{bmatrix} Z_{11} & |Z| \\ 1 & Z_{22} \end{bmatrix} \tag{9.36}$$

If

$$[Z]^{-1} = \begin{bmatrix} Y_{11} & Y_{12} \\ Y_{21} & Y_{22} \end{bmatrix} \quad [Z] = \frac{1}{|Y|} \begin{bmatrix} Y_{22} & -Y_{12} \\ -Y_{21} & Y_{11} \end{bmatrix}$$

$$[A] = \frac{-1}{Y_{21}} \begin{bmatrix} Y_{22} & 1 \\ |Y| & Y_{11} \end{bmatrix} \tag{9.37}$$

The reader is reminded that the conventions for the $[Z]$ matrix differ from those of the $[A]$ matrix with respect to the direction of I_2 but this does not affect the subsequent proof.

9.7.2 Series

If two networks $[A_a]$ and $[A_b]$ are connected in series, Fig. 9.7,

$$\begin{array}{l} I_1 = I_{1a} = I_{1b} \\ I_2 = I_{2a} = I_{2b} \end{array} \quad \text{or} \quad \begin{bmatrix} I_1 \\ I_2 \end{bmatrix} = \begin{bmatrix} I_1 \\ I_2 \end{bmatrix}_a = \begin{bmatrix} I_1 \\ I_2 \end{bmatrix}_b$$

also

$$\begin{array}{l} V_1 = V_{1a} + V_{1b} \\ V_2 = V_{2a} + V_{2b} \end{array} \quad \text{or} \quad \begin{bmatrix} V_1 \\ V_2 \end{bmatrix} = \begin{bmatrix} V_1 \\ V_2 \end{bmatrix}_a + \begin{bmatrix} V_1 \\ V_2 \end{bmatrix}_b$$

Since

$$\begin{bmatrix} V_1 \\ V_2 \end{bmatrix}_a = [Z_a] \cdot \begin{bmatrix} I_1 \\ I_2 \end{bmatrix}$$

and

$$\begin{bmatrix} V_1 \\ V_2 \end{bmatrix}_b = [Z_b] \cdot \begin{bmatrix} I_1 \\ I_2 \end{bmatrix}$$

$$\begin{bmatrix} V_1 \\ V_2 \end{bmatrix} = \left[[Z_a] + [Z_b] \right] \cdot \begin{bmatrix} I_1 \\ I_2 \end{bmatrix} \tag{9.38}$$

Thus to determine the $[A]$ matrix equivalent to the system, commencing with $[A_a]$ and $[A_b]$ the corresponding matrices $[Z_a]$ and $[Z_b]$ should be found from (9.35). $[Z_a]$ and $[Z_b]$ need to be added and converted back to $[A]$ with the aid of (9.36).

9.7.3 Parallel

If two networks $[A_a]$ and $[A_b]$ are connected in parallel, Fig. 9.8,

$$\begin{aligned} V_1 &= V_{1a} = V_{1b} \\ V_2 &= V_{2a} = V_{2b} \end{aligned} \quad \text{or} \quad \begin{bmatrix} V_1 \\ V_2 \end{bmatrix} = \begin{bmatrix} V_1 \\ V_2 \end{bmatrix}_a = \begin{bmatrix} V_1 \\ V_2 \end{bmatrix}_b$$

and

$$\begin{aligned} I_1 &= I_{1a} + I_{1b} \\ I_2 &= I_{2a} + I_{2b} \end{aligned} \quad \text{or} \quad \begin{bmatrix} I_1 \\ I_2 \end{bmatrix} = \begin{bmatrix} I_1 \\ I_2 \end{bmatrix}_a + \begin{bmatrix} I_1 \\ I_2 \end{bmatrix}_b$$

Hence

$$\begin{bmatrix} I_1 \\ I_2 \end{bmatrix}_a = [Z_a]^{-1} \cdot \begin{bmatrix} V_1 \\ V_2 \end{bmatrix}$$

and

$$\begin{bmatrix} I_1 \\ I_2 \end{bmatrix}_b = [Z_b]^{-1} \cdot \begin{bmatrix} V_1 \\ V_2 \end{bmatrix}$$

and thus

$$\begin{bmatrix} I_1 \\ I_2 \end{bmatrix} = \left[[Z_a]^{-1} + [Z_b]^{-1} \right] \cdot \begin{bmatrix} V_1 \\ V_2 \end{bmatrix} \tag{9.39}$$

Again $[A_a]$ and $[A_b]$ must be converted, this time to $[Z_a]^{-1}$ and $[Z_b]^{-1}$ which are then added and converted back to $[A]$.

It is thus interesting to note how the familiar rules 'add impedances in series circuits and admittances in parallel circuits' continue to be applicable in appropriate matrix problems.

9.8 The transmission line

9.8.1 Line parameters

A transmission line, whether overhead line or cable, is an excellent example of a two-port network. Its sole function is to receive energy at the input port and deliver it at the output port. It is also an example of distributed parameters and an exercise in the skill required in ensuring a satisfactory lumping technique. For the purpose of this text we will consider only the single-phase line with two conductors under steady-state a.c. conditions.

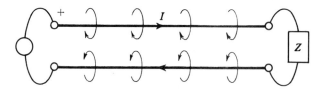

FIG. 9.9 A single-phase transmission line showing the magnetic field due to current. The line has therefore a loop inductance

Assuming in the first instance direct current flowing in the line (Fig. 9.9) and considering the magnetic field due to this current, we note that the line is in the form of a long loop with magnetic flux lines entering the plane of the loop. We can thus speak of the *loop inductance* of the line and divide this up according to the length of the line and so obtain a figure for the *loop inductance per kilometre* of the circuit.*

Examining the line under open-circuit conditions with a potential difference established between the conductors, the capacitance of the system is seen to depend on the length and spacing of the conductors. Both L and C are distributed values and so are the corresponding impedance and admittance parameters

$$Z = j\omega L \tag{9.40}$$

$$Y = j\omega C \tag{9.40}$$

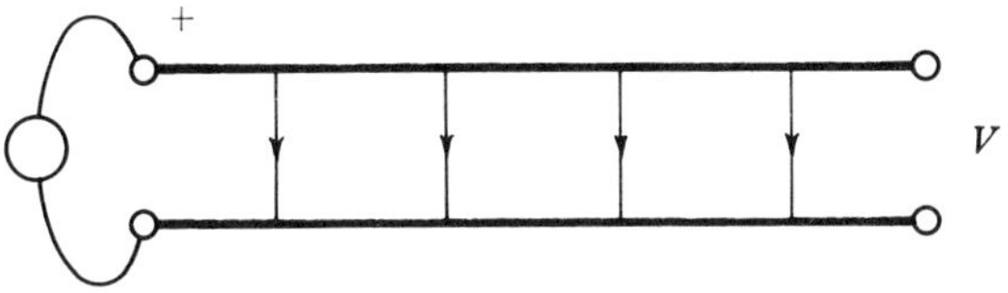

FIG. 9.10 A single phase transmission line showing the electric field due to voltage. The line has therefore distributed capacitance

For relatively short lines there are two possible methods of lumping these parameters (*a*) to divide the inductance into two equal parts and consider the capacitance concentrated at the centre of the line or (*b*) to divide the capacitance into two parts placing these at the ends of the line. These two suggestions are depicted in Fig. 9.11.

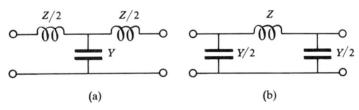

(a)   (b)

FIG. 9.11 The equivalent circuit of a short transmission line represented by lumped parameters (a) nominal T connection, (b) nominal Π connection

It should be appreciated that these two suggestions are not exact equivalents of each other nor do they make the same approximations. The difference between them is noted in the table on page 135, lines 7 and 8 where the corresponding $A\ B\ C\ D$ constants are compared.

---

* This figure is sometimes divided by two to give a figure for the 'inductance per conductor-kilometre' but this practice may lead to ambiguity.

Inductance is a property of the loop and not of the individual conductor.

The length of the line is the distance between the two ports and not the total length of conductor used.

These approximations have been used successfully in the past for relatively short lines but if computers are available to cope with more complicated calculations these methods are no longer needed.

It is more accurate and ultimately more straightforward to use the method explained in the next section.

9.8.2 The *A B C D* constants of a transmission line

Figure 9.12 shows a short section of a transmission line for which we assume the following parameters. We consider the steady-state a.c. operation of the line with angular frequency ω.

l = length of the line
x = distance of point **P** from the output port
z = series complex impedance per unit length
Z = zl = total series complex impedance
y = shunt complex admittance per unit length
Y = yl = total shunt complex admittance
dx = elementary length of line at point **P**
V = voltage phasor between lines at **P**
I = current phasor in the line at **P**

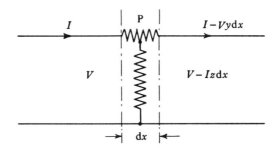

FIG. 9.12 An elementary section of a transmission line showing shunt admittance and series impedance

Considering the elementary section itself as a two-port network, its series impedance is $z\,dx$ and its shunt admittance $y\,dx$.

If the input quantities are V and I, the voltage is reduced in the section by the drop in series impedance $I z\,dx$ and current is reduced by $V y\,dx$. The output quantities for the section are therefore

$$V - Iz\,dx \quad \text{and} \quad I - Vy\,dx$$

In other words the changes in the voltage and current phasors within the section are given by

$$dV = -Iz\,dx \tag{9.41}$$

$$dI = -Vy\,dx \tag{9.42}$$

or
$$\frac{dV}{dx} = -Iz \tag{9.43}$$

and
$$\frac{dI}{dx} = -Vy \tag{9.44}$$

Differentiating the expression for dV/dx again

$$\frac{d^2V}{dx^2} = -\frac{dI}{dx} \cdot z = Vyz \tag{9.45}$$

This expression is of the form

$$\frac{d^2\beta}{dx^2} = \alpha^2\beta \tag{9.46}$$

which has the general solution

$$\beta = P \cosh \alpha x + Q \sinh \alpha x \tag{9.47}$$

and the solution for Eq. 9.45 becomes

$$V = P \cosh (\sqrt{(yz)}x) + Q \sinh (\sqrt{(yz)}x) \tag{9.48}$$

The values of the arbitrary constants P and Q depend on the conditions at the ends of the line. These are determined in the following manner.

If the zero for x is taken to be at the output port where the voltage and current are V_2 and I_2, substituting for $x = 0$ in Eq. 9.48

$$V_2 = P + 0 \tag{9.49}$$

Differentiating Eq. (9.48)

$$\frac{dV}{dx} = \sqrt{(yz)} \cdot P \sinh (\sqrt{(yz)}x) + \sqrt{(yz)}Q \cosh (\sqrt{(yz)}x) \tag{9.50}$$

and by dividing by z using Eq. (9.43)

$$I = -\sqrt{(y/z)}P \cdot \sinh (\sqrt{(yz)}x) - \sqrt{(y/z)}Q \cosh (\sqrt{(yz)}x) \tag{9.51}$$

Substituting for $x = 0$ in Eq. (9.51)

$$I_2 = 0 - \sqrt{(y/z)}Q \tag{9.52}$$

We also note that if the length of the line is l, at the input port we have

$$x = -l \tag{9.53}$$
$$\cosh (\sqrt{(yz)}x) = \cosh (-\sqrt{(yz)}l) = \cosh \sqrt{(YZ)} \tag{9.54}$$
$$\sinh (\sqrt{(yz)}x) = \sinh(-\sqrt{(yz)}l) = -\sinh\sqrt{(YZ)} \tag{9.55}$$

Substituting (9.49), (9.52), (9.54), (9.55) in Eqs. (9.48) and (9.51) we have for the input port

$$V_1 = V_2 \cosh \sqrt{(YZ)} + I_2 \sqrt{(Z/Y)} \sinh \sqrt{(YZ)} \quad \cdot \tag{9.56}$$
$$I_1 = V_2 \sqrt{(Y/Z)} \sinh \sqrt{(YZ)} + I_2 \cosh \sqrt{(YZ)} \tag{9.57}$$

In matrix form

$$\begin{bmatrix} V_1 \\ I_1 \end{bmatrix} = \begin{bmatrix} \cosh \sqrt{(YZ)} & \sqrt{(Z/Y)} \sinh \sqrt{(YZ)} \\ \sqrt{(Y/Z)} \sinh \sqrt{(YZ)} & \cosh \sqrt{(YZ)} \end{bmatrix} \cdot \begin{bmatrix} V_2 \\ I_2 \end{bmatrix} \quad (9.58)$$

In this form we have the transmission constants $A\,B\,C\,D$ for the line.
    Y and Z are complex numbers

$$Y = G + j\omega C$$
$$Z = R + j\omega L \quad (9.59)$$

to allow for resistance and leakage of the line
The complex constants are

$$A = D = \cosh \sqrt{(YZ)} = M \quad (9.60)$$
$$B = \sqrt{(Z/Y)} \sinh \sqrt{(YZ)} = ZN \quad (9.61)$$
$$\dot{C} = \sqrt{(Y/Z)} \sinh \sqrt{(YZ)} = YN \quad (9.62)$$

where
$$M = \cosh \sqrt{(YZ)} \quad (9.63)$$
$$N = \frac{\sinh \sqrt{(YZ)}}{\sqrt{(YZ)}} \quad (9.64)$$

Expanding $\cosh \alpha$ and $\sinh \alpha$ by infinite series

$$\cosh \alpha = 1 + \frac{\alpha^2}{2!} + \frac{\alpha^4}{4!} + \frac{\alpha^6}{6!} \quad (9.65)$$

$$\sinh \alpha = \alpha + \frac{\alpha^3}{3!} + \frac{\alpha^5}{5!} + \frac{\alpha^7}{7!} \quad (9.66)$$

with sufficient accuracy for all practical purposes. Hence by substitution

$$M = 1 + \frac{(YZ)}{2} + \frac{(YZ)^2}{24} + \frac{(YZ)^3}{720} \quad (9.67)$$

$$N = 1 + \frac{YZ}{6} + \frac{(YZ)^2}{120} + \frac{(YZ)^3}{5040} \quad (9.68)$$

Summarising the above analysis we observe that a transmission line is regarded as a two-port device with a transmission equation for steady-state a.c. working given by

$$\begin{bmatrix} V_1 \\ I_1 \end{bmatrix} = \begin{bmatrix} M & ZN \\ YN & M \end{bmatrix} \cdot \begin{bmatrix} V_2 \\ I_2 \end{bmatrix} \quad (9.69)$$

The $A\,B\,C\,D$ constants are simply 1, Z, Y and 1 respectively with M and N regarded as correcting factors, depending on the product $Y\,Z$.
    Table 16 in the Appendix gives values of M and N which have been calculated from Eqs. (9.67) and (9.68) over a range of complex values of the product $Y\,Z$ normally met in practice. When M and N are determined from the tables, Eqs. (9.69) can be quickly obtained.

CHAPTER 10

Network theorems

Numerous theorems exist leading to the simplification of complicated networks, pointing out matters of symmetry, or introducing transforms making for a more complete understanding of a system.

This chapter deals with some of the more important of these aids.

10.1 The principle of superposition

This theorem states that if a network has a number of voltage sources, the current at any point is the sum of the currents that would flow at that point due to each of the sources acting separately. When considering the sources acting individually, the other sources should be removed and the terminals to which they were connected should be short-circuited.* In other words the e.m.f. of the source should be removed but the circuit should not be broken.

The theorem of superposition is simply a restatement of the linearity principle (page 15) but it does introduce an alternative point of view, even for simple circuits. In the simple circuit of Fig. 10.1 the current is obviously 0·5 amps due to a total voltage of 5 V applied to the resistance of 10 Ω. On the other hand, this is seen to be equal to the sum of the currents 0·2 A and 0·3 A due to the two sources acting separately.

The proof of the theorem as applied to networks is as follows:
If the loop currents and voltages are given by

$$[V] = [Z][I],\qquad(10.1)$$

this impedance matrix can be partitioned to give the port voltages and currents (see Sec. 9.1), yielding

$$[V'] = [Z'][I'],\qquad(10.2)$$

---

* This statement assumes that the sources are ideal ones. A practical source which has internal impedance is considered to be an ideal source in series with impedance. Under these circumstances the ideal source is removed but the impedance must remain.

the inverse of which is

$$[I'] = [Y'][V'] \qquad (10.3)$$

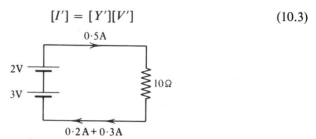

FIG. 10.1 A simple circuit illustrating the principle of superposition

For the kth port we see from the kth row of Eq. (10.3),

$$I'_k = Y'_{k1}V'_1 + Y'_{k2}V'_2 + Y'_{k3}V'_3 + \cdots Y'_{kk}V'_k + \cdots \qquad (10.4)$$

The current in the kth port is thus seen to be the superposition of components, one due to each of the port voltages as stated by the theorem.

10.2 Thevenin's theorem

This theorem (which should really be attributed to Helmholtz) is defined in several ways. Possibly the most useful is the following one where it is expressed as a set of instructions for finding a current.

'To determine the current at any point in a network:

(a) break the circuit at this point and determine (measure) the open-circuit voltage across the break (V_0);
(b) replace all source voltages by their internal impedances and determine the internal impedance as measured at the break (Z_0);
(c) then the original current I is given by

$$I = V_0/Z_0, ' \qquad (10.5)$$

Figure 10.2 illustrates these steps.

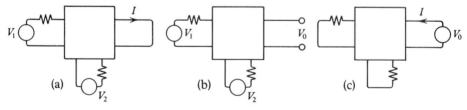

FIG. 10.2 (a) A circuit in which it is required to find the current I. (b) The same circuit with a break at the point where current is to be determined. (c) Voltage V_0 applied at the break

Figure 10.2(*a*) shows the original current flowing in part of a network where the source voltages are V_1 and V_2.

In Fig. 10.2(*b*) the break has been made, V_0 appearing across the break and in Fig. 10.2(*c*) the sources have been short-circuited to enable the impedance to be measured.

The proof also arises from Eq. (10.4).

Since the current in the kth port of a network is given by

$$I_k = Y_{k1}V_1 + Y_{k2}V_2 + Y_{k3}V_3 \cdots Y_{kk}V_k + \cdots \qquad (10.4)$$

if the break is made at this port the voltage V_0 is subtracted from the applied voltage V_k at the port and I_k vanishes.

Thus

$$0 = Y_{k1}V_1 + Y_{k2}V_2 + Y_{k3}V_3 \cdots Y_{kk}(V_k - V_0) \qquad (10.6)$$

Subtracting Eq. (10.6) from Eq. (10.4)

$$I_k = Y_{kk}V_0 \qquad (10.7)$$

This agrees with Eq. (10.5) since Y_{kk} is equal to $1/Z_0$ by definition.

An alternative version of Thevenin's theorem is as follows:

'Viewed from a given port, a linear network can be replaced by a two terminal device comprising a voltage source V_0 and an impedance Z_0 in series.' This is illustrated in Fig. 10.3(*a*).

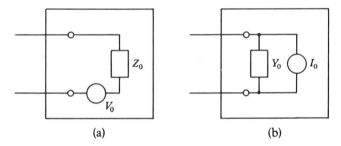

(a)   (b)

FIG. 10.3 (a) Illustrating Thevenin's Theorem. (b) Illustrating Norton's Theorem

The following example illustrates the use of Thevenin's theorem.

10.2.1 Example

Determine the current in the 5 Ω resistor in Fig. 10.4(*a*).

If we imagine a break in series with this resistor, the network is immediately simplified. The voltage V_{cd} is divided at **a** in the ratio 3:7, and at **b** in the ratio 8:2. Thus by inspection we can write

$$V_0 = V_{ab} = +5 \ V$$

When the source is removed and terminals **cd** short-circuited, the network viewed from the break is shown in Fig. 10.4(*c*).

Thus the internal impedance is given by

$$Z_0 = 5 + \frac{2 \times 8}{2+8} + \frac{3 \times 7}{3+7}$$

$$= 5 + 1\cdot6 + 2\cdot1$$

$$= 8\cdot7$$

and therefore in the original circuit

$$I = V_0/Z_0 = 5/8\cdot7 \text{ amperes.}$$

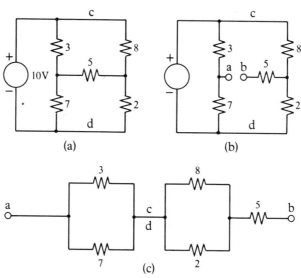

(a)

(b)

(c)

FIG. 10.4 (a) A circuit in which it is required to find the current in the 5Ω resistor. (b) The same circuit with a Thevenin break. (c) The Thevenin impedance across the break

10.2.2 Example

Another example of the use of Thevenin's Theorem is the method for determining the necessary rupturing capacity required by a circuit breaker installed at a given point in a power network. Figure 10.5 shows a power system consisting of three generating stations, **A**, **B** and **C**.

A duplicate feeder interconnects **A** and **B**, and single inter-connectors join **B.C** and **C.A**. The diagram is a single line diagram with each line representing the three conductors of a three-phase system. Circuit breakers are installed at various places in the system and the problem in hand is to determine the rating of the breaker at **D**.

In a problem of this type, resistance is usually ignored, and each unit (generator, transformer or line) is represented by its reactance. It is preferable for each reactance to be expressed as a per-unit value with reference to an entirely arbitrary common base load.*

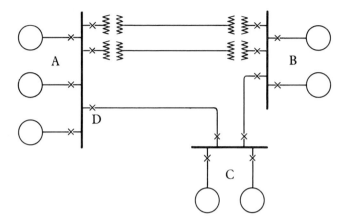

FIG. 10.5 A single-line diagram of a power system comprising generators, transmission lines and transformers

The equivalent circuit corresponding to one phase of the system is shown in Fig. 10.6 with the neutral conductor (actual or virtual) represented by a dotted line.

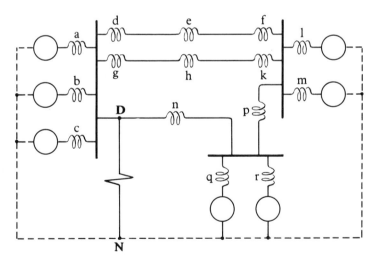

FIG. 10.6 The equivalent circuit of one phase of the power system in Fig. 10.5

The normal rated voltage of the system is produced between **D** and **N** at the circuit breaker.

The most severe conditions with which the breaker has to contend is due to

* See section 16.7 and *Electrical Machines*, section 1.3.

a three-phase short circuit occurring on the line in close proximity to the breaker. Therefore, two conditions must be investigated.

(a) with the fault on the line side,

(b) with the fault on the bus bar side.

In both cases the fault short circuits points **D** and **N** and so the fault current will be the same though the currents in the breaker will differ.

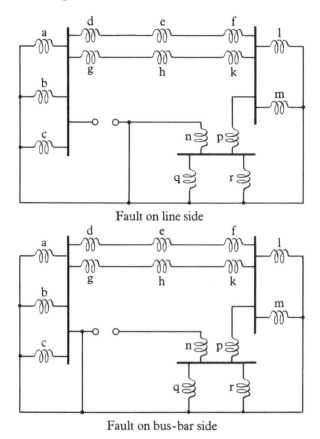

Fault on line side

Fault on bus-bar side

FIG. 10.7 The effect of a three-phase fault is a short-circuit between points *D* and *N*

To determine this fault current using Thevenin's Theorem, we first note that the Thevenin open-circuit voltage is the voltage appearing between **D** and **N** before the incidence of the fault and that is the normal system voltage. Thus $V_0 = V$.

Next the internal e.m.f.'s of all the generators are removed, giving the equivalent circuits of Fig. 10.7, and the value of Z_0 is measured at the fault terminals **D** and **N**.

Figure 10.8 shows the voltage V applied at the fault terminals to measure the circuit impedance.

The current I_1 in this circuit is equal to the fault current, I_2 is the current to be interrupted by the breaker when the fault is on the line side, and I_3 is the breaker current with the fault on the busbar side.

This circuit can be solved analytically or, alternatively, can be set up on a d.c. network analyser, consisting of variable resistance elements representing the reactances.

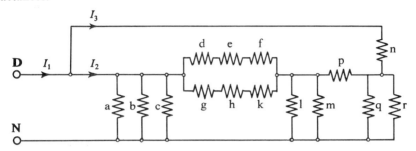

FIG. 10.8 The Thevenin impedance between D and N

The values of these resistances are set to correspond to the problem and are interconnected accordingly, thus enabling the equivalent impedance and also the equivalent fault current to be measured.

10.3 Norton's theorem

The dual form of Thevenin's theorem is Norton's theorem. 'Viewed from a given port a linear network can be replaced by a two-terminal device comprising a current source I_0 and an admittance Y_0 in parallel.' (Fig. 10.3(b)).

The two diagrams 10.3(a) and 10.3(b) are equivalent, provided

$$I_0 = V_0/Z_0$$

and

$$V_0 = I_0/Y_0 \tag{10.8}$$

Norton's theorem is used in nodal analysis in the same way that Thevenin's theorem is helpful to mesh analysis.

10.4 Star-mesh transform

Figure 10.9 shows two possible three-terminal networks, the first consisting of a star or T system of three impedances, Z_a, Z_b, Z_c and the second a mesh or Π system of three impedances, Z_1, Z_2, Z_3.

If the values of Z_a, Z_b, Z_c are known, it is possible to choose values of Z_1, Z_2 and Z_3, which make the two circuits indistinguishable to external measurement. In such circumstances, one system could replace the other as part of a larger network without altering external currents and voltages.

According to Chap. 9, page 135, the transmission matrices corresponding to these networks are given by

$$[A]_{\text{star}} = \begin{bmatrix} 1+Z_a/Z_c & Z_a+Z_b+Z_aZ_b/Z_c \\ 1/Z_c & 1+Z_b/Z_c \end{bmatrix} \qquad (10.9)$$

$$[A]_{\text{mesh}} = \begin{bmatrix} 1+Z_3/Z_1 & Z_3 \\ 1/Z_1+1/Z_2+Z_3/Z_1Z_2 & 1+Z_3/Z_2 \end{bmatrix} \qquad (10.10)$$

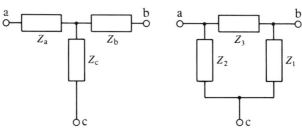

FIG. 10.9 Star/mesh or T/Π equivalents

If the two networks are to be equivalent, the transmission matrices must be identical

$$[A]_{\text{star}} = [A]_{\text{mesh}}$$

Thus by equating the B terms it is noted

$$Z_3 = Z_a+Z_b+Z_aZ_b/Z_c$$

and similarly

$$Z_1 = Z_b+Z_c+Z_bZ_c/Z_a \qquad \left.\right\} \qquad (10.11)$$

$$Z_2 = Z_c+Z_a+Z_cZ_a/Z_b$$

Equating the C terms

$$1/Z_c = 1/Z_1+1/Z_2+Z_3/Z_1Z_2$$

whence

$$Z_c = Z_1Z_2/(Z_1+Z_2+Z_3)$$

and similarly

$$Z_a = Z_2Z_3/(Z_1+Z_2+Z_3) \qquad \left.\right\} \qquad (10.12)$$

$$Z_b = Z_3Z_1/(Z_1+Z_2+Z_3)$$

Equations (10.11) and (10.12) can be used to determine the mesh elements corresponding to a given star and vice versa.

10.4.1 Reduction of networks by star-mesh substitution

Figure 10.10(a) is an example of a three-loop network, one loop consisting of the three impedances Z_3, Z_4 and Z_5 connected in mesh. The only connections to this mesh are made at the nodes **a**, **b** and **c**. If the mesh is replaced by its equivalent star, Fig. 10.10(b), the simplification of the network is apparent. It can now be solved as a simple series–parallel circuit. It should be noted that the introduction of a mesh-star transform reduces the number of loops of a network by one. The process can be repeated if more mesh-connected elements exist, and a complicated network can be considerably reduced in this way.*

The process is equivalent to the method of partitioning a matrix described in Sec. 3.8. This can be demonstrated with reference to the following example. The impedance matrix corresponding to the network of Fig. 10.10(a) is given by

$$\begin{bmatrix} V_1 \\ 0 \\ 0 \end{bmatrix} \begin{bmatrix} Z_1+Z_4 & -Z_1 & -Z_4 \\ -Z_1 & Z_1+Z_2+Z_3 & -Z_3 \\ -Z_4 & -Z_3 & Z_3+Z_4+Z_5 \end{bmatrix} \cdot \begin{bmatrix} I_1 \\ I_2 \\ I_3 \end{bmatrix}$$ (10.13)

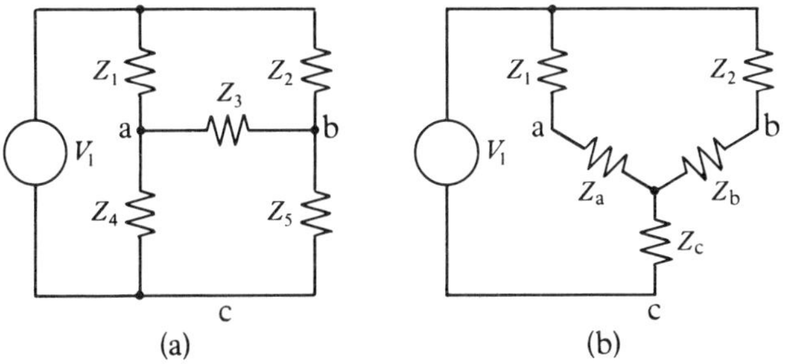

(a)   (b)

FIG. 10.10 The mesh formed by $Z_3Z_4Z_5$ is replaced by the star $Z_aZ_bZ_c$

Eliminating the last row and column by partitioning,

$$\begin{bmatrix} V_1 \\ 0 \end{bmatrix} = [Z'] \cdot \begin{bmatrix} I_1 \\ I_2 \end{bmatrix}$$

where

$$[Z'] = \begin{bmatrix} Z_1+Z_4 & -Z_1 \\ -Z_1 & Z_1+Z_2+Z_3 \end{bmatrix} - \frac{1}{Z_3+Z_4+Z_5} \begin{bmatrix} -Z_4 \\ -Z_3 \end{bmatrix} [-Z_4 \quad -Z_3]$$

$$= \begin{bmatrix} Z_1+Z_4 & -Z_1 \\ -Z_1 & Z_1+Z_2+Z_3 \end{bmatrix} - \frac{1}{Z_3+Z_4+Z_5} \begin{bmatrix} Z_4{}^2 & Z_3Z_4 \\ Z_3Z_4 & Z_3{}^2 \end{bmatrix}$$ (10.14)

The corresponding equations relating to Fig. 10.10(b) after the mesh-star substitution has been made are

$$[Z'] = \begin{bmatrix} Z_1+Z_a+Z_c & -Z_1-Z_a \\ -Z_1-Z_a & Z_1+Z_2+Z_a+Z_b \end{bmatrix}$$ (10.15)

where

$$Z_a = Z_3Z_4/(Z_3+Z_4+Z_5)$$

$$Z_b = Z_5Z_3/(Z_3+Z_4+Z_5)$$

$$Z_c = Z_4Z_5/(Z_3+Z_4+Z_5)$$

* If the principle of duality has been thoroughly assimilated it will be appreciated that the number of *nodes* in a network can be reduced by the progressive replacement of *stars* by *meshes* thus facilitating nodal analysis.

Equations (10.14) and (10.15) both reduce to

$$[Z'] = \frac{1}{Z_3+Z_4+Z_5} \begin{bmatrix} Z_1Z_3+Z_1Z_4+Z_1Z_5 \\ +Z_3Z_4+Z_4Z_5 \\ \\ -(Z_1Z_3+Z_1Z_4 \\ +Z_1Z_5+Z_3Z_4) \end{bmatrix} \begin{matrix} -(Z_1Z_3+Z_1Z_4 \\ +Z_1Z_5+Z_3Z_4) \\ \\ Z_1Z_3+Z_1Z_4+Z_1Z_5 \\ +Z_2Z_3+Z_2Z_4+Z_2Z_5 \\ +Z_3Z_4+Z_3Z_5 \end{matrix} \end{bmatrix}$$

$$(10.16)$$

10.5 Reciprocity

The reciprocity theorem states that if a current is produced at point **a** in a network by a source acting at point **b**, then the same current would be produced at point **b** by the source acting at point **a**.

Thus, in Fig. 10.11, the network shown is a two-port device, and if a voltage V is applied at one port and the current is I at the other port short-circuited, the same current would be produced by the same voltage with the ports interchanged.

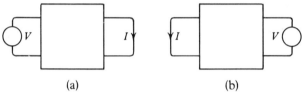

(a)   (b)

FIG. 10.11 Illustrating reciprocity

Simple proof follows from the equations of a two-port network since

$$\begin{bmatrix} V_1 \\ I_1 \end{bmatrix} = \begin{bmatrix} A & B \\ C & D \end{bmatrix} \cdot \begin{bmatrix} V_2 \\ I_2\dagger \end{bmatrix} \qquad (9.10)$$

when V_2 is equal to zero (Fig. 10.11(a))

$$V_1 = BI_2\dagger \qquad (10.17)$$

But

$$\begin{bmatrix} V_2 \\ I_2\dagger \end{bmatrix} = \begin{bmatrix} D & -B \\ -C & A \end{bmatrix} \cdot \begin{bmatrix} V_1 \\ I_1 \end{bmatrix} \qquad (9.16a)$$

so that when V_1 is equal to zero (Fig. 10.11(b)),

$$V_2 = B(-I_1) \qquad (10.18)$$

The network thus presents the same transfer impedance in the two directions.

10.6 The parallel generator theorem

Generators in parallel (Fig. 10.12) form a simple two-node network readily solved by nodal analysis. Eq. (6.9) applied to node **a**, yields

$$(E_1 Y_1 + E_2 Y_2 + E_3 Y_3) = (Y + Y_1 + Y_2 + Y_3)V_{ab} \qquad (10.19)$$

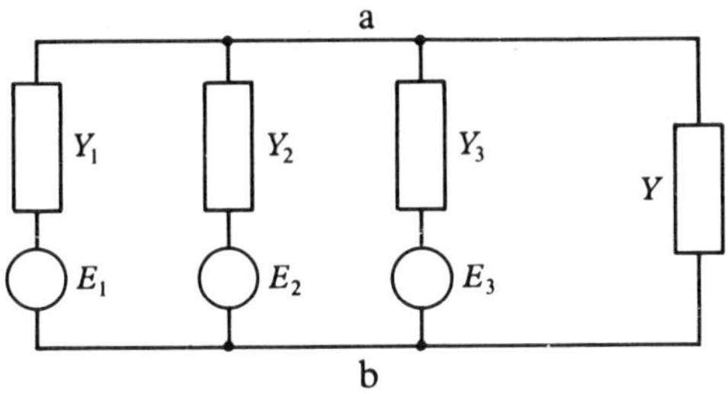

FIG. 10.12 Three generators operating in parallel with a common load

Thus the parallel generator theorem states:
'The common terminal voltage of a number of generators operating in parallel is equal to the sum of their short-circuit currents divided by the total admittance of the system including the load.'

An example of the parallel generator theorem applied to transformers is given in *Electrical Machines*, p. 42.

CHAPTER 11

The two-winding transformer

11.1 Mutual inductance

Mutual inductance exists between two coils if a voltage is set up in one of the coils as the result of a current change in the other coil. The value of the mutual inductance in henries is defined as the value of the voltage in volts produced in coil **1** by current changing at the rate of one ampere per second in coil **2**.

In Chap. 3 the symbol M_{12} is ascribed to this parameter. Similarly M_{21} is the voltage produced in coil **2**, by unit rate of change of current in coil **1**.

It is usually assumed that M_{12} and M_{21} are identical, hence the term mutual

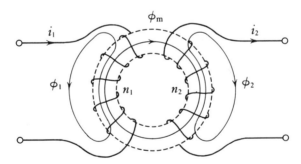

FIG. 11.1 A simplified diagram of the mutal and leakage flux produced by two coils

inductance, one single parameter which is a common property of the two coils. This singularity is not immediately obvious, and it is therefore suitable for further investigation.

Consider two coils shown in Fig. 11.1 having a number of turns n_1 and n_2 respectively. The resistance of the coils will be neglected.

Figure 11.1 shows two coils in close proximity, the first coil having n_1 turns carrying a current i_1 and the second having n_2 turns carrying a current i_2.

Three magnetic circuits are superimposed:

1. a magnetic field consisting of flux ϕ_1 generated by and linking with the n_1 turns of i_1;
2. a magnetic field consisting of flux ϕ_2 generated by and linking with the n_2 turns of i_2;
3. a magnetic field consisting of flux ϕ_m generated by the combined magneto-motive-force of both coils and linking with both.

The permeances of these magnetic circuits are Λ_1, Λ_2 and Λ_m respectively. Provided these permeances are constant, that is to say there is a linear relation-ship between flux and current and that no relative motion occurs between the coils or with respect to any other magnetic field, we can apply the principle of superposition and consider the effect of these fields acting separately. This is

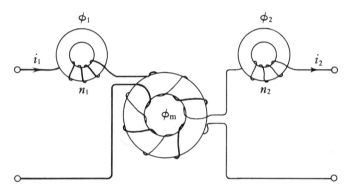

FIG. 11.2 Applying the principle of superposition to separate mutual and leakage flux

shown diagrammatically in Fig. 11.2, where the flux-turns $\phi_1 n_1$ and $\phi_2 n_2$ have been separated.

The values of the flux in the three magnetic circuits are

$$\phi_1 = \Lambda_1 n_1 i_1 \tag{11.1}$$

$$\phi_2 = \Lambda_2 n_2 i_2 \tag{11.2}$$

$$\phi_m = \Lambda_m(n_1 i_1 + n_2 i_2) \tag{11.3}$$

and the total flux linkages with the two electrical circuits are

$$\text{flux linkages with } i_1 = (\phi_1 + \phi_m)n_1 \tag{11.4}$$

$$\text{flux linkages with } i_2 = (\phi_2 + \phi_m)n_2 \tag{11.5}$$

In all real mutual inductances the linking flux is not necessarily the same for each of the n turns of a single coil as the Eqs. (11.4) and (11.5) suggest. It is justifiable and convenient to assume a value of ϕ common to all the turns by modifying Λ accordingly.

Equations (11.4) and (11.5) are presented below in matrix notation

$$\begin{bmatrix} \text{flux linkages with } i_1 \\ \text{flux linkages with } i_2 \end{bmatrix} = \begin{bmatrix} (\phi_1 + \phi_m)n_1 \\ (\phi_2 + \phi_m)n_2 \end{bmatrix} \tag{11.6}$$

$$= \begin{bmatrix} (\Lambda_1 + \Lambda_m)n_1{}^2 i_1 + \Lambda_m n_2 n_1 i_2 \\ \Lambda_m n_1 n_2 i_1 + (\Lambda_2 + \Lambda_m)n_2{}^2 i_2 \end{bmatrix}$$

$$= \begin{bmatrix} (\Lambda_1 + \Lambda_m)n_1{}^2 & \Lambda_m n_1 n_2 \\ \Lambda_m n_1 n_2 & (\Lambda_2 + \Lambda_m)n_2{}^2 \end{bmatrix} \cdot \begin{bmatrix} i_1 \\ i_2 \end{bmatrix} \tag{11.7}$$

Since voltage is equal to rate of change of flux linkages

$$\begin{bmatrix} v_1 \\ v_2 \end{bmatrix} = p \begin{bmatrix} (\Lambda_1 + \Lambda_m)n_1{}^2 & \Lambda_m n_1 n_2 \\ \Lambda_m n_1 n_2 & (\Lambda_2 + \Lambda_m)n_2{}^2 \end{bmatrix} \cdot \begin{bmatrix} i_1 \\ i_2 \end{bmatrix} \tag{11.8}$$

$$= p \begin{bmatrix} L_1 & M_{12} \\ M_{21} & L_2 \end{bmatrix} \cdot \begin{bmatrix} i_1 \\ i_2 \end{bmatrix} \tag{11.9}$$

Equation (11.9) is derived from Eq. (4.6) for two primitive coils. Thus

$$L_1 = (\Lambda_1 + \Lambda_m)n_1{}^2 \tag{11.10}$$

$$L_2 = (\Lambda_2 + \Lambda_m)n_2{}^2 \tag{11.11}$$

and

$$M_{12} = M_{21} = \Lambda_m n_1 n_2 \tag{11.12}$$

Provided the coils are not in relative motion with respect to each other or to any other magnetically coupled circuit, Eq. (11.12) applies. The effect of relative movement will be considered later in Chap. 16.

11.2 Coupling factor

Now that it has been proved that M_{12} is equal to M_{21}, the suffixes can be dropped and the relationship between M, L_1 and L_2 can be determined.

From Eqs. (11.10), (11.11) and (11.12)

$$M^2 = \Lambda_m{}^2 n_1{}^2 n_2{}^2$$

$$= \frac{\Lambda_m{}^2}{(\Lambda_1 + \Lambda_m)(\Lambda_2 + \Lambda_m)} L_1 L_2 \tag{11.13}$$

and

$$M = k \sqrt{(L_1 L_2)} \tag{11.14}$$

where

$$k^2 = \frac{\Lambda_m{}^2}{(\Lambda_1 + \Lambda_m)(\Lambda_2 + \Lambda_m)} \tag{11.15}$$

and k is known as the coupling factor.

The values of k depends on the spacing of the two coils. It is equal to unity in the ideal case of two closely coupled coils when all the flux generated by one coil links with the other (i.e. when $\phi_1 = \phi_m = \phi_2$). This is the ideal relationship referring to a perfect transformer.

11.3 Impedance matrix of a transformer

Ignoring core loss and the effects of capacitance either between windings, between windings and core or between turns, a two winding transformer is simply two circuits coupled by mutual inductance and, therefore, the general impedance matrix (Eqs. (4.2) and (4.6)) of this device is given by

$$\begin{bmatrix} V_1 \\ V_2 \end{bmatrix} = \begin{bmatrix} Z_{11} & Z_{12} \\ Z_{21} & Z_{22} \end{bmatrix} \cdot \begin{bmatrix} I_1 \\ I_2 \end{bmatrix} \tag{11.16}$$

or

$$\begin{bmatrix} V_1 \\ V_2 \end{bmatrix} = \begin{bmatrix} Z_1 & Z_m \\ Z_m & Z_2 \end{bmatrix} \cdot \begin{bmatrix} I_1 \\ I_2 \end{bmatrix} \tag{11.17}$$

where

$$Z_1 = R_1 + pL_1 \tag{11.18}$$

$$Z_2 = R_2 + pL_2 \tag{11.19}$$

$$Z_m = pM \tag{11.20}$$

Values of Z_1, Z_2 and Z_m (for a given transformer) can all be determined from open-circuit tests.

11.4 Equivalent T circuit

The equivalent **T** circuit shown in Fig. 11.3 could be used to represent a transformer, since this circuit has an impedance matrix identical with Eq. (11.17). It is not often used, however, since it does not offer any simplification.

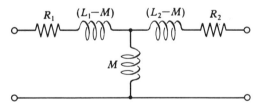

FIG. 11.3 An equivalent T circuit representing two coils with self inductances L_1 and L_2 and mutual inductance M

11.5 Equivalent [A] matrix

Since a transformer is invariably used as a two-port device, the [A] matrix is probably the best way of expressing its parameters. It is of particular value since, as we have seen in Chap. 10, the elements of the $[A]$ matrix are obtained directly from open-circuit and short-circuit tests. From Eq. (9.28)

$$\begin{bmatrix} V_1 \\ I_1 \end{bmatrix} = \begin{bmatrix} A & Z_{10}Z_{1\mathbf{x}}/A(Z_{10}-Z_{1\mathbf{x}}) \\ A/Z_{10} & Z_{10}/A(Z_{10}-Z_{1\mathbf{x}}) \end{bmatrix} \cdot \begin{bmatrix} V_2\dagger \\ I_2\dagger \end{bmatrix} \tag{11.21}$$

where A is the open-circuit voltage ratio V_{10}/V_{20}

Z_{10} is the open-circuit impedance V_{10}/I_{10}

$Z_{1\mathbf{x}}$ is the short-circuit impedance $V_{1\mathbf{x}}/I_{1\mathbf{x}}$

For a power transformer Z_{10} is much greater than $Z_{1\mathbf{x}}$ hence

$$Z_{10}/(Z_{10}-Z_{1\mathbf{x}}) \simeq 1$$

and Eq. (11.21) reduces to

$$\begin{bmatrix} V_1 \\ I_1 \end{bmatrix} = \begin{bmatrix} A & Z_{1\mathbf{x}}/A \\ A/Z_{10} & 1/A \end{bmatrix} \cdot \begin{bmatrix} V_2\dagger \\ I_2\dagger \end{bmatrix} \tag{11.22}$$

11.6 The hybrid matrix

It is customary for a power transformer rating to be specified in terms of the primary voltage V_1 and the secondary current I_2. Thus a 500 kVA, 1000/400–V transformer will be expected to give 400 V across the L.V. terminals on an open-circuit test when the applied voltage to the H.V. winding is 1000, but the L.V. terminal voltage on load will differ slightly due to the *inherent voltage regulation*.* The rated output current is 1250 but the corresponding input current will not be exactly 500 A since magnetising current must be included.

The $[A]$ matrix gives $[V_1 I_1]$ in terms of $[V_2\dagger I_2\dagger]$. For a power transformer we wish to know $[V_2\dagger I_1]$ in terms of $[V_1 I_2\dagger]$ and the matrix connecting these is known as the hybrid matrix $[H]$.
Thus if

$$[A] = \begin{bmatrix} A & B \\ C & D \end{bmatrix} \tag{11.23}$$

where A, B, C and D have the meanings given in Sec. 9.5

$$V_1 = AV_2\dagger + BI_2\dagger$$
$$I_1 = CV_2\dagger + DI_2\dagger$$
$$V_2\dagger = (1/A)V_1 + (-B/A)I_2\dagger$$
$$I_1 = (C/A)V_1 + (-BC/A)I_2\dagger + DI_2\dagger$$
$$= (C/A)V_1 + (1/A)I_2\dagger$$

or

$$\begin{bmatrix} V_2\dagger \\ I_1 \end{bmatrix} = \begin{bmatrix} 1/A & -B/A \\ C/A & 1/A \end{bmatrix} \cdot \begin{bmatrix} V_1 \\ I_2\dagger \end{bmatrix} \tag{11.24}$$

* *Electrical Machines*, p. 32.

$$= [H] \cdot \begin{bmatrix} V_1 \\ I_2\dagger \end{bmatrix} \qquad\qquad (11.25)$$

The elements of the hybrid matrix are obtained directly from open-circuit and short-circuit tests. The open-circuit is usually taken by measurements on the L.V. side giving Z_{10} and $A = V_{10}/V_2$ and the short-circuit test is made on the H.V. side with the L.V. winding short-circuited, thus measuring Z_{2x}. From Eq. (9.29) Z_{2x} was found to be equal to B/A. Hence

$$\begin{bmatrix} V_2\dagger \\ I_1 \end{bmatrix} = \begin{bmatrix} 1/A & -Z_{2x} \\ 1/Z_{10} & 1/A \end{bmatrix} \cdot \begin{bmatrix} V_1 \\ I_2\dagger \end{bmatrix} \qquad\qquad (11.26)$$

11.6.1 Equivalent circuit

It should be observed that although in the general case the element A is a complex ratio, for a power transformer it is assumed to have a negligible imaginary component and is known as the turns ratio.

The equivalent circuit representing Eqs. (11.22) and (11.26) is shown in Fig. 11.4.

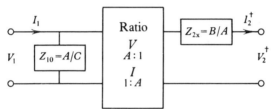

FIG. 11.4 Block diagram representing Equation 11.26

CHAPTER 12

The three-winding transformer

12.1 Introduction

The three-winding transformer is an excellent example of a three-port network. It is used to couple three circuits together, usually for the purpose of transferring energy from one to the other two. In these circumstances the first winding becomes the primary winding and the others are the secondary windings. Usually the secondaries are wound for different voltages and have different ratings.

When such a transformer is loaded, the secondary terminal voltages will be subject to leakage impedance drops in a somewhat similar manner to those occurring in a two-winding transformer, but to some extent the drop in voltage at one of the secondary terminals will depend on the load on the other secondary.

12.2 The impedance matrix

Since in the three-winding transformer we have three independent circuits linked with mutual inductance, the impedance matrix is given by

$$[Z] = \begin{bmatrix} Z_{11} & Z_{12} & Z_{13} \\ Z_{21} & Z_{22} & Z_{23} \\ Z_{31} & Z_{32} & Z_{33} \end{bmatrix} \tag{4.2}$$

which can be simplified to give

$$[Z] = \begin{bmatrix} Z_1 & Z_{21} & Z_{13} \\ Z_{21} & Z_2 & Z_{32} \\ Z_{13} & Z_{32} & Z_3 \end{bmatrix} \tag{12.1}$$

where

$$Z_1 = R_1 + pL_1$$
$$Z_2 = R_2 + pL_2$$

$$Z_3 = R_3 + pL_3$$
$$Z_{21} = pM_{21}$$
$$Z_{13} = pM_{13}$$
$$Z_{32} = pM_{32}$$

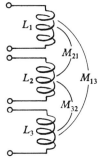

FIG. 12.1 Three coils with self and mutual inductance

and since it is known that the matrix must be symmetrical about the diagonal (from Eq. (11.12)).

The elements in this matrix can be determined from open-circuit tests. The results of open-circuit and short-circuit tests obtained by taking the windings in pairs and leaving the third winding open-circuited are given below:

(a) Winding (1) primary
  Winding (2) secondary
  Winding (3) open circuit

$$\begin{bmatrix} V_1 \\ V_2 \end{bmatrix} = \begin{bmatrix} Z_1 & Z_{21} \\ Z_{21} & Z_2 \end{bmatrix} \cdot \begin{bmatrix} I_1 \\ I_2 \end{bmatrix} \tag{12.2}$$

Open circuit test

$$Z_{10} = V_{10}/I_{10} = Z_1 \tag{12.3}$$
$$V_{20}/I_{10} = Z_{21} \tag{12.4}$$

Short circuit test

$$* \quad Z_{12x} = V_{1x}/I_{1x} = Z_1 - Z_{21}^2/Z_2 \tag{12.5}$$

(b) Winding (1) open circuit
  Winding (2) primary
  Winding (3) secondary

$$\begin{bmatrix} V_2 \\ V_3 \end{bmatrix} = \begin{bmatrix} Z_2 & Z_{32} \\ Z_{32} & Z_3 \end{bmatrix} \cdot \begin{bmatrix} I_2 \\ I_3 \end{bmatrix} \tag{12.6}$$

* This is obtained by putting V_2 equal to zero and eliminating the last row and column using the method of section 3.8.1.

Open-circuit test

$$Z_{20} = V_{20}/I_{20} = Z_2 \tag{12.7}$$

$$V_{30}/I_{20} = Z_{32} \tag{12.8}$$

Short-circuit test

$$Z_{23x} = V_{2x}/I_{2x} = Z_2 - Z_{32}^2/Z_3 \tag{12.9}$$

(c) Winding (1) secondary
    Winding (2) open circuit
    Winding (3) primary

$$\begin{bmatrix} V_3 \\ V_1 \end{bmatrix} = \begin{bmatrix} Z_3 & Z_{13} \\ Z_{13} & Z_1 \end{bmatrix} \cdot \begin{bmatrix} I_3 \\ I_1 \end{bmatrix} \tag{12.10}$$

Open-circuit test

$$Z_{30} = V_{30}/I_{30} = Z_3 \tag{12.11}$$

$$V_{10}/I_{30} = Z_{13} \tag{12.12}$$

Short-circuit test

$$Z_{31x} = V_{3x}I_{3x} = Z_3 - Z_{13}^2 Z_1 \tag{12.13}$$

12.3 The hybrid matrix

Although the impedance matrix is correct for three air-cored coils coupled by mutual inductance, substantial inaccuracies arise if it is used in the above form to represent a three-winding power transformer.

1. The impedance matrix does not include the effects of core loss.
2. When the impedance matrix is inverted, many terms involve the difference between two quantities which are nearly alike. Normal errors in measurement completely obscure these differences.

A much more suitable hybrid matrix may be devised which summarises the equations for the unknown voltages and currents in a form which does not require inversion.

If winding (1) is the primary and windings (2) and (3) are loaded, values of V_1, $I_2\dagger$ and $I_3\dagger$ will be known, and V_2, V_3 and I_1 will have to be determined. Thus

$$\begin{bmatrix} V_2 \\ V_3 \\ I_1 \end{bmatrix} = \begin{bmatrix} L & M & N \\ P & Q & R \\ S & T & U \end{bmatrix} \cdot \begin{bmatrix} V_1 \\ I_2\dagger \\ I_3\dagger \end{bmatrix} \tag{12.14}$$

The elements of this hybrid can be determined directly from open-circuit and short-circuit tests in the following manner.

1. Open-circuit test on winding (1)

$$I_2 = 0$$

$$I_3 = 0$$

Thence

$$\begin{bmatrix} V_{20} \\ V_{30} \\ I_{10} \end{bmatrix} = \begin{bmatrix} L \\ P \\ S \end{bmatrix} \cdot [V_{10}]$$

or

$$L = V_{20}/V_{10} = n_2/n_1 = k_{21} \qquad (12.15)$$

$$P = V_{30}/V_{10} = n_3/n_1 = k_{31} \qquad (12.16)$$

$$S = I_{10}/V_{10} = 1/Z_{10} \qquad (12.17)$$

2. Short-circuit test with winding (1) short-circuited, winding (2) excited and winding (3) open-circuited.

Thus

$$V_1 = 0$$

$$I_3 = 0$$

The supply to winding (2) will be V_{2x} and $I_{2x} = (-I_2\dagger)$.

Thus

$$\begin{bmatrix} V_{2x} \\ V_{3x} \\ -I_{1x} \end{bmatrix} = \begin{bmatrix} M \\ Q \\ T \end{bmatrix} \cdot [-I_{2x}]$$

or

$$M = V_{2x}/(-I_{2x}) = -Z_{2x} \qquad (12.18)$$

$$Q = V_{3x}/(-I_{2x}) = -Z_{32x} \qquad (12.19)$$

$$T = (-I_{1x})/(-I_{2x}) = n_2/n_1 = k_{21} \qquad (12.20)$$

3. Short-circuit test with winding (1) short-circuited, winding (2) open-circuited, and winding (3) excited.

$$V_1 = 0$$

$$I_2 = 0$$

The supply to winding (3) will be V_{3y} and $I_{3y} = (-I_3\dagger)$.

Thus

$$\begin{bmatrix} V_{2y} \\ V_{3y} \\ -I_{1y} \end{bmatrix} = \begin{bmatrix} N \\ R \\ U \end{bmatrix} \cdot [-I_{3y}]$$

$$N = V_{2y}/(-I_{3y}) = -Z_{23y} \qquad (12.21)$$

$$R = V_{3y}/(-I_{3y}) = -Z_{3y} \qquad (12.22)$$

$$U = (-I_{1y})/(-I_{3y}) = n_3/n_1 = k_{31} \qquad (12.23)$$

Matrix Eq. (12.14) can be rewritten

$$\begin{bmatrix} V_2 \\ V_3 \\ I_1 \end{bmatrix} = \begin{bmatrix} k_{21} & -Z_{2x} & -Z_{23y} \\ k_{31} & -Z_{32x} & -Z_{3y} \\ 1/Z_{10} & k_{21} & k_{31} \end{bmatrix} \cdot \begin{bmatrix} V_1 \\ I_2\dagger \\ I_3\dagger \end{bmatrix} \qquad (12.24)$$

CHAPTER 13

Symmetrical components

13.1 Introduction

This chapter deals specifically with three-phase circuits. In many elementary problems, three-phase circuits are assumed to be balanced, that is to say conditions are similar in the three phases, impedances are equal, and geometrical symmetry exists between the line conductors and between the windings in machines. Generated voltages are equal in magnitude and each is equally related to the other two voltages, having one a third of a cycle before it and the other a third of a cycle behind it, in time. This also is a type of symmetry, since the phasors representing these voltages form an equilateral triangle.

If such symmetrical voltages are applied to a symmetrical system, it is reasonable to assume that the currents will also be symmetrical and on this assumption, the established technique is to perform calculations relating to only one of the phases, finding the currents in this phase and deducing the others therefrom.

This simplified technique cannot be used when impedances are unequal or when the voltages are out of balance.

To deal with unbalanced problems in general, and to assist in defining the relationships between phase voltages and also phase currents, we shall make use of a sequence operator.

13.2 The sequence operator

The operator j which turns a phasor through 90° is widely used in electrical engineering. A similar operator, which will turn a phasor through 120°, can also be of value, particularly in three-phase problems. This is the operator h.

In Fig. 13.1, if the phasor **OA** represents a voltage V, then **OC** which is equal to **OA** advanced by 120° can be designated hV and similarly **OB** advanced by a further 120° is equal to hhV or h^2V. The sequence operator h has the following properties

$$h = \exp j(2\pi/3) \quad \text{by definition} \tag{13.1}$$
$$h^2 = \exp j(4\pi/3) \tag{13.2}$$
$$h^3 = 1 \tag{13.3}$$

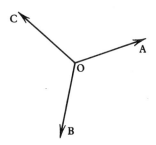

FIG. 13.1 A balanced three-phase system of phasors

If three phasors V, hV and h^2V are added together (Fig. 13.2) they form an equilateral triangle. Hence

$$1+h+h^2 = 0 \tag{13.4}$$
$$-h^2 = 1+h \tag{13.5}$$
$$-h = 1+h^2 \tag{13.6}$$

FIG. 13.2 The geometrical interpretation of the equation $V+hV+h^2V = 0$

Expressing Eqs. (13.1) and (13.2) in cartesian coordinates, we have

$$h = -0\cdot5+j\sqrt{3}/2 \tag{13.7}$$
$$h^2 = -0\cdot5-j\sqrt{3}/2 \tag{13.8}$$

The geometrical interpretation of Eqs. (13.7) and (13.8) is given in Fig. 13.3.

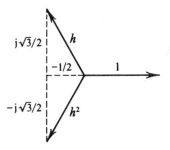

FIG. 13.3 The transformation of h and h^2 to complex operators

13.3 Symmetrical systems

For a system of three phasors, three different types of symmetry are possible:

(a) They may be of equal magnitude and in phase with each other. Since the alternating quantities they represent will each reach its maximum value at the same instant, they are said to have *zero sequence*.

(b) They may be of equal magnitude but may be spaced 120° with respect to each other. Now if they are labelled **a**, **b** and **c**, respectively, and if the alternating quantities they represent reach their maximum values in the same order, **a** followed by **b** followed by **c**, they are said to have *positive sequence*.

(c) Again they may be of equal magnitude, spaced by 120° and labelled **a**, **b** and **c**. However, if this time they follow each other in the reverse order, **a** followed by **c** followed by **b**, they are said to have *negative sequence*.*

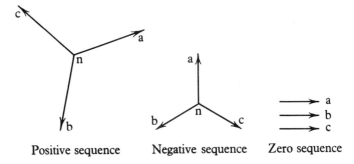

Positive sequence   Negative sequence   Zero sequence

FIG. 13.4 Positive sequence, negative sequence, zero sequence groups of phasors

Groups of phasors representing these three types of symmetry are shown in Fig. 13.4.

Figure 13.5 shows the same groups with the **a** phasors labelled V_1, V_2 and V_0, respectively. These are sometimes known as the reference phasors since the **b** and **c** phasors can be obtained by operating on them by h or h^2, as shown.

The phasor groups are also conveniently represented by the following matrices:

Positive sequence system $[V_{a1} \quad V_{b1} \quad V_{c1}] = [1 \quad h^2 \quad h]V_1$ (13.9)

Negative sequence system $[V_{a2} \quad V_{b2} \quad V_{c2}] = [1 \quad h \quad h_2]V_2$ (13.10)

Zero sequence system $[V_{a0} \quad V_{b0} \quad V_{c0}] = [1 \quad 1 \quad 1]V_0$ (13.11)

* The reader should take care not to be confused if he meets elsewhere the rather ambiguous term 'phase rotation'.

Phasors do not rotate. They are complex numbers which can be displayed geometrically as in Section 7.1.2, and the instantaneous values of the a.c. quantities they represent can be obtained by multiplying by exp $j\omega t$ and taking the real part.

The term, *phase sequence* applied to phasors, refers to the order in which the alternating quantities represented reach their maximum values.

13.4 Synthesis of an unbalanced system

If the nine component phasors shown in Fig. 13.5 are assembled by adding all the **a** phasors together and treating the **b** and **c** phasors similarly, we have the

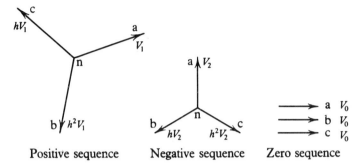

Positive sequence	Negative sequence	Zero sequence

FIG. 13.5 Positive sequence, negative sequence, zero sequence groups of phasors in terms of reference phasors $V_1 V_2$ and V_0

result shown in Fig. 13.6. Obviously these nine components produce a set of three resultant phasors which are unbalanced in magnitude and phase angle.

A useful laboratory demonstration of this fact can be made if we connect an induction regulator in series with the secondary windings of a three-phase

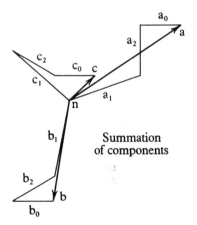

Summation of components

FIG. 13.6 The summation of the nine symmetrical components to form an unbalanced three-phase system

transformer. With normal connection (Fig. 13.7(a)) each phase of the transformer receives the same boost and the resultant line-to-neutral voltages are balanced. Both transformer and regulator are supplying positive sequence voltages.

If, however, the regulator connections are changed (Fig. 13.7(b)), the regulator now supplies voltage in negative sequence, and the resultant voltages will be unbalanced.

So far the system contains only positive and negative sequence components, but if the secondary of a single-phase transformer is connected to the neutral point **n** (Fig. 13.7(c)), zero sequence voltage is added to all three phases. Figure

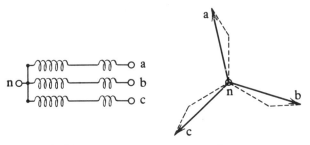

(a) Positive sequence components only

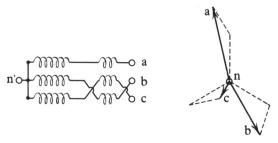

(b) Positive and negative sequence components

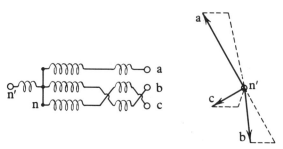

(c) Positive, negative and zero sequence components

FIG. 13.7 Synthesising an unbalanced system from components using a transformer and an induction regulator: (a) positive sequence components only; (b) positive and negative sequence components; (c) positive, negative and zero-sequence components

13.7(c) thus shows how each of the line-to-neutral voltages is built up of three components.

If an unbalanced system of three phasors can be synthesised from nine components in this way, it follows that any unbalanced system (of three phasors) can be analysed into its nine components.

13.5 Symmetrical component theorem

The symmetrical component theorem states that any unbalanced system of three phasors is the resultant of three sets of three balanced phasors, a positive sequence set, a negative sequence set, and a zero sequence set.*

The nine component phasors are known as the symmetrical components of the unbalanced system.

Referring again to Fig. 13.6, we see that

$$V_a = V_{a1} + V_{a2} + V_{a0} = V_1 + V_2 + V_0 \qquad (13.12)$$

where V_1, V_2 and V_0 are the reference phasors for the positive, negative and zero sequence components,

also

$$V_b = V_{b1} + V_{b2} + V_{b0} = h^2 V_1 + h V_2 + V_0 \qquad (13.13)$$

and

$$V_c = V_{c1} + V_{c2} + V_{c0} = h V_1 + h^2 V_2 + V_0 \qquad (13.14)$$

These equations can be written in matrix form.

Thus

$$\begin{bmatrix} V_a \\ V_b \\ V_c \end{bmatrix} = \begin{bmatrix} 1 & 1 & 1 \\ h^2 & h & 1 \\ h & h^2 & 1 \end{bmatrix} \cdot \begin{bmatrix} V_1 \\ V_2 \\ V_0 \end{bmatrix} \qquad (13.15)$$

or

$$[V] = [h][V_s] \qquad (13.16)$$

where

$$[h] = \begin{bmatrix} 1 & 1 & 1 \\ h^2 & h & 1 \\ h & h^2 & 1 \end{bmatrix} \qquad (13.17)$$

This is a very convenient display, since the rows are the groups of components forming the phase voltages and the columns are the groups forming the sequence sets.

Inverting Eqs. (13.15) we obtain

$$\begin{bmatrix} V_1 \\ V_2 \\ V_0 \end{bmatrix} = \frac{1}{3} \begin{bmatrix} 1 & h & h^2 \\ 1 & h^2 & h \\ 1 & 1 & 1 \end{bmatrix} \cdot \begin{bmatrix} V_a \\ V_b \\ V_c \end{bmatrix} \qquad (13.18)$$

---

* The symmetrical component theorem first demonstrated by C. L. Fortescue relates to any general system of m phasors. It proves that each of the m phasors can be subdivided into m components so that the total $m \times m$ components will form m symmetrical systems. The first of the symmetrical systems will simply consist of m equal phasors and will be known as the zero-sequence system. Each of the remaining $(m-1)$ systems will be a set of m phasors equal to each other in length and each being separated from the next by the angle $(2\pi/m)$. The phase sequences in each of the $(m-1)$ symmetrical systems will be different.

The most important practical use of this general theorem arises when it is applied to three-phase power systems as discussed in this chapter.

or

$$[V_s] = [h]^{-1}[V] \qquad (13.19)$$

where

$$[h]^{-1} = \frac{1}{3}\begin{bmatrix} 1 & h & h^2 \\ 1 & h^2 & h \\ 1 & 1 & 1 \end{bmatrix} \qquad (13.20)$$

Equation (13.19) is the transform equation showing how a set of phasors $[V]$ is transformed to its symmetrical components $[V_s]$.

The matrix $[h]$ is known as the symmetrical component transform.*

13.6 Analysis of an unbalanced system into components

The use of Eq. (13.18) to determine the symmetrical components of a given unbalanced system is shown in the following example. Analytical or geometrical methods of addition of phasors can be used.

Example:

Determine the symmetrical components corresponding to the following system of three phasors.

$$\begin{bmatrix} V_a \\ V_b \\ V_c \end{bmatrix} = \begin{bmatrix} 3+j4 \\ 10+j0 \\ 0+j5 \end{bmatrix}$$

This system is shown in Fig. 13.8(a).

From Eq. (13.18) the positive sequence components:

$$\begin{aligned}
V_1 &= \tfrac{1}{3}(V_a + hV_b + h^2V_c) \\
&= \tfrac{1}{3}[(3+j4) + (-0\cdot5+j0\cdot866)10 + (-0\cdot5-j0\cdot866)j5] \\
&= \tfrac{1}{3}[3+j4-5+j8\cdot66-j2\cdot5+4\cdot33] \\
&= \tfrac{1}{3}(2\cdot33+j10\cdot16) \\
&= 0\cdot78+j3\cdot39
\end{aligned}$$

Geometrically, this is shown in Fig. 13.8(b), commencing with V_a, hV_b is added and then h^2V_c. The resultant divided by three gives V_1.

To find the negative sequence component Eq. (13.18) gives

$$\begin{aligned}
V_2 &= \tfrac{1}{3}(V_a + h^2V_b + hV_c) \\
&= \tfrac{1}{3}[(3+j4) + (-0\cdot5-j0\cdot866)10 + (-0\cdot5+j0\cdot866)j5] \\
&= \tfrac{1}{3}(3+j4-5-j8\cdot66-j2\cdot5-4\cdot33) \\
&= \tfrac{1}{3}(-6\cdot33-j7.16) \\
&= -2\cdot11-j2.39
\end{aligned}$$

* Equation (13.19) may be taken as proof of the symmetrical component theorem. In other words since the inverse transform $[h]^{-1}$ is shown to exist, it follows that the symmetrical components $[V_s]$ can be found corresponding to *any* system of three unbalanced phasors $[V]$.

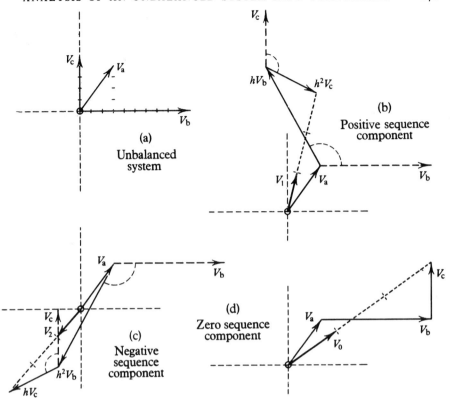

FIG. 13.8 Analysis of an unbalanced system into symmetrical components

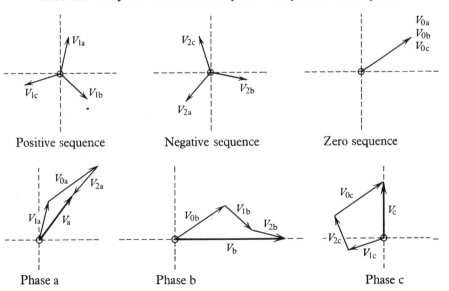

FIG. 13.9 Re-synthesis of the symmetrical components to form the original system

In Fig. 13.8(c) V_a, h^2V_b and hV_c are added geometrically and divided by three.

The zero sequence component is obtained from

$$V_0 = \tfrac{1}{3}(V_a + V_b + V_c)$$
$$= \tfrac{1}{3}(3 + j4 + 10 + j5)$$
$$= \tfrac{1}{3}(13 + j9)$$
$$= 4{\cdot}33 + j3$$

The addition of V_a, V_b and V_c is shown in Fig. 13.8(d).

In order to check the accuracy of this computation, the nine symmetrical components are shown in Fig. 13.9. Adding the corresponding positive, negative and zero sequence components, the original set of three unbalanced phasors is produced.

13.7 Three-phase circuits

13.7.1 Three-phase four-wire circuit

Consider a relatively simple three-phase four-wire system (Fig. 13.10(a)) consisting of a three-phase source supplying three loads Z_a, Z_b and Z_c.

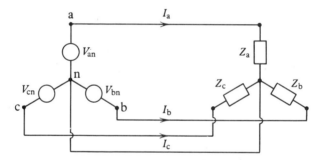

FIG. 13.10a A three-phase four-wire system

The line and neutral conductors are assumed to be without impedance but mutual inductance exists between the loads. The system is a network with three loops, the line currents I_a, I_b and I_c being the loop currents.
Hence

$$\begin{bmatrix} V_{an} \\ V_{bn} \\ V_{cn} \end{bmatrix} = \begin{bmatrix} Z_a & Z_{ab} & Z_{ac} \\ Z_{ba} & Z_b & Z_{bc} \\ Z_{ca} & Z_{cb} & Z_c \end{bmatrix} \cdot \begin{bmatrix} I_a \\ I_b \\ I_c \end{bmatrix} \tag{13.21}$$

To determine the currents, Eq. (13.21) must be inverted. This is a difficult procedure since the impedances are all complex numbers and the matrix is virtually a 6×6 square. No simplification is possible in the general case but if mutual inductance can be neglected, Eq. (13.21) reduces to

$$\begin{bmatrix} V_{an} \\ V_{bn} \\ V_{cn} \end{bmatrix} = \begin{bmatrix} Z_a & 0 & 0 \\ 0 & Z_b & 0 \\ 0 & 0 & Z_c \end{bmatrix} \cdot \begin{bmatrix} I_a \\ I_b \\ I_c \end{bmatrix} \tag{13.22}$$

Now the equations can be solved independently with much less trouble. Moreover, if voltages and impedances are balanced, that is

$$Z_a = Z_b = Z_c \tag{13.23}$$

and V_{an}, V_{bn} and V_{cn} form a symmetrical positive sequence system, solving for one phase gives an answer equally applicable to the other two.

13.7.2 Balanced self and mutual impedances

Again simplification is possible if a degree of symmetry exists. If, for example, all the mutual impedances are alike, and all the self impedances also are equal

$$\begin{bmatrix} V_a \\ V_b \\ V_c \end{bmatrix} = \begin{bmatrix} Z & Z_m & Z_m \\ Z_m & Z & Z_m \\ Z_m & Z_m & Z \end{bmatrix} \cdot \begin{bmatrix} I_a \\ I_b \\ I_c \end{bmatrix} \tag{13.24}$$

Now if the currents can be assumed to be a symmetrical set of positive sequence

$$\begin{bmatrix} I_a \\ I_b \\ I_c \end{bmatrix} = \begin{bmatrix} 1 \\ h^2 \\ h \end{bmatrix} \cdot [I_a] = \begin{bmatrix} h \\ 1 \\ h^2 \end{bmatrix} [I_b] = \begin{bmatrix} h^2 \\ h \\ 1 \end{bmatrix} [I_c] \tag{13.25}$$

Thus substituting in Eq. (13.24)

$$\left. \begin{array}{l} V_a = (Z + h^2 Z_m + h Z_m) I_a \\ V_b = (h Z_m + Z + h^2 Z_m) I_b \\ V_c = (h^2 Z_m + h Z_m + Z) I_c \end{array} \right\} \tag{13.26}$$

or

$$\begin{bmatrix} V_a \\ V_b \\ V_c \end{bmatrix} = \begin{bmatrix} Z_t & 0 & 0 \\ 0 & Z_t & 0 \\ 0 & 0 & Z_t \end{bmatrix} \cdot \begin{bmatrix} I_a \\ I_b \\ I_c \end{bmatrix} \tag{13.27}$$

where

$$Z_t = Z + Z_m(h + h^2) \tag{13.28}$$
$$= Z - Z_m$$

This shows that under these conditions the equations once again can be solved independently, and that the voltages will be balanced.

We can, therefore, infer that if we start with balanced generated voltages applied to a system of balanced impedances even though mutual impedances are included, the currents will be balanced and we can still use single-phase techniques for our calculations obtaining the currents by dividing their respective line-to-neutral voltages, by an equivalent impedance Z_t which allows for the mutual impedances.

13.7.3 Three-phase three-wire system

A common problem involving unbalanced currents arises when three unequal impedances are connected in star across a three-wire system, the voltages of which may or may not be balanced. Mutual inductance may exist between the loads.

Such a problem is shown in Fig. 13.10(b), where impedances Z_a, Z_b, Z_c are connected to supply lines **a**, **b**, **c**. The line currents are designated I_a, I_b and I_c.

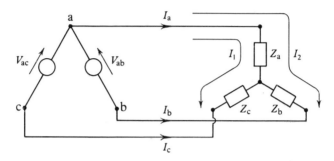

FIG. 13.10b A three-phase three-wire system

The circuit has only two loops. If the loop currents are I_1 and I_2, flowing in the paths shown in the diagram, we have for the connection matrix

$$\begin{bmatrix} I_a \\ I_b \\ I_c \end{bmatrix} = \begin{bmatrix} 1 & 1 \\ 0 & -1 \\ -1 & 0 \end{bmatrix} \cdot \begin{bmatrix} I_1 \\ I_2 \end{bmatrix}$$

(13.29)

The unconnected primitive impedance matrix

$$[Z'] = \begin{bmatrix} Z_a & Z_{ab} & Z_{ac} \\ Z_{ba} & Z_b & Z_{bc} \\ Z_{ca} & Z_{cb} & Z_c \end{bmatrix}$$

(13.30)

Hence, multiplying by the connection matrix and its transpose,

$$\begin{bmatrix} V_{ac} \\ V_{ab} \end{bmatrix} = \begin{bmatrix} 1 & 0 & -1 \\ 1 & -1 & 0 \end{bmatrix} \cdot \begin{bmatrix} Z_a & Z_{ab} & Z_{ac} \\ Z_{ab} & Z_b & Z_{bc} \\ Z_{ca} & Z_{cb} & Z_c \end{bmatrix} \cdot \begin{bmatrix} 1 & 1 \\ 0 & -1 \\ -1 & 0 \end{bmatrix} \cdot \begin{bmatrix} I_1 \\ I_2 \end{bmatrix}$$

$$= \begin{bmatrix} Z_{11} & Z_{12} \\ Z_{21} & Z_{22} \end{bmatrix} \cdot \begin{bmatrix} I_1 \\ I_2 \end{bmatrix}$$

(13.31)

where

$$Z_{11} = Z_a + Z_c - (Z_{ac} + Z_{ca})$$
$$Z_{12} = Z_a + Z_{cb} - (Z_{ab} + Z_{ca})$$
$$Z_{21} = Z_a + Z_{bc} - (Z_{ba} + Z_{ac})$$
$$Z_{22} = Z_a + Z_b - (Z_{ab} + Z_{ba})$$
(13.32)

Inverting

$$\begin{bmatrix} I_1 \\ I_2 \end{bmatrix} = \frac{1}{\Delta} \begin{bmatrix} Z_{22} & -Z_{12} \\ -Z_{21} & Z_{11} \end{bmatrix} \cdot \begin{bmatrix} V_{ac} \\ V_{ab} \end{bmatrix}$$
(13.33)

where

$$\Delta = Z_{11} Z_{22} - Z_{12} Z_{21}$$
(13.34)

Thus

$$I_a = I_1 + I_2 = \frac{(Z_{22} - Z_{21}) V_{ac} + (Z_{11} - Z_{12}) V_{ab}}{Z_{11} Z_{22} - Z_{12} Z_{21}}$$
(13.35)

$$I_b = -I_2 = \frac{Z_{21} V_{ac} - Z_{11} V_{ab}}{Z_{11} Z_{22} - Z_{12} Z_{21}}$$
(13.36)

$$I_c = -I_1 = \frac{Z_{12} V_{ab} - Z_{22} V_{ac}}{Z_{11} Z_{22} - Z_{12} Z_{21}}$$
(13.37)

13.8 Power networks

In most power systems a considerable degree of symmetry exists in the phase and mutual impedances. Machine windings are completely symmetrical and although the self-impedance of the three conductors comprising a transmission line are not exactly alike, these differ only slightly and occasionally this difference is minimised by transposition of the conductors. In no case is there deliberately introduced a generated voltage with negative or zero sequence components.

Every attempt is made to balance the load across the phases. Machine loads are completely symmetrical; indeed it can be shown that an induction motor's load helps to balance a system.

But, in spite of this, large single-phase loads have to be accommodated, particularly those due to furnaces, also heating and lighting loads in remote districts for which it is uneconomic to supply a three-phase feeder.

The largest out-of-balance currents arise during fault conditions on the system as when, for example, a flash-over occurs on a line insulator.

It is extremely important to devise methods of calculation for the analysis of these problems. It will be shown that the method of symmetrical components has definite advantages.

If the topology of the network is at all complicated, the individual equations of Sec. 13.7.1 are difficult to set up and their solution as three simultaneous equations is much too involved.

It will be shown in the next section that when a certain degree of symmetry exists, it is possible to apply the method of symmetrical components, enabling a single-phase technique corresponding to Sec. 13.7.2. to be used. This reduces the computations to a tractable level, adopts parameters which are measurable, and also enables simpler model networks to be set up on network analysers.

13.9 Analysis of a simplified power network

We will now consider the elementary power network of Fig. 13.11. Here we have represented a generator whose line-to-neutral generated voltages are E_a, E_b and E_c connected by a transmission line to terminals **a**, **b**, **c**, **n**, the supply points for a three-phase four-wire load.

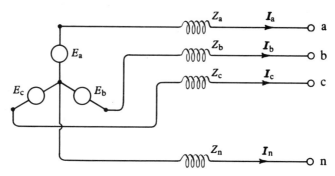

FIG. 13.11 An elementary three-phase power network

Self and mutual inductances exist in and between all four conductors, not only in the transmission line, but also in the generator. Lumping these impedances and treating them as the coils of a primitive network, the equations of the volt-drops due to the primitive currents, are given by:

$$\begin{bmatrix} V'_a \\ V'_b \\ V'_c \\ V'_n \end{bmatrix} = \begin{bmatrix} Z_a & Z_{ab} & Z_{ac} & Z_{an} \\ Z_{ba} & Z_b & Z_{bc} & Z_{bn} \\ Z_{ca} & Z_{cb} & Z_c & Z_{cn} \\ Z_{na} & Z_{nb} & Z_{nc} & Z_n \end{bmatrix} \cdot \begin{bmatrix} I'_a \\ I'_b \\ I'_c \\ I'_n \end{bmatrix} \qquad (13.38)$$

When these primitive coils are connected in accordance with Fig. 13.11

$$I'_n = -(I_a + I_b + I_c)$$

and thus the corresponding matrix is

$$\begin{bmatrix} I'_a \\ I'_b \\ I'_c \\ I'_n \end{bmatrix} = \begin{bmatrix} 1 & 0 & 0 \\ 0 & 1 & 0 \\ 0 & 0 & 1 \\ -1 & -1 & -1 \end{bmatrix} \cdot \begin{bmatrix} I_a \\ I_b \\ I_c \end{bmatrix} \qquad (13.39)$$

and thus

$$
\begin{bmatrix} V'_a \\ V'_b \\ V'_c \\ V'_n \end{bmatrix} = \begin{bmatrix} Z_a & Z_{ab} & Z_{ac} & Z_{an} \\ Z_{ba} & Z_b & Z_{bc} & Z_{bn} \\ Z_{ca} & Z_{cb} & Z_c & Z_{cn} \\ Z_{na} & Z_{nb} & Z_{nc} & Z_n \end{bmatrix} \begin{bmatrix} 1 & 0 & 0 \\ 0 & 1 & 0 \\ 0 & 0 & 1 \\ -1 & -1 & -1 \end{bmatrix} \cdot \begin{bmatrix} I_a \\ I_b \\ I_c \end{bmatrix}
$$

$$
= \begin{bmatrix} Z_a - Z_{an} & Z_{ab} - Z_{an} & Z_{ac} - Z_{an} \\ Z_{ba} - Z_{bn} & Z_b - Z_{bn} & Z_{bc} - Z_{bn} \\ Z_{ca} - Z_{cn} & Z_{cb} - Z_{cn} & Z_c - Z_{cn} \\ Z_{na} - Z_n & Z_{nb} - Z_n & Z_{nc} - Z_n \end{bmatrix} \cdot \begin{bmatrix} I_a \\ I_b \\ I_c \end{bmatrix} \tag{13.40}
$$

Equation (13.40) thus gives the voltage drops across individual coils. In each loop the voltage across the neutral coils is in the opposite direction to that across the line impedance, consequently the total impedance voltage per loop is given by

$$
[Z][I] = \begin{bmatrix} 1 & 0 & 0 & -1 \\ 0 & 1 & 0 & -1 \\ 0 & 0 & 1 & -1 \end{bmatrix} \cdot \begin{bmatrix} Z_a - Z_{an} & Z_{ab} - Z_{an} & Z_{ac} - Z_{an} \\ Z_{ba} - Z_{bn} & Z_b - Z_{bn} & Z_{bc} - Z_{bn} \\ Z_{ca} - Z_{cn} & Z_{cb} - Z_{cn} & Z_c - Z_{cn} \\ Z_{na} - Z_n & Z_{nb} - Z_n & Z_{nc} - Z_n \end{bmatrix} \cdot \begin{bmatrix} I_a \\ I_b \\ I_c \end{bmatrix}
$$

$$
= \begin{bmatrix} Z_a - Z_{an} - Z_{na} + Z_n & Z_{ab} - Z_{an} - Z_{nb} + Z_n & Z_{ac} - Z_{an} - Z_{nc} + Z_n \\ Z_{ba} - Z_{bn} - Z_{na} + Z_n & Z_b - Z_{bn} - Z_{nb} + Z_n & Z_{bc} - Z_{bn} - Z_{nc} + Z_n \\ Z_{ca} - Z_{cn} - Z_{na} + Z_n & Z_{cb} - Z_{cn} - Z_{nb} + Z_n & Z_c - Z_{cn} - Z_{nc} + Z_n \end{bmatrix}
$$

$$
\begin{bmatrix} I_a \\ I_b \\ I_c \end{bmatrix} \tag{13.41}
$$

and finally the relationship between the generated e.m.f.'s and terminal voltages is given by

$$
[E] = [V] + [Z][I] \tag{13.42}
$$

or

$$
[V] = [E] - [Z][I] \tag{13.43}
$$

Equation (13.43) shows that the relationship between the line-to-neutral voltages and the line currents is extremely complicated. They are linked by a 3×3 matrix containing 36 complex terms.

It should be especially noted that, although the mutual inductance Z_{ab} and Z_{ba} are alike in static circuits, this is no longer true in circuits such as machines, which include moving members (see Sec. 11.1).

We now examine the effect of transforming each of the sets of unbalanced phasors $[V]$, $[E]$ and $[I]$ into their symmetrical components using the symmetrical component transform of Sec. 13.5.

From Eqs. (13.16) and (13.19)

$$[V_s] = [h]^{-1}[V] \tag{13.44}$$
$$[E_s] = [h]^{-1}[E] \tag{13.45}$$

and

$$[I] = [h][I_s] \tag{13.46}$$

Multiplying both sides of Eq. (13.42) by $[h]^{-1}$

$$[h]^{-1}[E] = [h]^{-1}[V] + [h]^{-1}[Z][I]$$
$$= [h]^{-1}[V] + [h]^{-1}[Z][h][I_s]$$

or

$$[E_s] = [V_s] + [h]^{-1}[Z][h][I_s] \tag{13.47}$$
$$= [V_s] + [Z_s][I_s] \tag{13.48}$$

where

$$[Z_s] = [h]^{-1}[Z][h] \tag{13.49}$$

According to Eq. (13.49), it is possible to determine a sequence impedance matrix $[Z_s]$ which, when multiplied by the sequence currents and added to the sequence components of terminal voltage, is equal to the sequence components of generated voltage.

In other words, the same type of equation exists between sequence components of voltage and current as that showing the relationship between actual line voltages and currents.

There is no reason to suppose that the matrix $[Z_s]$ is any less complicated than $[Z]$ and, consequently, it is not to be expected that this transformation to symmetrical components will lead to numerical simplification. This is true for a completely general three-loop network, but if a certain degree of symmetry exists between the elements of $[Z]$ it can be shown that $[Z_s]$ is a simple diagonal matrix leading to considerable simplifications.

13.9.1 Simplification due to symmetry

For the majority of power networks, we may assume

$$Z_a = Z_b = Z_c$$
$$Z_{ab} = Z_{bc} = Z_{ca}$$
$$Z_{ba} = Z_{cb} = Z_{ac} \tag{13.50}$$
$$Z_{an} = Z_{bn} = Z_{cn} = Z_{na} = Z_{nb} = Z_{nc}$$

but

$$Z_{ab} \neq Z_{ba}.$$

The first four equations of (13.50) arise naturally where steps are taken to maintain equality between phases. In machine and transformer windings the phase impedances are obviously alike and transmission line conductors are approximately equally spaced and often transposed.

The fact that $Z_{ab} \neq Z_{ba}$ requires explanation. It has been shown (see Sec. 11.1) that the mutual inductance between two static coils is indeed mutual, i.e. $M_{ab} = M_{ba}$. However, with a rotating machine, the effective mutual inductance between two stator phase windings **a** and **b** is complicated by the presence of the moving rotor.

Since at any instant the rotor coils will be moving towards one of the stator coils but away from the other, the result is an effective increase in the mutual inductance from **a** to **b** but a decrease of the inductance from **b** to **a**. This will be discussed at length in Chap. 15.

Arising from the simplifications of Eqs. (13.50), Eq. (13.41) becomes

$$[Z] = \begin{bmatrix} p & q & r \\ r & p & q \\ q & r & p \end{bmatrix} \tag{13.51}$$

where

$$p = Z_a + (Z_n - 2Z_{an})$$
$$q = Z_{ab} + (Z_n - 2Z_{an})$$
$$r = Z_{ba} + (Z_n - 2Z_{an})$$

Then

$$[Z_s] = [h]^{-1}[Z][h] \tag{13.49}$$

$$= \frac{1}{3} \begin{bmatrix} 1 & h & h^2 \\ 1 & h^2 & h \\ 1 & 1 & 1 \end{bmatrix} \cdot \begin{bmatrix} p & q & r \\ r & p & q \\ q & r & p \end{bmatrix} \cdot \begin{bmatrix} 1 & 1 & 1 \\ h^2 & h & 1 \\ h & h^2 & 1 \end{bmatrix}$$

$$= \frac{1}{3} \begin{bmatrix} 1 & h & h^2 \\ 1 & h^2 & h \\ 1 & 1 & 1 \end{bmatrix} \cdot \begin{bmatrix} p + h^2 q + hr & p + hq + h^2 r & p + q + r \\ r + h^2 p + hq & r + hp + h^2 q & r + p + q \\ q + h^2 r + hp & q + hr + h^2 p & q + r + p \end{bmatrix} \tag{13.52}$$

Let

$$Z_1 = p + h^2 q + hr$$
$$Z_2 = p + hq + h^2 r \tag{13.53}$$
$$Z_0 = p + q + r$$

Then

$$[Z_s] = \frac{1}{3} \begin{bmatrix} 1 & h & h^2 \\ 1 & h^2 & h \\ 1 & 1 & 1 \end{bmatrix} \cdot \begin{bmatrix} Z_1 & Z_2 & Z_0 \\ h^2 Z_1 & hZ_2 & Z_0 \\ hZ_1 & h^2 Z_2 & Z_0 \end{bmatrix}$$

$$= \begin{bmatrix} Z_1 & 0 & 0 \\ 0 & Z_2 & 0 \\ 0 & 0 & Z_0 \end{bmatrix}$$

Matrix Eq. (13.48) thus becomes

$$
\begin{bmatrix} E_1 \\ E_2 \\ E_0 \end{bmatrix} = \begin{bmatrix} V_1 \\ V_2 \\ V_0 \end{bmatrix} + \begin{bmatrix} Z_1 & 0 & 0 \\ 0 & Z_2 & 0 \\ 0 & 0 & Z_0 \end{bmatrix} \cdot \begin{bmatrix} I_1 \\ I_2 \\ I_0 \end{bmatrix} \qquad (13.55)
$$

which reduces to three independent equations

$$
\left. \begin{aligned}
E_1 &= V_1 + Z_1 I_1 \\
E_2 &= V_2 + Z_2 I_2 \\
E_0 &= V_0 + Z_0 I_0
\end{aligned} \right\} \qquad (13.56)
$$

13.9.2 Sequence impedances

The independence of the three Eqs. (13.56) demonstrates that the three components of current can be regarded as flowing independently through their own impedances, and these alone are responsible for the difference between generated and load voltages of that sequence.

This profound simplification has three important advantages:

1. The three independent equations (13.56) each refer to balanced current, hence single-phase computation using line-to-neutral values is all that is necessary.

2. The impedance parameters Z_1, Z_2 and Z_0 are comparatively readily measurable in a practical problem, whereas the determination of the self and mutual impedances Z_a and Z_{ab} invariably presents measurement difficulties.
   Z_1 is known as the impedance of the positive sequence network (the positive sequence impedance).
   Z_2 is known as the impedance of the negative sequence network (the negative sequence impedance).
   Z_0 is known as the impedance of the zero sequence network (the zero sequence impedance).

3. Only three impedance parameters, Z_1, Z_2 and Z_0 are needed to specify a problem. It should also be noted that if the neutral is disconnected (as in a three-wire system), Z_0 becomes infinity.

Also if Z_{ab} is equal to Z_{ba} as for static networks (transmission lines and transformers) $Z_1 = Z_2$.

13.9.3 Typical values of Z_1, Z_2 and Z_0

Typical values of the three sequence impedances used in intricate power-system problems submitted to the Electrical Research Association for solution on the Electronic Network Analyser are given in the following table.

Table 13.1

APPARATUS	TYPICAL PER UNIT VALUES ON RATING		
	Z_1	Z_2	Z_0
TRANSFORMERS:			
(a) 132 kV, nominal	j0·10	j0·10	—
(b) 275 kV, nominal	j0·14	j0·14	—
(c) 400 kV, nominal	j0·16	j0·16	—
GENERATORS:			
Turbo-Alternators			
(a) 3600 r.p.m.	j2·00	j0·15	j0·10
(b) 1800 r.p.m.	j1·20	j0·25	j0·25
Water Wheel Generators			
(a) With dampers	j0·90	j0·20	j0·15
(b) Without dampers	j0·90	j0·30	j0·15
SYNCHRONOUS CONDENSERS	j1·50	j0·15	j0·05
INDUCTION MOTORS	3·40 + j0·80	0·15 + j0·80	—
TRANSMISSION LINES			
(a) 132 kV, nominal	0·0008 + j0·0017	0·0008 + j0·0017	—
(b) 275 kV, nominal	0·0004 + j0·0017	0·0004 + j0·0017	—
(c) 400 kV, nominal	0·0002 + j0·0017	0·0002 + j0·0017	—

Note:

(i) *Transformers.* Zero sequence reactance values depend largely on the type of connections and may vary from $6Z_1$ to $0·6Z_1$.

(ii) *Synchronous machines.* Positive sequence reactance values refer to the unsaturated synchronous reactance. Under normal operating conditions, these values are multiplied by a load adjustment factor which may vary from 0·6 to 0·8 for generators and from 0·4 to 0·6 for synchronous condensers.

(iii) *Induction motors.* Approximate values are given for an equivalent circuit operating at small values of slip, the large shunt magnetizing impedance being neglected.

(iv) *Transmission lines.* 'Per-mile' values are quoted. The line rating is taken as being the ratio of the square of the nominal voltage to the surge impedance $V^2/\sqrt{(L/C)}$. Zero sequence values depend largely on the circuit configuration and may vary from $2Z_1$ to $5Z_1$.

13.9.4 The method of symmetrical component analysis

Figure 13.12 summarises the method of symmetrical components for the solution of certain types of power system problem involving unbalanced circuits. Orthodox methods of general circuit analysis would require the formulation and inversion of the impedance matrix of the system, the elements of which are not easily determined from tests taken on the system. On the other hand the separate sequence impedance networks are more readily defined and these are either solved separately or if other constraints are introduced in the problem they are interconnected to form a single network.

When the sequence currents in any particular three-phase branch have been determined, the actual line currents are obtained by multiplying by the transform $[h]$.

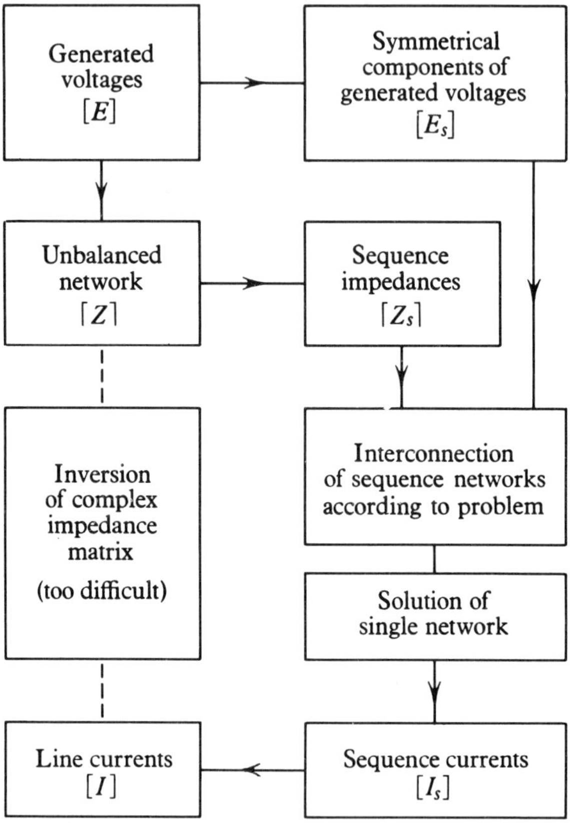

FIG. 13.12 **The transform method used for solving unbalanced circuits with the aid of symmetrical components**

The next section shows how the method is applied to problems arising in power systems where failures of insulation have occurred. Such conditions usually distort the system and large unbalanced currents are produced. Not only is it important for such currents to be measured but protective equipment must be arranged to isolate the faulty section quickly and with minimum disturbance to the remainder of the system. Symmetrical component analysis is particularly valuable in this type of problem.

13.10 Faults in power networks

13.10.1 Types of fault

Fault conditions arising in power networks may be placed in four categories:

(*a*) Single line-to-earth fault as for example when a single line insulator flashes over.

(b) Double line-to-earth fault when two insulators on two lines flash over to earth simultaneously, leaving the third line clear.

(c) Line-to-line fault, when two overhead conductors swing together and touch but remain insulated from earth.

(d) Symmetrical three-phase fault, which is the case when all three lines are short circuited.

The analysis of these conditions is conducted in the following manner.

13.10.2 Sequence networks

Equations

$$\left.\begin{aligned} E_1 &= V_1 + I_1 Z_1 \\ E_2 &= V_2 + I_2 Z_2 \\ E_0 &= V_0 + I_0 Z_0 \end{aligned}\right\} \tag{13.56}$$

are represented by the separate networks shown in Fig. 13.13(a).

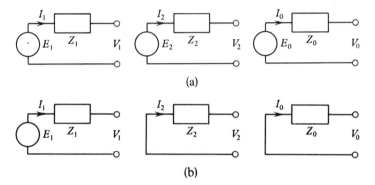

(a)

(b)

FIG. 13.13 (a) Circuits corresponding to Equation 13.56. (b) The same, when $E_2 = E_0 = 0$

In power circuits generated voltages are invariably balanced. No negative or zero sequence voltages are deliberately generated, hence $E_2 = E_0 = 0$ and the sequence networks are simplified to the circuits shown in Fig. 13.13(b).

When a fault occurs in the power network, special relationships arise between the sequence currents and the sequence line voltages according to the type of fault. This is due to the interconnections produced between the phases and comes about in the following manner.

13.10.3 Single line-to-earth fault

Assuming that the fault occurs on line **a**, if I_f is the fault current, it will be seen from Fig. 13.14 that

$$I_a = I_f \tag{13.57}$$
$$I_b = 0 \tag{13.58}$$
$$I_c = 0 \tag{13.59}$$

The voltage V_{an} will disappear but the voltages V_{bn} and V_{cn} will remain. Hence

$$V_a = 0 \qquad (13.60)$$

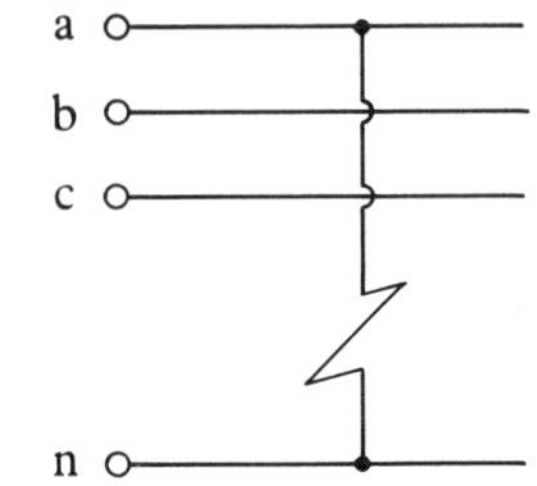

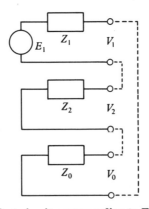

FIG. 13.14 A line-to-earth fault on line a

To obtain the symmetrical components of current

$$[I_s] = [h]^{-1} \quad [I] \qquad (13.61)$$

Therefore

$$\begin{bmatrix} I_1 \\ I_2 \\ I_0 \end{bmatrix} = \frac{1}{3} \begin{bmatrix} 1 & h & h^2 \\ 1 & h^2 & h \\ 1 & 1 & 1 \end{bmatrix} \cdot \begin{bmatrix} I_a \\ 0 \\ 0 \end{bmatrix} \qquad (13.62)$$

Thus

$$I_1 = I_2 = I_0 = \tfrac{1}{3} I_a = \tfrac{1}{3} I_t \qquad (13.63)$$

Also from (13.60)

$$V_a = V_1 + V_2 + V_0 = 0 \qquad (13.64)$$

Equations (13.63) and (13.64) therefore represent the particular characteristics of an earth fault. These conditions must be introduced as constraints on the general sequence networks of Fig. 13.13.

If the independent sequence networks are connected in series then the three components of current are bound to be equal. If the two remaining terminals are short-circuited, then $V_1 + V_2 + V_0$ is necessarily zero.

FIG. 13.15 The equivalent circuit corresponding to Equations 13.63 and 13.64

The equivalent circuit (Fig. 13.15) thus represents the power network under these fault conditions.

Solving this simple series circuit, we obtain the result

$$I_1 = I_2 = I_0 = E_1/(Z_1 + Z_2 + Z_0) \tag{13.65}$$

and

$$I_t = 3E_1/(Z_1 + Z_2 + Z_0) \tag{13.66}$$

remembering that E_1, Z_1, Z_2 and Z_0 are all phase values.

The sequence voltages can then be found since from Fig. 13.15

$$\left. \begin{aligned} V_1 &= E_1 - I_1 Z_1 \\ V_2 &= 0 - I_2 Z_2 \\ V_0 &= 0 - I_0 Z_0 \end{aligned} \right\} \tag{13.67}$$

hence the line voltages V_b and V_c are obtained from

$$\begin{bmatrix} V_a \\ V_b \\ V_c \end{bmatrix} = \begin{bmatrix} 1 & 1 & 1 \\ h^2 & h & 1 \\ h & h^2 & 1 \end{bmatrix} \cdot \begin{bmatrix} V_1 \\ V_2 \\ V_0 \end{bmatrix} \tag{13.15}$$

It is also worth noting that if the fault has impedance Z_t, the volt drop across the fault impedance is given by

$$\begin{aligned} I_t Z_t &= (3I_0) Z_t \\ &= I_0(3 Z_t) \end{aligned} \tag{13.68}$$

The effect of fault impedance may be thought of as increasing the zero sequence impedance of the system by $3 Z_t$ and Eq. (13.66) becomes

$$I_t = 3E_1/[Z_1 + Z_2 + (Z_0 + 3 Z_t)] \tag{13.69}$$

13.10.4 Double line-to-earth fault

If two lines are faulted to earth, simultaneously, leaving the third line sound (Fig. 13.16), the following constraints are incurred:

$$I_a = 0 \tag{13.70}$$

$$V_b = V_c = 0 \tag{13.71}$$

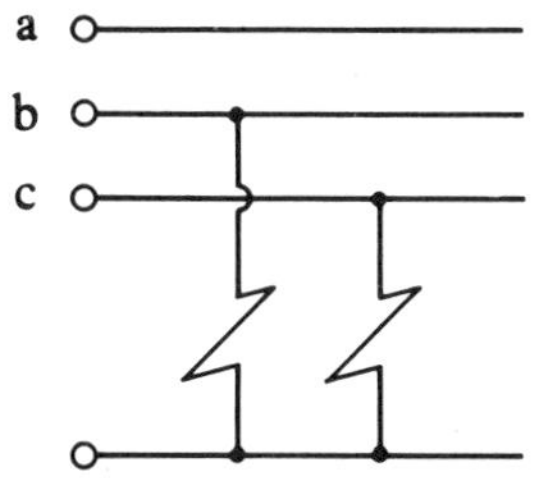

FIG. 13.16 A double line-to-earth fault on lines a and b

Thus

$$I_1 + I_2 + I_0 = 0 \qquad (13.72)$$

and

$$\begin{bmatrix} V_1 \\ V_2 \\ V_0 \end{bmatrix} = \frac{1}{3} \begin{bmatrix} 1 & h & h^2 \\ 1 & h^2 & h \\ 1 & 1 & 1 \end{bmatrix} \cdot \begin{bmatrix} V_a \\ 0 \\ 0 \end{bmatrix} \qquad (13.73)$$

hence

$$V_1 = V_2 = V_0 = \tfrac{1}{3} V_a \qquad (13.74)$$

To comply with Eqs. (13.72) and (13.74) new constraints must be applied to the sequence networks.

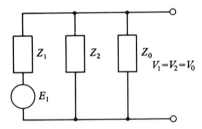

FIG. 13.17 The equivalent circuit corresponding to Equations 13.72 and 13.74

The parallel connections of Fig. 13.17 are seen to satisfy these conditions since this ensures that the branch voltages are equal and that the three currents add up to zero. Hence

$$I_1 = \frac{E_1}{Z_1 + \dfrac{Z_2 Z_0}{Z_2 + Z_0}}$$

$$= \frac{E_1(Z_2 + Z_0)}{Z_1 Z_2 + Z_2 Z_0 + Z_0 Z_1} \qquad (13.75)$$

and

$$I_2 = \frac{-Z_0}{(Z_0 + Z_2)} I_1 \qquad (13.76)$$

$$I_0 = \frac{-Z_2}{(Z_0 + Z_2)} I_1 \qquad (13.77)$$

and

$$V_1 = V_2 = V_0 = E_1 - I_1 Z_1$$
$$= -I_0 Z_0 \qquad (13.78)$$

13.10.5 Line-to-line fault (unearthed)

An example of this type of fault is shown in Fig. 13.18, where lines **b** and **c** are short circuited as could take place if two conductors of an overhead line swung together in abnormally stormy weather.

These conditions are similar to those imposed by a double line-to-earth fault with the additional constraint that $I_0 = 0$ since no path exists in which zero

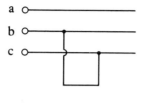

FIG. 13.18 A line-to-line fault between b and c unearthed

sequence currents can flow. The equivalent circuit is thus shown in Fig. 13.19 and the sequence currents will be as follows:

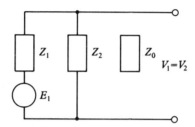

FIG. 13.19 A simplified form of Fig. 13.17 when $I_0 = 0$

$$Z_0 = \text{infinity} \tag{13.79}$$
$$I_1 = E_1/(Z_1 + Z_2) \tag{13.80}$$
$$I_2 = -I_1 \tag{13.81}$$
$$I_0 = 0 \tag{13.82}$$
$$V_1 = V_2 = E_1 - I_1 Z_1$$
$$= -I_2 Z_2 \tag{13.83}$$
$$V_0 = 0 \tag{13.84}$$

13.10.6 Symmetrical three-phase fault

Under these conditions (Fig. 13.20)

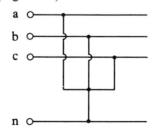

FIG. 13.20 A symmetrical three-phase fault on lines a, b and c

$$V_a = V_b = V_c = 0 \qquad (13.85)$$

and thus, if there is no output voltage the individual sequence networks must be shortcircuited (Fig. 13.21). No driving voltages exist in the negative and zero sequence networks, hence

$$I_2 = I_0 = 0 \qquad (13.86)$$
$$V_2 = V_0 = 0 \qquad (13.87)$$
$$I_1 = E_1/Z_1 \qquad (13.88)$$
and
$$V_1 = E_1 - I_1 Z_1 \qquad (13.89)$$

These results underline the obvious symmetry of the problem.

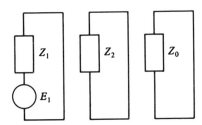

FIG. 13.21 Circuits corresponding to Equations 13.86, 13.87 and 13.88

The discussion on faults in power networks is continued in Sec. 22.3.

13.11 Analysis of power systems

In order to illustrate the procedure adopted in the determination of fault currents on a system, a model network has been constructed in the laboratories of the Lanchester Polytechnic with which the author was associated. The purpose is to enable students to use simple measurement techniques to determine the

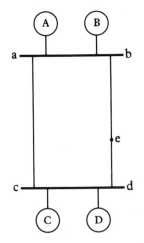

FIG. 13.22 A model network

sequence currents at various points in the system after the sequence networks have been interconnected to correspond to given fault conditions.

The model represents a system shown in Fig. 13.22 comprising four generators **A**, **B**, **C** and **D**. They are connected to busbars in pairs, the busbars being interconnected by duplicate transmission lines. Only the reactances of these elements are represented on the assumption that resistance is negligible in comparison.

Arrangements are made to simulate faults of the types previously described occurring at either of the busbars or at a point **e** part-way along the transmission line **bd**.

13.11.1 Construction

Three networks are arranged, representing the positive, negative and zero sequence reactances and are displayed in a pictorial manner on a 19-in. rack panel with wander plug sockets arranged at all nodes. As resistance in the

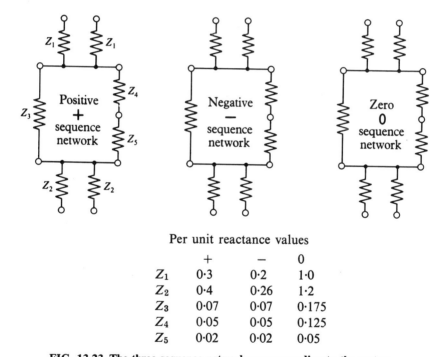

Per unit reactance values

	+	−	0
Z_1	0·3	0·2	1·0
Z_2	0·4	0·26	1·2
Z_3	0·07	0·07	0·175
Z_4	0·05	0·05	0·125
Z_5	0·02	0·02	0·05

FIG. 13.23 The three-sequence networks corresponding to the system

real system may be neglected, all currents are in quadrature with their driving voltages and their magnitudes depend on reactance values. In the model, resistances are chosen to represent each of the system reactances (Fig. 13.23). Although currents and voltages in the model will be in phase, the current magnitudes will represent those of the real system.

13.11.2 Measuring equipment

A supply of 50 V at 50 Hz is obtained from a transformer and applied to the network at the terminals representing the fault (Fig. 13.24). A potential divider

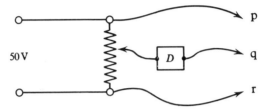

FIG. 13.24 The measuring circuit

with three decades is also connected across these terminals and a sensitive null detector connected to its slider. The other side of the detector is connected to the wander lead marked **q** in the figure. When this probe is attached to a given node in the network, and the bridge is balanced, the potential divider indicates the potential of that node with respect to terminal **r** in per unit notation.

13.11.3 Line-to-earth fault in the transmission line

If we wish to represent the conditions produced by a single line-to-earth fault on the system occurring at a point **e**, two-sevenths of the distance along the transmission line **bd** when all the generators are running, the model network is connected as shown in Fig. 13.25.

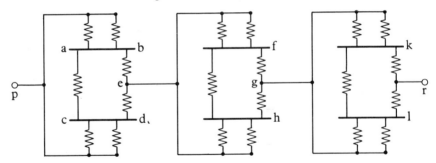

FIG. 13.25 Connecting the networks to correspond to a line-to-earth fault

The neutral terminals of the generators are connected together in each of the sequence networks and the networks are connected in series in accordance with the requirements of Sec. 13.10.3. The measuring voltage is applied to the terminals **p** and **r** and the detector terminal **q** is connected to **b, d, e, f, g, h, k** and **l** in turn. In this way the voltages V_{br}, V_{dr}, V_{er}, etc. are measured.

By subtraction, V_{pb}, V_{be}, V_{de}, V_{ef}, V_{eh}, V_{fg}, V_{hg}, V_{gk}, V_{gl}, V_{kr}, V_{lr}, are obtained and from these voltages, knowing the individual reactances, the currents in all the elements are determined.

For example, the negative sequence current (in per unit notation) in generator **B** is given by $V_{ef}/0.2$.

13.11.4 Analytical solution

The analytical solution of the problem described in the previous section is obtained in the following way.

In terms of the seven loop currents shown in Fig. 13.26, the impedance network is given by

	A	B	C	D	E	F	G
	1·02	−0·2	−0·02	−0·13	−0·02	−0·6	−0·05
	−0·2	0·42	−0·07	0	0	0	0
	−0·02	−0·07	0·14	0	0	0	0
$[Z] =$	−0·13	0	0	0·3	−0·07	0	0
	−0·02	0	0	−0·07	0·14	0	0
	−0·6	0	0	0	0	1·275	−0·175
	−0·05	0	0	0	0	−0·175	0·35

By inversion:

$$\begin{bmatrix} I_A \\ I_B \\ I_C \\ I_D \\ I_E \\ I_F \\ I_G \end{bmatrix} = \begin{bmatrix} 2\cdot042 \\ 1\cdot114 \\ 0\cdot849 \\ 1\cdot079 \\ 0\cdot831 \\ 1\cdot075 \\ 0\cdot829 \end{bmatrix} \cdot [V = 1\cdot0]$$

and from these loop currents the branch currents in the impedances shown in Fig. 13.23 can be found by substitution in the following equations:

Positive sequence currents

$$\begin{bmatrix} I_1 \\ I_2 \\ I_3 \\ I_4 \\ I_5 \end{bmatrix} = \begin{bmatrix} 0 & \frac{1}{2} & 0 \\ \frac{1}{2} & -\frac{1}{2} & 0 \\ 0 & 1 & -1 \\ 0 & 0 & 1 \\ 1 & 0 & -1 \end{bmatrix} \cdot \begin{bmatrix} I_A \\ I_B \\ I_C \end{bmatrix}$$

Negative sequence currents

$$\begin{bmatrix} I_1 \\ I_2 \\ I_3 \\ I_4 \\ I_5 \end{bmatrix} = \begin{bmatrix} 0 & \frac{1}{2} & 0 \\ \frac{1}{2} & -\frac{1}{2} & 0 \\ 0 & 1 & -1 \\ 0 & 0 & 1 \\ 1 & 0 & -1 \end{bmatrix} \cdot \begin{bmatrix} I_A \\ I_D \\ I_E \end{bmatrix}$$

Zero sequence
currents

$$
\begin{bmatrix} I_1 \\ I_2 \\ I_3 \\ I_4 \\ I_5 \end{bmatrix} = \begin{bmatrix} 0 & \frac{1}{2} & 0 \\ \frac{1}{2} & -\frac{1}{2} & 0 \\ 0 & 1 & -1 \\ 0 & 0 & 1 \\ 1 & 0 & -1 \end{bmatrix} \cdot \begin{bmatrix} I_A \\ I_F \\ I_G \end{bmatrix}
$$

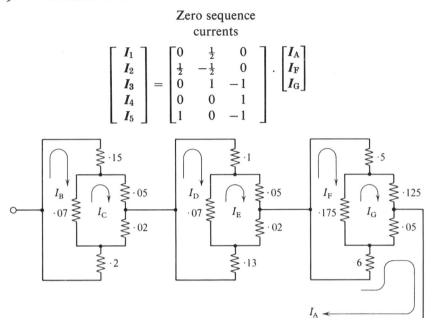

FIG. 13.26 The circuit of Fig. 13.25 is seen to have seven loops to be solved by mesh analysis

13.11.5 Alternative treatment

This particular network is ideally suitable for treatment by systematic reduction, using the star-mesh transformation suggested in Sec. 10.4.1.

If, in Fig. 13.26, the meshes **C**, **E** and **G** are replaced by their equivalent star-networks, the equivalent circuit is reduced to that of Fig. 13.27. It is now a simple series-parallel circuit which can be solved by elementary methods.

13.11.6 Earth fault on busbars

The conditions corresponding to an earth fault on the busbars **ab** are shown in Fig. 13.28, where the three sequence networks are connected in series, connections being made at the neutral and at the busbar **ab** in each case.

Once again, this is a simple series parallel system and mesh analysis is unnecessary.

The laboratory experiment described above shows the student how to use elementary measuring techniques to obtain solutions to relatively complicated problems with sufficient accuracy for practical purposes. It paves the way for an understanding of network analyser techniques such as are used by the Electrical Research Association, the Central Electricity Generating Board, and the electrical manufacturing industry, for dealing with the vastly complicated networks of modern power systems.

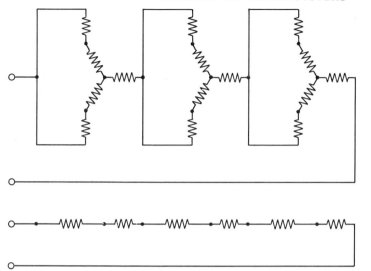

FIG. 13.27 The circuit in Fig. 13.26 can be simplified by star-mesh transformations

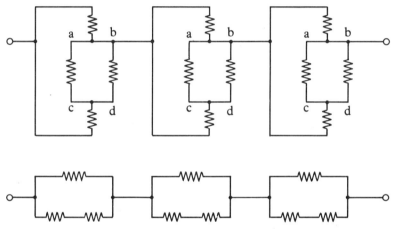

FIG. 13.28 The circuit corresponding to an earth fault at the busbars ab

13.12 Invariance of power

At this stage it should be noted that the symmetrical component transform $[h]$ does not conform to the conditions required for invariance of power. This is evident from the fact that its inverse and conjugate transpose are unequal.† Thus we find that if the complex power of a system is given by

$$S = V_a{}^*I_a + V_b{}^*I_b + V_c{}^*I_c \qquad (13.90)$$

then if the voltages and currents are transformed to their symmetrical components, the power

$$S' = V_1{}^*I_1{}^* + V_2{}^*I_2 + V_0{}^*I_0 \qquad (13.91)$$

† See Section 4.3.3.

and

$$S = 3S' \qquad (13.92)$$

The difference between S and S' can be explained thus: S is the total complex power of the system whereas S' obtained from the symmetrical components refers to only one phase and so must be multiplied by three to be equal to S.

It is sometimes suggested that if the symmetrical component transform is taken to be

$$[h'] = \frac{1}{\sqrt{3}} \begin{bmatrix} 1 & 1 & 1 \\ h^2 & h & 1 \\ h & h^2 & 1 \end{bmatrix} \qquad (13.93)$$

then

$$[h']^{-1} = \frac{1}{\sqrt{3}} \begin{bmatrix} 1 & h & h^2 \\ 1 & h^2 & h \\ 1 & 1 & 1 \end{bmatrix} \qquad (13.94)$$

$$= [h'_t{}^*]$$

The transform $[h']$ thus satisfies the condition for power invariance whereas $[h]$ does not.

For this reason, certain authorities favour the exclusive use of $[h']$ as the symmetrical component transform.

The mathematical basis of this reasoning is unquestioned but nevertheless the factor $1/\sqrt{3}$ is seldom employed in British practice. It is generally preferred to assume that

$$V_a = V_1 + V_2 + V_0 \qquad (13.12)$$

i.e., that the sum of the symmetrical components of voltage is equal to the voltage of the reference phase, rather than to introduce the factor $1/\sqrt{3}$ making

$$V_a = (1/\sqrt{3})(V'_1 + V'_2 + V'_0) \qquad (13.95)$$

CHAPTER 14

Mechanical analogues

14.1 Introduction

Mechanical systems have many properties which correspond to the characteristics of electrical systems. In the first place most of them operate under the action of a force stimulus which is a function of time ultimately producing an output response. For every system a system function can be determined which describes the relationship between the response and the stimulus. Examples range from simple levers, gears and pulley systems, hydraulic jacks to complicated servomechanisms for power assisted steering of motor vehicles and ships, hydraulic variable speed drives, and boring and tunnelling devices. As with electrical systems, a mechanical system is made up of components which either store or dissipate energy and similar laws govern this energy transfer. The laws of motion give rise to linear differential equations which correspond almost exactly with those relating to electrical circuits, and consequently the system function of a given mechanical device could be exactly the same mathematical expression as that which refers to the system function of an electrical circuit.

It is therefore possible to set up an electrical network to represent a mechanical system. To do this the response of the electrical system must be the same time function as that of the mechanical system when both systems are subjected to equivalent applied stimuli.

The electrical system is then said to be the analogue of the mechanical system.

The term *analogue* has two meanings when used in the comparison of two physical systems. Parts of the systems might be similar in some way such as mechanical force and electromotive force, or they might have similar mathematical relationships, e.g. force = mass × rate of change of velocity; current = capacitance × rate of change of voltage. We normally use the analogue which is most convenient for processing in an analogue computer not necessarily the one which has the greatest physical similarity to the original system.

Techniques of measurement and analysis which have been developed for use with electrical systems can, therefore, be applied (with care) to mechanical systems and this is of particular importance in the study of mechanical vibration and of system stability. Moreover, many electrical systems are combined with other systems—usually mechanical ones—and it is convenient to be able to express their behaviour by a single mathematical process.

We have seen the immense importance that is attached to the concept of equivalent electrical circuits. It is probably the most important tool employed by the electrical engineer. In this chapter we shall attempt to set up mechanical circuits and to analyse them in the same way that we have done with their electrical equivalents. First we consider only linear motion.

14.2 Mechanical elements

Mechanical systems can usually be reduced to a chain of 'two terminal elements', which are either energy sources, energy sinks or energy stores. In each case the two terminals are points moving relatively to each other, between which force is applied or between which relative velocity can be measured. Examples of these fall into four categories:

14.2.1 Force source

The combination of cylinder, piston and piston rod in an engine is an example of a force source. Force and relative motion are produced between two members, the cylinder head and the cross-head, which are regarded as the two terminals of the device. A cam and roller mechanism is another such example. The circuit symbol representing a force source is shown in Fig. 14.1.

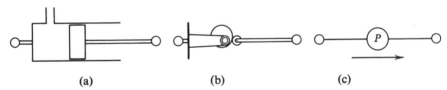

(a)   (b)   (c)

FIG. 14.1 Examples of force sources (a) pressure cylinder and piston, (b) cam and roller, (c) symbol for a force source

14.2.2 Energy sink

A dash-pot or other frictional device is an example of a force sink. When force is applied between the terminals, relative motion of the terminals takes place and energy is converted to heat. Two alternative symbols for an energy sink are shown in Fig. 14.2.

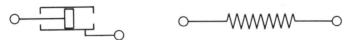

FIG. 14.2 Alternative symbols used for sinks

14.2.3 Energy store (potential)

A spring is readily seen to be an example of an energy store. As relative motion takes place between the terminals marking the ends of the spring, energy is either stored or returned according to the direction of the relative movement. The simple coil spring of Fig. 14.3 thus represents a potential energy store.

FIG. 14.3 Symbol for a store (potential energy)

A weight moving in the vertical direction is also represented by the same symbol.

14.2.4 Energy source (kinetic)

A moving mass stores kinetic energy, the amount of energy depending on its speed. It is a little difficult to appreciate a mass as a two-terminal element until it is realised that measurement of the stored energy requires the measurement of the velocity of the mass relative to a datum. The symbol adopted to show the mass and the datum is given in Fig. 14.4.

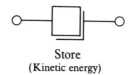

Store
(Kinetic energy)

FIG. 14.4 Symbol for a store (kinetic energy)

14.3 Mechanical circuits

A mechanical system is first split up into a series of two terminal elements according to the number of sources, stores and sinks present. These are then interconnected by first joining together all stationary points (this corresponds to an earthed busbar in an electric circuit). Similarly all points which move with the same velocity relative to earth are also joined together.

Figure 14.5 is an example illustrating a simple carriage suspension system. The elements included are the mass of the carriage, two suspension springs, and two dampers. A disturbing force causing displacement would be applied between the mass and earth. Since in this system all the elements are connected between the mass and earth, the corresponding mechanical circuit is shown in Fig. 14.5(c) where all the elements are seen to be in parallel.

Another mechanical system is shown in Fig. 14.6. A mass hangs from a spring, the upper end of which has applied to it an oscillating force by a crank/connecting-rod mechanism.

A dashpot is connected to the mass in an attempt to damp its movement.

The mechanical circuit is built up from the stationary frame **e**, to which one terminal of the force source, the mass and the sink are connected. Terminals **b**

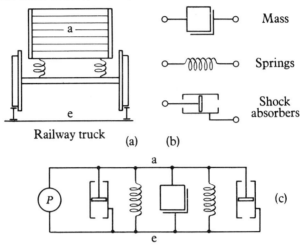

Mass

Springs

Shock absorbers

Railway truck   (a)   (b)

(c)

FIG. 14.5 Example of a mechanical system: (a) diagram of the system, (b) symbols used, (c) equivalent circuit

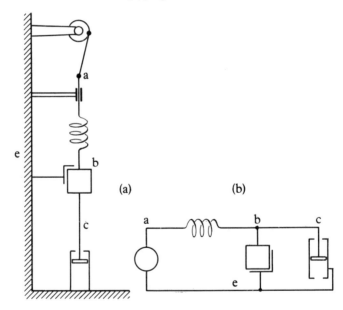

(a)   (b)

FIG. 14.6 (a) An oscillating-mass system. (b) Its equivalent circuit

and **c** move together and so are connected to each other and the spring lies between terminals **a** and **b**.

It will be noted that the 'mechanical circuit' is a type of skeleton outline of the system, and also that these circuits resemble the more familiar electrical circuits.

Furthermore in the example of Fig. 14.5 the disturbing force is applied to the elements in parallel, consequently the applied force is equal to the sum of the forces applied to each element and the velocities measured across each element are alike.

In symbols

$$P = P_1 + P_2 + P_3 + P_4 + P_5 \qquad (14.1)$$

and

$$v = v_1 = v_2 = v_3 = v_4 = v_5 \qquad (14.2)$$

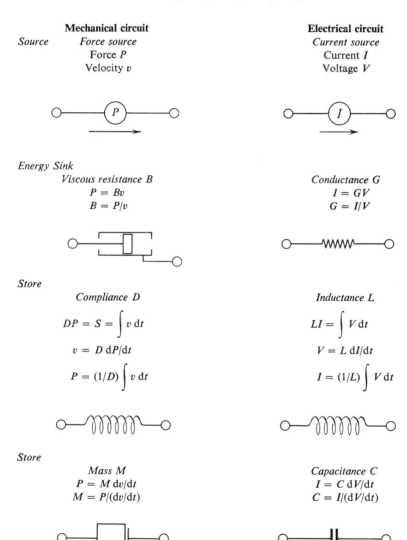

	Mechanical circuit	**Electrical circuit**
Source	*Force source* Force P Velocity v	*Current source* Current I Voltage V

Energy Sink

Viscous resistance B $P = Bv$ $B = P/v$	*Conductance G* $I = GV$ $G = I/V$

Store

Compliance D $DP = S = \displaystyle\int v\,dt$ $v = D\,dP/dt$ $P = (1/D)\displaystyle\int v\,dt$	*Inductance L* $LI = \displaystyle\int V\,dt$ $V = L\,dI/dt$ $I = (1/L)\displaystyle\int V\,dt$

Store

Mass M $P = M\,dv/dt$ $M = P/(dv/dt)$	*Capacitance C* $I = C\,dV/dt$ $C = I/(dV/dt)$

FIG. 14.7 Electrical analogues of mechanical circuits

These equations correspond precisely with those for an electrical circuit of similar parallel formation
where

$$I = I_1 + I_2 + I_3 + I_4 + I_5 \tag{14.3}$$

and

$$V = V_1 = V_2 = V_3 = V_4 = V_5 \tag{14.4}$$

Thus in the two circuits the relations between force and velocity in one correspond to the relations between current and voltage in the other.

We express this by saying that the electrical analogue of force is current and that the analogue of velocity is voltage.

Beginning with these relationships, we now examine the elements of electrical and mechanical circuits side by side in order to determine which elements are analogous pairs. These relationships are shown in the following table, together with the defining equations corresponding to each of the terms.

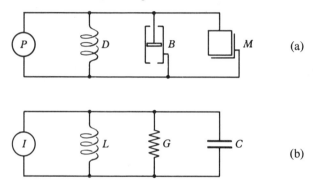

FIG. 14.8 (a) A mechanical circuit. (b) Its electrical analogue

Where mechanical resistance is concerned it will be noticed that only viscous resistance (force proportional to velocity) has an electrical analogue. Any other form of resistance would be represented by a non-linear conductance.

Returning to the mechanical problem of Fig. 14.5, in which an oscillatory force $P(t)$ is applied to a combination of M, D and B. It is required to determine an expression for the velocity of the mass. This can be obtained by first drawing the mechanical circuit and transforming it into its electrical analogue (Fig. 14.8). Consistent systems of units must be used in each of the two circuits and the numerical values of the analogue pairs must be equal or in constant proportion. The expression obtained by solving the electrical circuit for the voltage across the capacitor will, therefore, apply to the velocity of the mass in the mechanical analogue.

The unit systems must be consistent, but need not be the same. For example, if the ft-lb-sec system is used on the mechanical side, a force of 10 lb becomes a current of 10 A, a mass of 5 slugs transforms to a capacitance of 5 farads and so on. There is, however, a very strong case for using the SI system of units throughout, and so reducing the risk of error.

LINEAR MOTION			ANGULAR MOTION			DIRECT ELECTRICAL ANALOGUE		
QUANTITY	SYMBOL	SI UNIT	QUANTITY	SYMBOL	SI UNIT	QUANTITY	SYMBOL	SI UNIT
Displacement	x	metre	Displacement	θ	radian	Volt sec	$\int V\,dt$	Vs
Velocity	$v = dx/dt$	m/s	Velocity	$\omega = d\theta/dt$	rad/s	Voltage	V	Volt
Acceleration	$a = d^2x/dt^2$	m/s²	Acceleration	$d^2\theta/dt^2$	rad/s²	Rate of change of Voltage	dV/dt	V/s
Force	P	newton	Torque	f	N m	Current	I	A
Mass	M	kg	Moment of inertia	J	kg m²	Capacitance	C	F
Resistance (viscous)	$B = P/v$	N s/m	Resistance (viscous)	$B = f/\omega$	Nms/rad	Conductance	G	A/V
Compliance	$D = x/P$	m/N	Compliance	$D = \theta/f$	rad/Nm	Inductance	L	H

FIG. 14.9 Table of equivalence between quantities relating to linear motion, angular motion and electric circuit phenomena

14.4 Rotational systems

The equations of rotary motion are similar to those of linear motion. A rotary system has therefore a linear analogue and, consequently, electrical analogues of both systems can be obtained.

Corresponding terms are shown in the table (Fig. 14.9) where linear displacement becomes angular displacement, force becomes torque, mass becomes moment of inertia and so on. The corresponding quantities in the electrical analogue are also shown.

Figure 14.10 shows an example of a flexible shaft with a system of flywheels together with its mechanical circuit and electrical analogue.

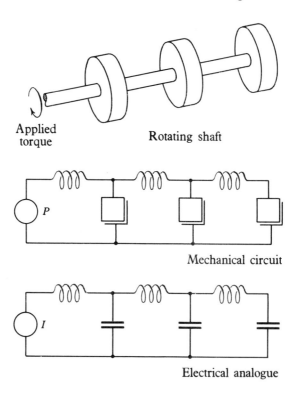

FIG. 14.10 Example of a rotating shaft with a series of flywheels

14.5 Dual electrical circuits

Since most electrical circuits can be transformed to their dual circuits (see Chap. 8), it follows that a mechanical circuit can be represented not only by its direct electrical analogue but by its dual electrical analogue.

This is shown in Fig. 14.11 where a mechanical circuit is given and its direct electrical analogue is obtained by using the same configuration and replacing the components, element by element, according to the table Fig. 14.9.

The direct electrical analogue is then converted to the dual electrical analogue using the methods of Chap. 8.

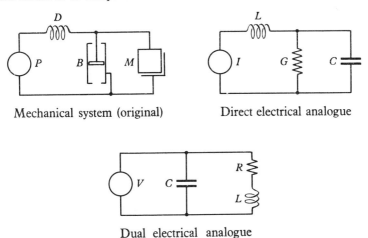

Mechanical system (original)   Direct electrical analogue

Dual electrical analogue

FIG. 14.11 The comparison of a mechanical circuit with its direct electrical analogue and its dual electric analogue

The voltage in the dual circuit thus represents force in the original and similarly current represents velocity.

This pairing of terms seems logical; voltage being looked upon as 'electrical force' and current being related to velocity of charge. It should be noted, however, that the similarity in appearance of the two circuits has been lost.

Problems can be solved equally well by using either the direct analogue or its dual.

14.6 Electro-mechanical transducers

The term transducer can be applied to any device which converts mechanical energy into electrical energy or vice versa. Apart from conventional electrical generators and motors which obviously conform to this definition, the term may be applied to many kinds of equipment such as record player pick-ups, loud-speakers, microphones and telephone receivers, pressure pick-ups, solenoid and plunger mechanisms and vibration generators.

These are divided into two categories, electro-magnetic transducers and electro-static transducers.

As an example of the former, consider a lossless d.c. motor operating under constant field conditions. The torque produced by the machine is directly proportional to the armature current and the terminal voltage is proportional to speed.

Thus

$$f = k_1 I \qquad (14.5)$$
$$v = k_2 V \qquad (14.6)$$

This kind of transducer is a link between a mechanical system and an electrical network in which current is transformed to torque (or force) and voltage to velocity. The combined network of an electro-mechanical system containing this type of transducer, is shown in Fig. 14.12.

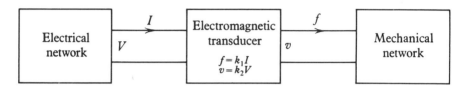

FIG. 14.12 Block diagram of an electromagnetic transducer system

With a transducer operating on an electrostatic principle, such as a capacitor microphone, force is proportional to voltage and velocity to current. Such a transducer links the electrical network with the dual of the mechanical network system, and not with the actual mechanical network as shown in Fig. 14.13.

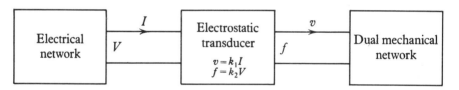

FIG. 14.13 Block diagram of an electrostatic transducer system

CHAPTER 15

Machine
conventions

15.1 Rotating electrical machines

Detailed study of the characteristics, performance and design of individual electrical machines invariably leads to emphasis on individual aspects of the many different types, with the result that many similarities which exist between them remain unobserved.

It is thus easy for a student to assume that a.c. and d.c. windings are quite different in character, or that an induction motor is designed according to rules that have no application to synchronous machines. This attitude can well be understood especially when he finds industrial design departments divided into separate sections, employing widely divergent design techniques.

It is, therefore, a useful exercise to consider a unified or generalised theory of electrical machines which places the emphasis on common characteristics and shows how the various parameters of the different machines are related to each other. It will be soon observed that such a treatment may not necessarily lead to the simplification of established routines for the design or determination of the performance of a given machine, and sometimes the results obtained may not be sufficiently accurate for practical purposes. Unified theory is essentially linear theory; non-linear effects such as saturation of magnetic circuits have to be ignored. In spite of this, a deeper understanding of fundamental principles may be gained, particularly in those applications which involve transient behaviour.

15.2 Types of rotating electrical machines

Although electrical machines sometimes have unusual shapes such as with disc or linear accelerators, the conventional machine consists of two members, a stator and a rotor which are in the form of concentric (more strictly coaxial) cylinders (Fig. 15.1).

Windings are distributed in slots on the inner circumference of the stator and the outer circumference of the rotor. Together they produce magnetic field patterns in the air gap between the cylinders, and it is due to this magnetic field

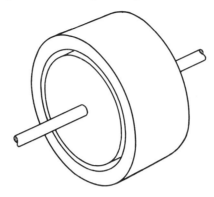

FIG. 15.1 A uniform-air-gap system

that torque is produced and energy transfer is made possible between electrical and mechanical systems. Machines are classified in two types according to the air gap.

15.2.1 Uniform air gap

This is the standard arrangement for induction motors and cylindrical-rotor alternators. Since air-gap permeance is uniform around the circumference, the flux density in the gap is proportional to the combined magneto-motive force of the windings. The special manner in which the windings are distributed determines the number of alternate north and south magnetic poles in the flux pattern. When polyphase alternating currents flow in a suitable winding, the field consists of a fundamental system of poles moving relative to the winding with a superimposed harmonic system of much smaller amplitude, also moving each at its own speed and direction relative to the winding.

15.2.2 Salient poles

If either the stator winding or the rotor winding is excited by direct current, the field pattern is stationary with respect to it, as for example in an alternator with a cylindrical rotor. The designer may wish to define the field polarities more strongly by dividing the air gap into alternate zones of high and low permeance, thus establishing salient poles.

Figure 15.2(*a*) shows a four-pole salient-pole rotor and Fig. 15.2(*b*), a similar stator. The d.c. windings can now be arranged around the poles instead of in slots and this is usually a more efficient system from the points of view of both the designer and the production engineer.

Figure 15.2(*c*) shows the composite arrangement of stator slotting and salient poles employed in typical cross-field d.c. generator construction.

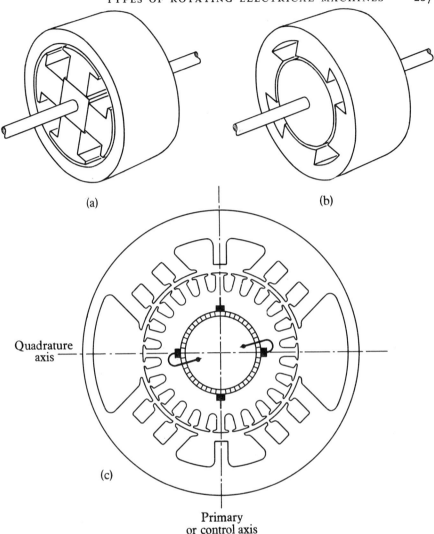

(a)

(b)

Quadrature axis

(c)

Primary
or control axis

**FIG. 15.2 Salient-pole systems (a) salient-pole rotor, (b) salient-pole stator, (c) slotting
for a cross-field generator with salient-pole stator**

15.2.3 Magnetic axes

A salient-pole machine has two axes of geometrical symmetry. Independent
of the number of poles, the machine is symmetrical about an axis through the
centre of a pole, and also about an axis mid-way between poles. These axes
are termed the direct axis (or the pole axis) and the quadrature axis respectively.
This terminology arises from the fact that in the two-pole machine the quadrature axis is at right angles to the pole axis (Fig. 15.3). The difference in permeance
of the gap in the direction of the direct axis from that in the quadrature axis
gives rise to the special features of salient-pole machines.

A coil in the position **ab**, which embraces the direct axis, produces more flux than a similar coil **cd** about the quadrature axis. It will be observed that coil **ab** has thus a greater inductance than **cd**. The inductance of a rotor coil which is

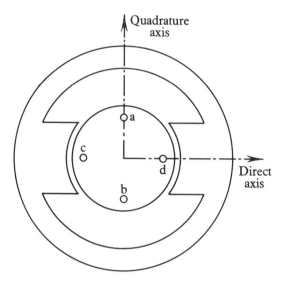

FIG. 15.3 The conventional axes of a salient-pole machine

continually moving from one axis to the other is continually changing from maximum to minimum value. Such a variation does not take place with a uniform-air-gap machine.

15.3 Windings

Detailed arrangements of rotor and stator windings have been dealt with elsewhere.* For the present purpose, only a few of the more general properties of windings need be re-stated. Two forms of windings are used, *concentrated windings* and *distributed windings*.

Field windings for both d.c. machines and salient pole alternators are examples of concentrated windings. All other windings are usually distributed. There are three types of distributed winding:

(*a*) cage windings, as with induction-motor rotors and alternator damper windings,
(*b*) phase windings, and
(*c*) commutator windings.

Apart from the cage-type windings, distributed windings are composed of coils, each having a spread of approximately a pole pitch, the coils being con-

* *Electrical Machines*, Chapter 4.

nected in series–parallel arrangements according to the purpose of the winding. The ends of the phase windings are brought out to terminals to which connections can be made. Connections may be made to a rotor winding by insulated slip rings on the rotor and fixed brushgear.

Commutator windings are invariably rotor windings where the coils are connected in series forming a closed circuit. Each of the junctions between the coils is connected to a commutator segment and connection to the winding is made by brushes sliding on the surface of the commutator.

15.4 Conventions for machine diagrams

The following conventions will be observed in later sections where sketches are used showing directions of current in windings, magnetic fluxes, and polarities. This is essential in order to avoid ambiguity.

1. It will be assumed that all field coils are wound as right-hand helices (Fig. 15.4).

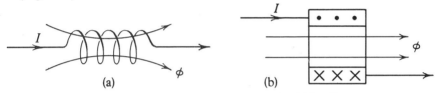

FIG. 15.4 The Fleming convention for field coils

The direction of the flux through the coil is, therefore, in the same direction as the current in the leads. If the coil is shown in cross-section, the relationship between flux and current will be as shown in Fig. 15.4(*b*).

2. Arrow heads will be used to show the positive direction of current.

3. Terminals will be marked + and − to show polarities corresponding to the positive direction of current or to positive rate of change of current.

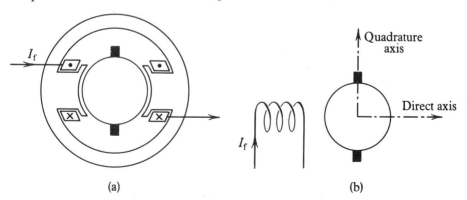

FIG. 15.5 Machine conventions (a) pictorial diagram of a commutator machine, (b) symbolic diagram for the same machine

4. Sketches of salient-pole machines will be drawn showing salient-pole stators with the direct axis horizontally to the right and the quadrature axis vertically upwards.

5. Rotation will be considered to be positive when the armature coil rotates from the direct axis to the quadrature axis. This corresponds to anti-clockwise rotation of the armature, Fig. 15.5.

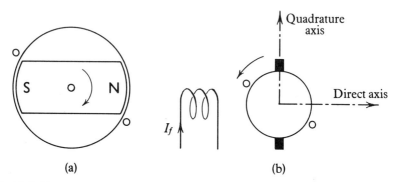

(a)   (b)

FIG. 15.6 (a) A machine with salient-pole rotor showing a stator coil. (b) Symbolic diagram for the corresponding salient-pole stator machine

It will be noted from Fig. 15.6 that a machine with a salient-pole rotor, rotating clockwise, has the same relative motion between field and armature coils as that of the salient pole-stator machine whose rotor rotation is anti-clockwise.

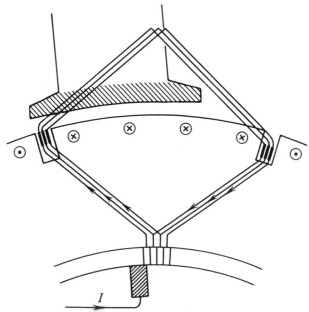

FIG. 15.7 The relative positions of field poles, and brush gear in an actual machine

6. Particular attention must be paid to the representation of commutator windings. Part of such a winding is sketched in Fig. 15.7, where adjacent coils in a slot and their connections to the commutator are shown. The winding is a progressive lap winding, termed progressive if the coils are connected in series with the left-hand commutator segment connected to the left-hand conductor and similarly on the right-hand side. The brush shown is placed on the centre line of the pole and if current enters the brush, the conductor currents under this pole will be downward (into the paper).

Although individual conductors move forward with the armature as it revolves, others take their place and the current pattern remains nearly constant in space, corresponding to a current sheet with downward direction underneath this particular pole.

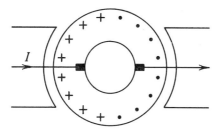

FIG. 15.8 Sketch of an actual two-pole d.c. machine showing position of poles and brush gear and the direction of current in the armature conductors

Figure 15.8 shows this current pattern for a two-pole machine with current downward under the left-hand pole and upwards under the right-hand one. The fact that the brush is physically on the centre line of the pole is due to

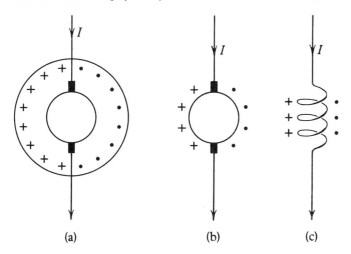

FIG. 15.9 (a) The conventional diagram for an armature and brush gear. (b) The Fleming convention for armature current. (c) The Fleming convention for field current

connections between the top and bottom conductors and the commutator segments being of equal length. If the bottom conductors were connected to the
nearest segments and the top connection lengthened accordingly, the brushes
would be moved through 90 degrees, giving the diagram of Fig. 15.9(*a*).

The winding is not so arranged in practical machines but it is a convenient
convention to assume it, since this convention leads to a simple diagram in
which the direction of the magneto-motive-force with respect to the current
corresponds to a right-hand helix whose axis is the brush axis, consequently,
this convention agrees with that relating to field coils (Fig. 15.9(*b*) and 15.9(*c*)).

It will always be assumed subsequently that progressive windings have been
used and that the m.m.f. of a commutator winding is directed along the brush
axis.

CHAPTER 16

The primitive machine

16.1 Introduction

All electrical machines consist of systems of interconnected coils each having resistance and inductance with mutual inductance between them. It seems reasonable to expect that they can be treated as networks and analysed by the methods of preceding chapters. So they can, but before too many assumptions are made, it must be remembered that the problem is complicated by a new factor, namely that there is relative motion between rotor and stator coils. We must first consider the effect of this relative motion on mutual inductance.

16.2 The primitive machine

In Chap. 4, our study of networks began with the consideration of unconnected coils having resistance inductance and capacitance, together with mutual inductance between them. This was known as the primitive system, later to be constrained by interconnections between the coils to form a network. The connection matrix $[c]$ transforming network currents to primitive currents was established and this enabled the impedance matrix for the network to be evaluated

$$[Z'] = [c_t][Z][c] \tag{4.29}$$

Making the same approach to machine theory, we again begin with a system of unconnected coils and construct the machine from them. We commence with a special type of machine which has several windings on both stator and rotor unconnected to each other except by mutual inductance. This we call a *primitive machine*. We shall first study the impedance matrix for this machine, or system, and then form various machines from it by suitable interconnections.

Fig. 16.1(*a*) shows diagrammatically such a machine. The stator is a two-pole salient-pole type with well defined direct and quadrature magnetic axes.*

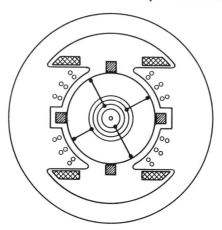

FIG. 16.1 (a) Diagram of a primitive machine.

Windings are arranged about each of these axes:

1. $\mathbf{k}_f$ a concentrated winding about the direct axis, known as the field winding;
2. $\mathbf{k}_{ds}$ a distributed winding about the direct axis known as the direct-axis stator winding;
3. $\mathbf{k}_{qs}$ a distributed winding about the quadrature axis known as the quadrature-axis stator winding.

$\mathbf{k}_f$ corresponds to the main field winding of an alternator or d.c. machine, and $\mathbf{k}_{ds}$ and $\mathbf{k}_{qs}$ to the stator windings of a two-phase induction motor.†

The rotor has a simple two-pole progressive commutator winding, either lap or wave, with two pairs of brushes on the commutator in the direct and quadrature axes respectively, and also four symmetrical tappings brought out to slip rings. The stator poles are divided at the centre to make a commutation zone for the direct-axis brushes.

We have seen in Chap. 15 that a rotating commutator winding creates the same m.m.f. pattern as a fixed solenoid coincident with the brush axis. The fact that the rotor is moving does not affect the magnetic field pattern though of course it is related to the voltage developed between the brushes. We therefore

---

* From a practical stand-point, salient-pole machines and uniform air-gap machines are essentially dissimilar in construction and a single machine cannot be converted from one type to the other just by changing interconnections.

The salient-pole machine is really the more general machine for fundamental study, since, *theoretically* it can be converted to a uniform air-gap machine by equating the parameters relating to the direct and quadrature axes.

† The suffixes d, q, s and r stand for direct-axis, quadrature-axis, stator and rotor respectively.

The current I_{dr} is thus the current in the direct-axis rotor circuit and the direction of this current is defined in terms of the fig. 16.1(*b*).

This type of double-subscript notation must not be confused with other forms.

assume that the rotor can be replaced by two *apparently stationary* windings $\mathbf{k}_{dr}$ and $\mathbf{k}_{qr}$ whose axes are in the direct axis and quadrature axis respectively as

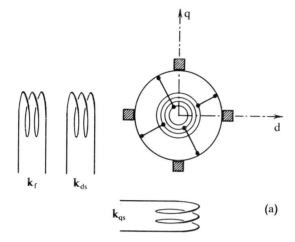

FIG. 16.1 (b) Symbolic representation of a primitive machine.

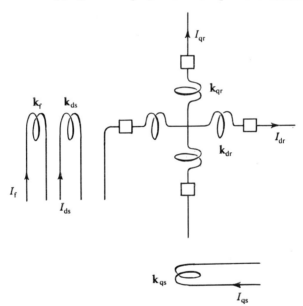

FIG. 16.1 (c) Conventional directions of current in a primitive machine showing the apparently stationary coils representing the commutator winding

shown in Fig. 16.1(c). Since the coils are at right angles, no mutual inductance exists between them and the two coils can be considered independent.

The primitive machine thus consists of *five independent windings* in a primitive system, but before we can write down the impedance matrix of this system we must learn more about the voltages which are induced in rotating coils and also in commutator windings.

16.3 Mutual inductance between stator and rotor coil

Figure 16.2. shows a stator coil carrying a current i and a rotor coil whose axis is inclined at an angle θ to the direct axis. It is assumed in the first place that the distribution of flux due to the stator coil is sinusoidal. Any other

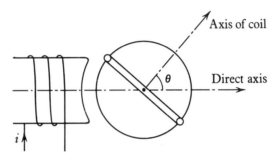

FIG. 16.2 A single armature coil of a primitive machine inclined at an angle θ to the direct axis

distribution can be accounted for by assuming appropriate harmonic components.

When θ is equal to zero let the mutual inductance between the two coils be M. This will be the maximum value of $M(\theta)$ since flux linkages will diminish as θ increases from 0° to 90°.

For sinusoidal flux distribution

$$M(\theta) = M \cos \theta \tag{16.1}$$

The voltage generated in the rotor coil

$$v = \frac{\mathrm{d}\phi}{\mathrm{d}t} = \frac{\mathrm{d}[M(\theta)i]}{\mathrm{d}t}$$

$$= \frac{\mathrm{d}(M \cos \theta)i}{\mathrm{d}t}$$

$$= M \cos \theta \frac{\mathrm{d}i}{\mathrm{d}t} - iM \sin \theta \frac{\mathrm{d}\theta}{\mathrm{d}t} \tag{16.2}$$

The voltage has thus two components, one dependent on the instantaneous position of the coil, together with the rate of change of primary current, and the other dependent on the position of the coil, the instantaneous current and the speed of rotation. The former may be termed the 'transformer' voltage and the second component the 'rotational' voltage. It is the second voltage component that makes the significant difference between machine circuits and the networks of preceding chapters.

16.4 Commutator windings

A commutator winding consists of a number of coils such as those described in the previous section distributed around the rotor and connected in series to form an endless loop. .As the armature makes one revolution, each coil passes in sequence through all the positions occupied by the other coils. The voltage picked off at the commutator between a pair of brushes at any instant

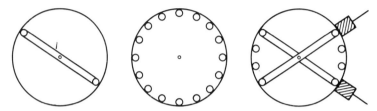

FIG. 16.3 (a) A single armature coil. (b) Several armature coils. (c) Those armature coils between a pair of brushes

is equal to the sum of the instantaneous voltages of the coils between them. For example, in Fig. 16.3, there are always four coils between the brushes shown in the diagram hence the voltage between these brushes is equal to that of these four coils. The voltage between brushes can be determined by finding the average of all the coil voltages at that instant within the brush zone and multiplying by the number of coils.

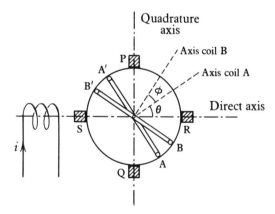

FIG. 16.4 A commutator winding with brushes in the direct and quadrature axes

In Fig. 16.4 brushes are placed diametrically on the commutator with brushes **P** and **Q** lying on the quadrature axis, and brushes **R** and **S** in the direct axis.

The voltage v_{QP} is thus the summation of the voltage of conductors whose axes are in the zone $0 < \theta < \pi$ and v_{SR} is the summation of the voltages of the conductors for which $-\pi/2 < \theta < \pi/2$.

The voltage in a given armature coil **A** (which we will designate as the reference coil) is given by Eq. (16.3)

$$v_A = M \cos \theta \frac{di}{dt} - iM \sin \theta \frac{d\theta}{dt} \qquad (16.3)$$

and the voltage in any other coil **B** displaced on the armature by angle ϕ with reference to **A** is

$$v_B = M \cos (\theta + \phi) \frac{di}{dt} - iM \sin (\theta + \phi) \frac{d\theta}{dt} \qquad (16.4)$$

At the instant t when the reference coil is at the angle θ (Fig. 16.3), the angles of the coils contributing to v_{QP} lie between $(0 - \theta)$ and $(\pi - \theta)$.

v_{QP} is obtained by finding the average value of these coil voltages at that instant, that is by integrating Eq. (16.4) between these limits, dividing by π and multiplying by the number of coils.

As this is a space average, θ, i, di/dt and $d\theta/dt$ may be treated as constants in the integration.

Thus
$$v_{QP} = \frac{g}{\pi} \int_{\varphi=0-\theta}^{\varphi=\pi-\theta} v_B \, d\phi \qquad (16.5)$$

$$= \frac{g}{\pi} \int_{0-\theta}^{\pi-\theta} \left[M \cos (\theta + \phi) \frac{di}{dt} - iM \sin (\theta + \phi) \, d\theta/dt \right] d\phi$$

$$= \frac{g}{\pi} [0 - 2iM \, d\theta/dt]$$

$$= -\frac{2}{\pi} gMi \, d\theta/dt \qquad (16.6)$$

where g is the number of coils in series between the brushes. Similarly v_{SR} is g times the average voltage of coils for which ϕ lies between $(-\pi/2 - \theta)$ and $(\pi/2 - \theta)$.

Thus
$$v_{SR} = \frac{g}{\pi} \int_{-\pi/2-\theta}^{\pi/2-\theta} v_B \, d\phi \qquad (16.7)$$

$$= \frac{g}{\pi} \left[2M \frac{di}{dt} + 0 \right]$$

$$= \frac{2g}{\pi} M \frac{di}{dt} \qquad (16.8)$$

The significance of the negative sign attributed to v_{QP} will be developed in Sec. 16.5.4.

From Eqs. (16.6) and (16.8) we see that the two commutator voltages picked out by the brush pairs at right angles are also the separate components of the generated voltage in the armature coils due to di/dt and $d\theta/dt$ respectively.

v_{SR} is known as the direct axis brush voltage and is the 'transformer' voltage.

v_{QP} is known as the quadrature axis brush voltage and is the 'rotational' voltage.

It is highly significant that the commutator and brushgear possess the property of separating out the two components of voltage in this manner and it is also important to recognise that the coefficients of $i\, d\theta/dt$ and di/dt in Eq. (16.6) and (16.8) are the same namely $(2/\pi)gM$.

16.4.1 Non-sinusoidal flux distribution

The above analysis assumes sinusoidal flux distribution where

$$M(\theta) = M \cos \theta \qquad (16.1)$$

Extending this to the more general case of non-sinusoidal distribution, $M(\theta)$ will contain harmonics, i.e.

$$M(\theta) = \Sigma\, M_n \cos (n\theta + \phi_n) \qquad (16.9)$$

where n is odd.

The equation to the coil voltage corresponding to the nth flux harmonic is derived in the manner of Eq. 16.2, and is

$$v_n = M_n \cos (n\theta + \phi_n)\frac{di}{dt} - iM_n n \sin (n\theta + \phi_n)\frac{d\theta}{dt} \qquad (16.10)$$

and the corresponding harmonic components in the commutator voltages become

$$(v_{QP})_n = -\frac{2g}{\pi} M_n ni\, (d\theta/dt) \qquad (16.11)$$

and

$$(v_{SR})_n = \frac{2g}{\pi} M_n\, (di/dt) \qquad (16.12)$$

It should be noticed that the coefficient $(2/\pi)gM_n$ is multiplied by n in Eq. (16.11) but not in (16.12).

We, therefore, infer that if the flux distribution is not sinusoidal, which is usually the case for a salient pole machine, the two commutator voltages will have different waveforms.

Assuming constant speed, the waveform of v_{SR} will correspond to that of di/dt and will not be affected by the space distribution of the flux. This, of course, is normal in any transformer. On the other hand v_{QP}, the generated voltage, is dependent on the flux distribution and its waveform contains the same harmonics.

16.4.2 Sinusoidal excitation

If the stator coil is excited by alternating current of r.m.s. value I and frequency

ω and the armature is rotating at the corresponding synchronous speed

$$i = \sqrt{2I} \sin \omega t \tag{16.13}$$

and
$$d\theta/dt = \omega \tag{16.14}$$

Thus
$$v_{\mathrm{PQ}} = \frac{2}{\pi} gMi \, d\theta/dt = \frac{2}{\pi} gM \sqrt{2\omega I} \sin \omega t \tag{16.15}$$

$$v_{\mathrm{SR}} = \frac{2}{\pi} gM \, di/dt = \frac{2}{\pi} gM \sqrt{2\omega I} \cos \omega t \tag{16.16}$$

The r.m.s. values of the commutator voltages V_{PQ} and V_{SR} are thus equal under these conditions.

If the armature rotates at a speed such that

$$d\theta/dt = \alpha\omega \tag{16.17}$$

then the ratio

$$V_{\mathrm{PQ}}/V_{\mathrm{SR}} = \alpha \tag{16.18}$$

16.4.3 Excitation in the quadrature axis

If the current i flows in a quadrature-axis stator winding (such as k_{qs} in Fig. 16.1(b) instead of in a direct-axis coil, which we have previously considered, the equation to the voltage in the armature coil **A** becomes

$$v'_{\mathrm{A}} = M \cos(\theta - \pi/2) \frac{di}{dt} - iM \sin(\theta - \pi/2) \frac{d\theta}{dt}$$

$$= M \sin \theta \cdot \frac{di}{dt} + iM \cos \theta \frac{d\theta}{dt} \tag{16.19}$$

instead of (16.3), and it can therefore be deduced that

$$v'_{\mathrm{B}} = iM \cos(\theta + \phi) \frac{d\theta}{dt} + M \sin(\theta + \phi) \cdot \frac{di}{dt}$$

and

$$v'_{\mathrm{QP}} = \frac{g}{\pi} \int_{\varphi=0-\theta}^{\varphi=\pi-\theta} v'_{\mathrm{B}} \, d\phi$$

$$= \frac{2}{\pi} g \cdot M \cdot \frac{di}{dt} \tag{16.20}$$

and

$$v'_{\mathrm{SR}} = \frac{g}{\pi} \int_{-\pi/2-\theta}^{\pi/2-\theta} v'_{\mathrm{B}} \cdot d\phi$$

$$= \frac{2}{\pi} gMi \, d\theta/dt \tag{16.21}$$

The fact that Eqs. (16.8) and (16.20) are similar and so are (16.6) and (16.21) *without the negative sign* should be noted for future reference.

16.5 The impedance matrix of a primitive machine

We are now in a position to determine the voltage equations for the five independent windings of the primitive machine in terms of impedance coefficients and currents in the windings. The assembled coefficients form the unconnected impedance matrix of the machine giving

$$[V] = [Z] \cdot [I] \tag{16.22}$$

The conventional order for the sequence of the windings is given by Eq. (16.23)

$$
\begin{bmatrix} V_f \\ V_{ds} \\ V_{dr} \\ V_{qr} \\ V_{qs} \end{bmatrix}
= [Z] \cdot
\begin{bmatrix} I_f \\ I_{ds} \\ I_{dr} \\ I_{qr} \\ I_{qs} \end{bmatrix}
\tag{16.23}
$$

where the first three lines refer to voltage in the direct-axis coils and lines four and five refer to the quadrature-axis coils. There is an advantage in putting the rotor coils adjacent in lines three and four.

The impedance matrix is built up by the superposition of (*a*) a resistance matrix, (*b*) a self inductance matrix and (*c*) a mutual inductance matrix as was done in Sec. 4.1 for stationary coils. Now we must include (*d*) a matrix of coefficients of generated voltages caused by rotation.

16.5.1 The resistance matrix

$$
[R] =
\begin{matrix} k_f \\ k_{ds} \\ k_{dr} \\ k_{qr} \\ k_{qs} \end{matrix}
\begin{bmatrix}
R_f & \cdot & \cdot & \cdot & \cdot \\
\cdot & R_{ds} & \cdot & \cdot & \cdot \\
\cdot & \cdot & R_r & \cdot & \cdot \\
\cdot & \cdot & \cdot & R_r & \cdot \\
\cdot & \cdot & \cdot & \cdot & R_{qs}
\end{bmatrix}
\tag{16.24}
$$
$$\text{where}\quad R_r = R_{dr} = R_{qr}$$

The resistance between the brush pairs R_{dr} and R_{qr} are necessarily the same by reason of symmetry.

16.5.2 The self-inductance matrix

$$
[L]p =
\begin{bmatrix}
L_f & \cdot & \cdot & \cdot & \cdot \\
\cdot & L_{ds} & \cdot & \cdot & \cdot \\
\cdot & \cdot & L_{dr} & \cdot & \cdot \\
\cdot & \cdot & \cdot & L_{qr} & \cdot \\
\cdot & \cdot & \cdot & \cdot & L_{qs}
\end{bmatrix} p
\tag{16.25}
$$

L_{dr} and L_{qr} differ in a salient-pole machine and so do L_{ds} and L_{qs} but they are equal in a uniform air gap machine.

16.5.3 The mutual-inductance matrix

$$[M]p = \begin{bmatrix} . & M_{fd} & M_{fr} & . & . \\ M_{fd} & . & M_d & . & . \\ M_{fr} & M_d & . & . & . \\ . & . & . & . & M_q \\ . & . & . & M_q & . \end{bmatrix} p$$

(16.26)

Mutual inductance occurs between windings on the same axis but there is no mutual inductance between windings on axes in quadrature. The mutual coefficients occur in the matrix in pairs about the diagonal (see Sec. 4.6). M_d and M_q differ in a salient pole machine but are equal if the air gap is uniform.

M_d and M_q refer to the 'transformer' voltages in the direct-axis and quadrature-axis coils and are derived from equations such as (16.8) and (16.20).

16.5.4 The coefficients of generated voltage

According to Eqs. (16.6) and (16.21) rotational voltages are given by expressions such as $(2g/\pi)Mi\,d\theta/dt$ or $Gi(p\theta)$ where G is equal to $(2g/\pi)M$ if the flux is sinusoidally distributed but is slightly modified for other distributions.

These voltages can only be produced in the rotor windings and are due to currents in the opposite axis. There are thus only five such terms, three contributing to V_{qr} and two to V_{dr}.

Thus $[V \text{(rotational components)}] = [G](p\theta)[I]$

and

$$[G](p\theta) = \begin{bmatrix} . & . & . & . & . \\ . & . & . & . & . \\ . & . & . & . & . \\ . & . & . & G_{qr} & G_q \\ -G_f & -G_d & -G_{dr} & . & . \\ . & . & . & . & . \end{bmatrix} (p\theta)$$

(16.27)

The reason for the negative signs preceding all terms in the fourth line will be seen if Eq. (16.6) is compared with (16.21). The difference between the two equations can also be deduced by observing the polarities of the brushes as shown in Fig. 16.5.

Figure 16.5(a) shows the polarities of the transformer component of voltage in V_{dr} due to positive rate of change of I_{ds}.

The corresponding diagram for the transformer voltage in the quadrature axis is given in Fig. 16.5(b). This diagram is obviously similar to 16.5(a) but turned through a right angle.

Figure 16.5(c) shows the rotational component of voltage in the direct axis due to rotation through flux set up by I_{qs}. Note that the polarity of the voltage agrees with that of the transformer voltage in 16.5(a).

Positive values are, therefore, attributed to the G coefficients relating to components of V_{dr} in order to correspond with the L and M coefficients.

In Fig. 16.5(d), which shows rotational voltage in the flux of I_{ds} and is similar

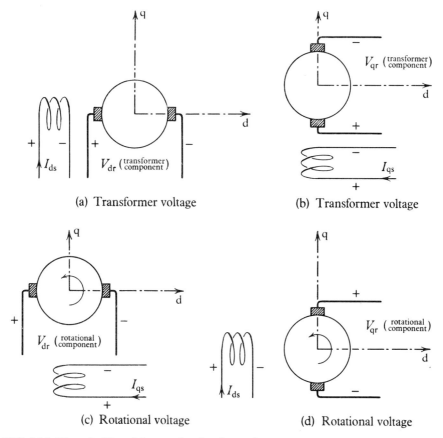

(a) Transformer voltage

(b) Transformer voltage

(c) Rotational voltage

(d) Rotational voltage

FIG. 16.5 The polarities of the rotational and transformer voltages in a primitive machine

to Fig. 16.5(c), rotated clockwise through ninety degrees, this component is seen to be in the opposite direction to the corresponding transformer voltage. Negative signs are, therefore, applied to all such terms in the fourth row of the matrix.

16.5.5 The G and M coefficients

G_d and G_q differ due to saliency and so do G_{dr} and G_{qr} for the same reason that M_d and M_q are unequal. For a *uniform air-gap machine* this difference does not exist when we may assume

$$G_d = G_q \qquad G_{dr} = G_{qr} \qquad M_d = M_q$$

Furthermore if we compare the coefficients in Eqs. (16.6) and (16.8) we observe that if we may assume *sinusoidally distributed flux* then all G coefficients will be equal in value to the M coefficients with the same subscripts.

$$G_d = M_d \qquad G_{dr} = M_{dr} \qquad G_f = M_f$$
$$G_q = M_q \qquad G_{qr} = M_{qr}$$

This approximation is frequently made for simplicity without involving serious error even in examples such as a salient pole generator where the flux waveform is known to be non-sinusoidal. The degree of approximation is equivalent to ignoring harmonics in a system where harmonics are known to be present though relatively small in amplitude.

16.5.6 The complete impedance matrix

Adding these component matrices, we obtain

$$[Z] = [R] + [L]p + [M]p + [G](p\theta) \tag{16.28}$$

$$= \begin{bmatrix} R_f + L_f p & M_{fd}p & M_{fr}p & 0 & 0 \\ M_{fd}p & R_{ds} + L_{ds}p & M_d p & 0 & 0 \\ M_{fr}p & M_d p & R_r + L_{dr}p & G_{qr}(p\theta) & G_q(p\theta) \\ -G_f(p\theta) & -G_d(p\theta) & -G_{dr}(p\theta) & R_r + L_{qr}p & M_q p \\ 0 & 0 & 0 & M_q p & R_{qs} + L_{qs}p \end{bmatrix} \tag{16.29}$$

The matrix 16.29 is known as the impedance matrix of the primitive machine.

It is the fundamental matrix from which the actual matrices of many types of machine can be derived by applying appropriate connection transforms.

It should be noted that the presence of the terms representing the generated voltages destroys the diagonal symmetry of the matrix.

A further point of practical importance is that all the coefficients in the impedance matrix of a given primitive machine can be measured in simple open-circuit tests. They can also be predetermined by standard design procedures.

16.6 Torque

For the study of any type of machine it is always important to be able to determine how the value of the gross torque is related to the electrical input or output of the machine and to the machine parameters.

The torque developed in a primitive machine is derived in the following manner.

Commencing with the generalised impedance matrix

$$[Z] = [R] + ([L + M]p + (p\theta)[G] \tag{16.28}$$

and multiplying by the current matrix $[I]$ we obtain the voltage matrix $[V]$

where

$$[V] = [Z][I]$$
$$= [R][I] + [L+M]p[I] + (p\theta)[G][I] \qquad (16.30)$$
$$= [R][I] + p[\Phi] + (p\theta)[B] \qquad (16.31)$$

and the new terms are defined by

$$[\Phi] = [L+M][I] \qquad (16.32)$$
$$[B] = [G] \cdot [I] \qquad (16.33)$$

These are flux terms which enable the transformer and rotational components of voltage to be identified as $p[\Phi]$ and $(p\theta)[B]$ respectively in Eq. (16.31). The matrix $[R][I]$ is the matrix of voltage-drop in resistance.

The electrical power input to the system is obtained by premultiplying the matrix $[V]$ by the transpose of $[I]$

Thus
$$P = [I_t] \cdot [V] \qquad (16.34)$$
$$= [I_t] \cdot [Z][I] \qquad (16.35)$$
$$= [I_t][R][I] + [I_t][L+M]p[I] + (p\theta)[I_t][G][I] \qquad (16.36)$$

The terms in this equation are seen to be

$$P = \text{the total electrical input power (instantaneous)}$$

$$[I_t][R][I] = \text{the electrical power dissipated in heat}$$

$$[I_t][L+M]p[I] = \text{the rate of storage of energy in the magnetic fields of the system}$$

$$(p\theta)[I_t][G][I] = \text{the rate of conversion of electrical energy to mechanical energy, i.e. mechanical power output.}$$

Since mechanical power is the product of torque f and speed

$$(p\theta)f = (p\theta)[I_t][G][I] \qquad (16.37)$$

and hence, dividing by $(p\theta)$, the torque of the primitive machine is given by

$$*f = [I_t][G][I] \qquad (16.38)$$

This equation is one of general application and can be used for all types of machine. When the voltage equations have been solved and the winding currents for a connected machine have been determined, this equation can be used to determine torque.

* If the currents are represented by phasors then according to section 4.3.3
$$P = \text{Re } [I*_t] [V] \qquad (16.41)$$
hence
$$f = \text{Re } [I*_t] [G] [I] \qquad (16.42)$$

Expanding the matrices of Eq. (16.34) the value of the torque is given by

$$f = [I_{dr} \quad I_{qr}] \cdot \begin{bmatrix} 0 & 0 & 0 & G_{qr} & G_q \\ -G_f & -G_d & -G_{dr} & 0 & 0 \end{bmatrix} \cdot \begin{bmatrix} I_f \\ I_{ds} \\ I_{dr} \\ I_{qr} \\ I_{qs} \end{bmatrix} \quad (16.39)$$

$$= I_{dr}(G_{qr}I_{qr} + G_q I_{qs}) - I_{qr}(G_f I_f + G_d I_{ds} + G_{dr} I_{dr}) \quad (16.40)$$

16.6.1 Components of torque

If for convenience we separate the terms in Eq. (16.35) into those components which are due to excitation from the stator and those due to rotor excitation,

$$f = [I_{dr} \quad I_{qr}] \begin{bmatrix} 0 & 0 & G_q \\ -G_f & -G_d & 0 \end{bmatrix} \begin{bmatrix} I_f \\ I_{ds} \\ I_{qs} \end{bmatrix}$$

$$+ [I_{dr} \quad I_{qr}] \begin{bmatrix} 0 & G_{qr} \\ -G_{dr} & 0 \end{bmatrix} \begin{bmatrix} I_{dr} \\ I_{qr} \end{bmatrix} \quad (16.43)$$

$$= f_1 + f_2$$

where

$$f_1 = I_{dr}G_q I_{qs} - I_{qr}(G_f I_f + G_d I_{ds}) \quad (16.44)$$
$$f_2 = I_{dr}I_{qr}(G_{qr} - G_{dr}) \quad (16.45)$$

The torque of the primitive machine is thus divided into two components. The first component, f_1, is the torque due to rotor currents in conductors situated at right angles to flux caused by stator currents and this is the major component. The second component, f_2, is due to the action of current in rotor conductors situated at right angles to flux set up by rotor currents in the opposite axis. It will be seen that this term vanishes for uniform air gap machines where G_{qr} and G_{dr} are equal.

The torque component f_2 is known as the reluctance torque introduced by saliency.

16.6.2 The torque equation of a d.c.-series motor

At this point the reader is reminded once again that we are using generalised impedance theory and consequently we are assuming a passive network which includes stores and sinks and to which external sources are connected and driving voltages applied.

The primitive machine is thus assessed as a network containing amongst other terms sinks which correspond to the energy converted from electrical to mechanical power. These are the terms with the G coefficients.

It follows therefore that we are considering the machine as a *motor* and the torque f in Eq. (16.37) is a *motoring torque*. In other words, if in a particular example the torque f obtained from Eq. (16.38) is found to have a positive value this means that the torque produced by the interaction of current and magnetic flux within the machine is in the positive direction.

The product of this torque and the speed (pθ) gives the power converted to mechanical work and the machine is acting as a motor. The following example is a simple illustration.

Suppose we connect a primitive machine to form a simple series d.c. motor. For this purpose we need the stator field winding k_f and the quadrature brushes only. For these two windings only the impedance matrix of the primitive machine is reduced to

$$\begin{bmatrix} V_f \\ V_{qr} \end{bmatrix} = \begin{bmatrix} R_f + L_f p & 0 \\ -G_f(p\theta) & R_r + L_{qr}p \end{bmatrix} \cdot \begin{bmatrix} I_f \\ I_{qr} \end{bmatrix} \tag{16.46}$$

For the unconnected state of these two windings we note that if (pθ) and I_f are positive the rotational component of voltage will be negative since

$$[V_{qr}] \text{ rotational} = -G_f(p\theta)I_f \tag{16.47}$$

The corresponding polarity of the brushes is shown in Fig. 16.6(a).

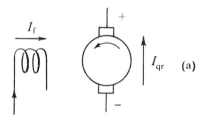

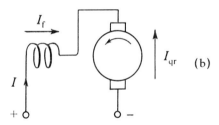

FIG. 16.6 (a) The conventional directions and polarities of a primitive machine with anti-clockwise rotation. (b) The connections of the windings to form a series d.c. motor

To connect the machine as a motor for this direction of rotation the windings must be in *series opposition* as shown in Fig. 16.6(b).

The single machine current I of the connected system is thus related to the winding currents by the connection matrix (c) where

$$\begin{bmatrix} I_f \\ I_{qr} \end{bmatrix} = \begin{bmatrix} 1 \\ -1 \end{bmatrix} I \tag{16.48}$$

$$[c] = \begin{bmatrix} 1 \\ -1 \end{bmatrix} \tag{16.49}$$

The impedance matrix for the connected machine is thus

$$V = \begin{bmatrix} 1 & -1 \end{bmatrix} \cdot \begin{bmatrix} R_\text{f} + L_\text{f}\text{p} & 0 \\ -G_\text{f}(\text{p}\theta) & R_\text{r} + L_\text{qr}\text{p} \end{bmatrix} \begin{bmatrix} 1 \\ -1 \end{bmatrix} I$$

$$= \begin{bmatrix} 1 & -1 \end{bmatrix} \cdot \begin{bmatrix} R_\text{f} + L_\text{f}\text{p} & 0 \\ -G_\text{f}(\text{p}\theta) & -R_\text{r} - L_\text{qr}\text{p} \end{bmatrix} I$$

$$= [(R_\text{f} + R_\text{r}) + (L_\text{f} + L_\text{qr})\text{p} + G_\text{f}(\text{p}\theta)] I \qquad (16.50)$$

Under steady-state conditions

$$V = [(R_\text{f} + R_\text{r}) + G_\text{f}(\text{p}\theta)] I \qquad (16.51)$$

from which speed is seen to be approximately inversely proportional to armature current.

Steady state torque is obtained from Eq. (16.38)

$$f = IGI = I^2 G \qquad (16.52)$$

16.7 The per-unit system

The per-unit system of values has been explained elsewhere* To recapitulate, the method consists of expressing all quantities as fractions of given base values.

It is customary with machines to use the normal rating as the base or unit value, so that a line current of $I = 0.5$ per unit means that the current I is equal to the full-load line current of the machine multiplied by 0.5. The voltage, current, power and speed, corresponding to a given load can all be expressed in terms of the full-load values in this way. The per-unit notation is particularly useful when relative size is more important than absolute value. It has particular significance when used in connection with transformers. To state that the voltage drop of a transformer on load is 100 volts can be most misleading unless it is clearly specified to which winding this drop refers. On the contrary there is no ambiguity in the statement that the voltage drop in a transformer is 0.1 per unit. This statement applies equally well to all windings.

In a power system comprising several machines, an arbitrary M.V.A. base may be chosen, the voltage (line-to-neutral) and the line current corresponding to this being taken as the unit values for the system. All other voltages and currents expressed in per-unit values are said to be referred to this base.

Per-unit values can also be attached to all resistances, reactances and impedances. The per-unit resistance of a device is defined as the value of the resistance voltage of the device when carrying a current of 1.0 per unit expressed as a fraction of the reference voltage of the system. Since

$$V = IR$$

$$\frac{V}{V_\text{base}} = \frac{I}{I_\text{base}} \left(R\frac{I_\text{base}}{V_\text{base}} \right) \qquad (16.53)$$

* *Electrical Machines*, Chapter 1, p. 5.

or $$V_{pu} = I_{pu}R_{pu} \tag{16.54}$$

where $$R_{pu} = R(I_{base}/V_{base}) \tag{16.55}$$

For example, a 50 kW, 500 V d.c. generator has an armature resistance of $0 \cdot 2\Omega$. Express R in per-unit values.

$$1 \cdot 0 \text{ pu current} \doteqdot 500000/500 = 100 \text{ A}$$

$$R_{pu} = 0 \cdot 2 \times 100/500 = 0 \cdot 04 \text{ pu}$$

Per-unit values of mutual inductance are determined in the following way.

Since $$V_2 = M \, dI_1/dt \tag{16.56}$$

$$\frac{V_2}{V_{2\,base}} = \frac{M \, dI_1/I_{1\,base}}{dt} \cdot \frac{I_{1\,base}}{V_{2\,base}} \tag{16.57}$$

also $$V_1 = M \, dI_2/dt \tag{16.58}$$

$$\frac{V_1}{V_{1\,base}} = M \frac{dI_2/I_{2\,base}}{dt} \cdot \frac{I_{2\,base}}{V_{1\,base}} \tag{16.59}$$

Thus $$M_{pu} = M\frac{I_{1\,base}}{V_{2\,base}} = M\frac{I_{2\,base}}{V_{1\,base}} \tag{16.60}$$

Different base values for current and voltage can, therefore, be used when two circuits are coupled by mutual induction provided

$$V_{1\,base}\,I_{1\,base} = V_{2\,base}\,I_{2\,base}, \tag{16.61}$$

that is to say the same volt-ampere base must be maintained.

16.7.1 Power formulae and per-unit values

Although impedance formulae apply equally well when per-unit values are used in place of volts, amperes and ohms, an additional factor has to be used in power equations. For example, the total power in a three-phase system (assuming unity power factor) is given by

$$P = V_a I_a + V_b I_b + V_c I_c \tag{16.62}$$

In an m phase system

$$P = \sum_{k=1}^{k=m} V_k I_k \tag{16.63}$$

where P is in watts, V in volts and I in amperes.

Thus $$\frac{P}{V_{base}\,I_{base}} = \Sigma \, V_{pu}I_{pu} \tag{16.64}$$

But $$P_{pu} = \frac{P}{P_{base}}$$

and $$P_{base} = mV_{base}\,I_{base}$$

if all the phases have the same base value.

Hence

$$P_{\text{pu}} = \frac{P}{mV_{\text{base}} I_{\text{base}}}$$

$$= \frac{1}{m} \sum V_{\text{pu}} I_{\text{pu}} \tag{16.65}$$

It is particularly to be noted that Eq. (16.65) contained the factor $1/m$ whereas Eq. (16.63) does not. It might be argued that the base has been changed in Eq. (16.65) and that this is wrong; it should have been maintained. Normal usage, however, demands that a machine operating on 1·0 per-unit voltage and 1·0 per-unit current delivers 1·0 per-unit power.

Care must be taken to include the factor $1/m$ in power formulae wherever per-unit values are involved or if transforms are used which effectively alter the base value in this way. An example of such a transform is the symmetrical component transform used in Sec. 13.12.

A convenient formula to determine per-unit reactance in a three-phase system is derived in the following way.

If

X = reactance in ohms per phase of a three-phase star-connected winding

I = nominal line current

V = nominal voltage line-to-line

S = apparent power = $\sqrt{3} \cdot V \cdot I$.

By definition

$$X_{\text{pu}} = \frac{IX}{\dfrac{V}{\sqrt{3}}} = \frac{\sqrt{3}VIX}{V^2} = \frac{SX}{V^2} \tag{16.66}$$

Thus

$$\text{per-unit reactance} = \frac{\text{rated apparent power in M.V.A.}}{\text{nominal line voltage in kV}} \cdot \left[\begin{array}{c} \text{equivalent star} \\ \text{reactance in ohms} \end{array} \right]$$

$$\tag{16.67}$$

16.7.2 Example using per-unit values

A three-phase power system consists of an alternator, transformer and transmission line having reactances as shown in the following table. Determine the overall per unit reactance of the system.

	Rating M.V.A.	Line voltage, kV	Connection	Reactance in ohms/phase
Alternator	30	33	star	4·0
Transformer	10	33/132	star/delta	1·0 winding (star) 48 winding (delta)
Transmission line		132		30

It will be convenient to use the alternator rating as the base for the system, viz., 30 M.V.A.

From Eq. (16.65)

$$\text{per-unit reactance (alternator)} \quad = \frac{30}{33^2} \cdot 4 \quad = 1 \cdot 102$$

$$\text{per-unit reactance} \left(\begin{array}{l}\text{transformer}\\\text{primary}\\\text{winding}\end{array}\right) = \frac{30}{33^2} \cdot 1 \quad = 0 \cdot 276$$

Equivalent star reactance corresponding to the mesh winding $= 48/3 \qquad = 16\Omega$
of transformer

$$\text{per-unit reactance} \left(\begin{array}{l}\text{transformer}\\\text{secondary}\\\text{winding}\end{array}\right) = \frac{30}{132^2} \cdot 16 = 0 \cdot 276$$

$$\text{per-unit reactance (line)} \qquad = \frac{30}{132^2} \cdot 30 = 0 \cdot 517$$

$$\begin{array}{ll}\text{Total per-unit reactance} & = 1 \cdot 102 \\ & 0 \cdot 276 \\ & 0 \cdot 276 \\ & 0 \cdot 517 \\ & \overline{ 2 \cdot 171}\end{array}$$

Note that the transformer has a per-unit reactance $0 \cdot 276 + 0 \cdot 276 = 0 \cdot 552$ on a basis of 30 M.V.A.

In terms of its own rating (10 M.V.A.) this is a per-unit reactance of $0 \cdot 552/3 = 0 \cdot 184$.

16.8 Transforms

In Sec. 16.6.2 we saw a simple example where a connection transform is used in conjunction with the impedance matrix of the primitive machine in order to form an actual machine, namely a series d.c. motor. Before we attempt to construct other machines in similar manner and determine the effect of their connection matrices we must introduce two other transforms.

16.8.1 Transforms which can be used for both voltage and current

If the transform [c] used to convert a set of currents into a related set, is also to be used for voltages, it follows in the first place that the matrix must be capable of inversion, that is it must be square.

If

$$[I] = [c][I']$$

then
$$[I'] = [c]^{-1}[I].$$

In the same way
$$[V] = [c][V']$$

and
$$[V'] = [c]^{-1}[V].$$

It also follows that if
$$[V] = [Z][I]$$

we have
$$[V'] = [c]^{-1}[Z][c][I'].$$

But for power invariance
$$[V'] = [c_t][Z][c] \cdot [I'].$$

Hence $[c]^{-1}$ and $[c_t]$ must be equal. *Only transforms which have this property can be used to transform the currents and voltages of a given practical system from one frame of reference to another.

The transform to rotating axes described in the next section has this property.

16.8.2 The transform to rotating axes

We have seen that currents I_d and I_q flowing into the rotor through the brushes and commutator of a primitive machine can be considered to be flowing in apparently stationary coils lying in the direct and quadrature axes.

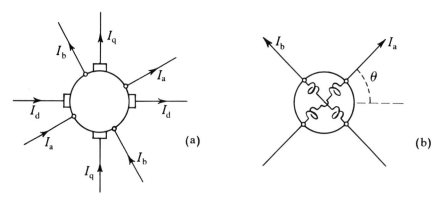

FIG. 16.7 (a) A commutator winding with fixed tappings on the armature. (b) The equivalent windings between diametrically opposite tappings

In the same manner two currents I_a and I_b flowing into the rotor through the slip rings as shown in Fig. 16.7(b) can be assumed to be flowing in two separate coils at right angles to each other (Fig. 16.7(b))
all the coils having the same number of turns. These coils are not stationary; they are attached to the rotor and rotate with it.

* Such matrices are known as *orthogonal matrices*, section 3.10.4.

We will examine conditions when the axis of coil **a** is inclined to the direct axis by angle θ.

The m.m.f.'s produced by currents I_a and I_b will be proportional to the values of the currents and will lie along the axes of the coils. If we resolve these m.m.f.'s along the direct axis and the quadrature axis we thus obtain the values of the currents I_d and I_q *which produce the same magnetic effect* as the currents I_a and I_b flowing in the inclined coils.

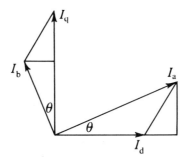

FIG. 16.8 Resolving the currents I_a and I_b along the direct and quadrature axes

From Fig. 16.8 we see that

$$I_d = I_a \cos \theta - I_b \sin \theta$$
$$I_q = I_a \sin \theta + I_b \cos \theta$$

or in matrix notation

$$\begin{bmatrix} I_d \\ I_q \end{bmatrix} = \begin{bmatrix} \cos \theta & -\sin \theta \\ \sin \theta & \cos \theta \end{bmatrix} \cdot \begin{bmatrix} I_a \\ I_b \end{bmatrix} \tag{16.68}$$

The matrix $\begin{bmatrix} \cos \theta & -\sin \theta \\ \sin \theta & \cos \theta \end{bmatrix}$ is known as the transform to inclined axes since it converts

$$\begin{bmatrix} I_a \\ I_b \end{bmatrix} \quad \text{to} \quad \begin{bmatrix} I_d \\ I_q \end{bmatrix}$$

By inversion

$$\begin{bmatrix} I_a \\ I_b \end{bmatrix} = \begin{bmatrix} \cos \theta & \sin \theta \\ -\sin \theta & \cos \theta \end{bmatrix} \cdot \begin{bmatrix} I_d \\ I_q \end{bmatrix} \tag{16.69}$$

Since this inverse matrix is seen to be the transpose of the original the equations can also be used for voltage.

$$\begin{bmatrix} V_d \\ V_q \end{bmatrix} = \begin{bmatrix} \cos \theta & -\sin \theta \\ \sin \theta & \cos \theta \end{bmatrix} \cdot \begin{bmatrix} V_a \\ V_b \end{bmatrix} \tag{16.70}$$

and
$$\begin{bmatrix} V_a \\ V_b \end{bmatrix} = \begin{bmatrix} \cos \theta & \sin \theta \\ -\sin \theta & \cos \theta \end{bmatrix} \cdot \begin{bmatrix} V_d \\ V_q \end{bmatrix} \tag{16.71}$$

If the rotor is in continuous rotation with constant speed ω we can write

$$\theta = \omega t$$

Using this transform (Eqs. (16.68)–(16.71)) currents and voltages at the slip rings of the primitive machine can be converted to the corresponding axis quantities (or commutator values) to be used in the primitive voltage equations of the machine. Using these transforms we can always replace one set of currents and voltages by the other.

In a machine which has only slip rings and is without commutator and brushgear, the equivalent axis values can still be determined and used in calculation even though no such actual currents and voltages exist.

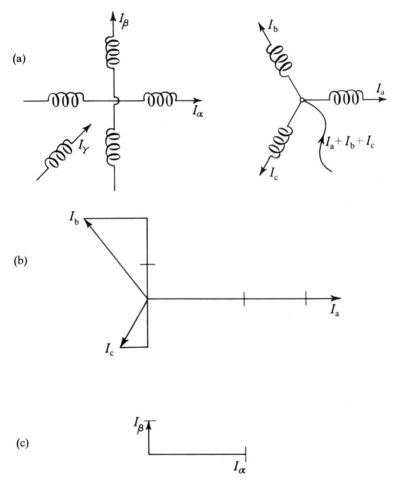

FIG. 16.9 (a) Diagrammatic representation of a two-phase system of coils and a corresponding three-phase system producing similar m.m.f. patterns at the centre of the coils. (b) Components of m.m.f. due to currents I_a I_b and I_c in the three-phase system. (c) M.m.f. due to currents I_α and I_β

16.8.3 The three-phase to two-phase transform

The primitive machine we have been discussing has two stator windings which form an adequate two-phase a.c. system. We must now consider if this is sufficient to represent three-phase systems since obviously the majority of practical power systems have three phases. It is well known that the magnetic field pattern of a balanced system of three-phase currents flowing in a balanced system of three stator coils, is precisely the same as that produced by a similar balanced system of two-phase currents flowing in a balanced system of two stator coils disregarding complications due to the use of iron cores.*

Both systems produce a rotating field pattern. We must now determine if this equivalence also applies to all current conditions balanced or unbalanced and find out how to transform from one system to the other.

We do this with reference to Fig. 16.9 where three currents I_a, I_b and I_c are flowing in three coils inclined at 120 degrees to each other. On the assumption that all coils have the same number of turns (hence m.m.f. is proportional to current), these currents are shown in Fig. 16.9(a) and have been resolved in the direction of the two axis coils α and β. It is obvious that the two currents I_α and I_β of Fig. 16.9(c) have the same m.m.f. as the system, I_a, I_b, I_c.

From the geometry of the diagram

$$I_\alpha = I_a - \frac{I_b}{2} - \frac{I_c}{2}$$

$$I_\beta = 0 + \frac{\sqrt{3}I_b}{2} - \frac{\sqrt{3}I_c}{2} \tag{16.72}$$

$$\begin{bmatrix} I_\alpha \\ I_\beta \end{bmatrix} = \begin{bmatrix} 1 & -\dfrac{1}{2} & -\dfrac{1}{2} \\ 0 & \dfrac{\sqrt{3}}{2} & -\dfrac{\sqrt{3}}{2} \end{bmatrix} \cdot \begin{bmatrix} I_a \\ I_b \\ I_b \end{bmatrix} \tag{16.73}$$

The matrix in Eq. (16.73) can be said to transform the three-phase set of currents into the two-phase set. So it does, but it is not capable of inversion, hence we learn that we cannot transform a two-phase system into a *unique* three-phase set.

This is because if each of the currents I_a, I_b, I_c in Fig. 16.9(b) have I_y added to them, the resultant two-phase system is still I_α and I_β of Fig. 16.9(c).

This we see because the three currents I_y in themselves form a balanced three-phase system with *no resultant m.m.f.* (Fig. 16.10).

Including I_y as part of each of the three-phase system $I_a I_b I_c$ we obtain

$$\begin{bmatrix} I_\alpha \\ I_\beta \\ I_y \end{bmatrix} = \begin{bmatrix} 1 & -\dfrac{1}{2} & -\dfrac{1}{2} \\ 0 & \dfrac{\sqrt{3}}{2} & -\dfrac{\sqrt{3}}{2} \\ 1 & 1 & 1 \end{bmatrix} \cdot \begin{bmatrix} I_a \\ I_b \\ I_c \end{bmatrix} \tag{16.74}$$

* *Electrical Technology*, E. Hughes, Longman, Sec. 16.2/3.

giving us a matrix that can be inverted and suggests that the three-phase system

$$\begin{bmatrix} I_a \\ I_b \\ I_c \end{bmatrix}$$

can be satisfactorily represented by an alternative system

$$\begin{bmatrix} I_\alpha \\ I_\beta \\ I_\gamma \end{bmatrix}$$

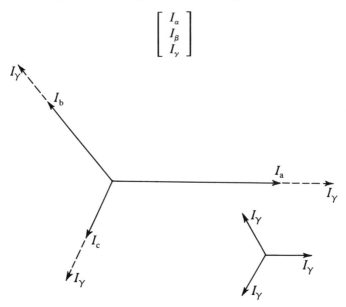

FIG. 16.10 Adding three equal currents I_γ to I_a I_b and I_c produces no additional m.m.f.

The current I_y does not contribute to the m.m.f. in the $\alpha\beta$ plane of Fig. 16.9 and some authorities infer that it acts in a third dimension. This is a mathematical device to remind us not to leave it out of our calculations entirely.

If this transform is to have satisfactory application with real electrical machines it should also apply to voltage and so $[c]^{-1}$ should be equal to $[c_t]$ in accordance with Sec. 16.8.1.

Introducing multipliers a and b in Eq. (16.74)

$$\begin{bmatrix} I_\alpha \\ I_\beta \\ I_\gamma \end{bmatrix} = \begin{bmatrix} a & \dfrac{a}{2} & -\dfrac{a}{2} \\ 0 & \left(\dfrac{\sqrt{3}}{2}\right)a & -\left(\dfrac{\sqrt{3}}{2}\right)a \\ b & b & b \end{bmatrix} \cdot \begin{bmatrix} I_a \\ I_b \\ I_c \end{bmatrix} = [c]\begin{bmatrix} I_c \\ I_b \\ I_c \end{bmatrix} \tag{16.75}$$

If this transform is to apply also to voltage

that is
$$\begin{bmatrix} V_\alpha \\ V_\beta \\ V_\gamma \end{bmatrix} = [c]\begin{bmatrix} V_a \\ V_b \\ V_c \end{bmatrix} \tag{16.76}$$

we must ensure that

$$[c]^{-1} = [c_t]$$

or

$$[c][c_t] = [1] \tag{16.77}$$

Thus

$$\begin{bmatrix} a & -a/2 & -a/2 \\ 0 & -3a/2 & -\sqrt{3}a/2 \\ b & b & b \end{bmatrix} \cdot \begin{bmatrix} a & 0 & b \\ -a/2 & \sqrt{3}a/2 & b \\ -a/2 & -\sqrt{3}a/2 & b \end{bmatrix} = \begin{bmatrix} 1 & 0 & 0 \\ 0 & 1 & 0 \\ 0 & 0 & 1 \end{bmatrix}$$

$$\begin{bmatrix} \dfrac{3a^2}{2} & 0 & 0 \\ 0 & \dfrac{3a^2}{2} & 0 \\ 0 & 0 & 3b^2 \end{bmatrix} = \begin{bmatrix} 1 & 0 & 0 \\ 0 & 1 & 0 \\ 0 & 0 & 1 \end{bmatrix}$$

$$a = \sqrt{(2/3)} \tag{16.78}$$
$$b = \sqrt{(1/3)} \tag{16.79}$$

Therefore

$$[c] = \sqrt{\left(\frac{2}{3}\right)} \begin{bmatrix} 1 & -\dfrac{1}{2} & -\dfrac{1}{2} \\ 0 & \dfrac{\sqrt{3}}{2} & -\dfrac{\sqrt{3}}{2} \\ \dfrac{1}{\sqrt{2}} & \dfrac{1}{\sqrt{2}} & \dfrac{1}{\sqrt{2}} \end{bmatrix}$$

We may thus recapitulate by stating that using this transform, currents (and voltages) in three coils forming a three-phase system **a b c** are changed to the corresponding quantities, in three other coils α β and γ. Coils α and β form a two-phase system with a separate coil γ carrying I_y.

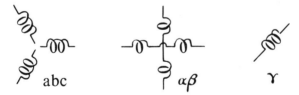

abc   αβ   γ

FIG. 16.11 Current $I\gamma$ produces flux in a separate system, but the resultant fluxes in the abc system and the $\alpha\beta$ system are the same

Current I_y produces flux in a separate system but the resultant fluxes of the **a b c** system and the α β system are identical.

CHAPTER 17

The polyphase
induction motor

17.1 Introduction

The polyphase induction machine is derived directly from the primitive machine without introducing any connection transforms but merely by observing the following simplifications. Figure 17.1 shows a primitive machine connected as a two-phase induction machine. The field winding $\mathbf{k}_f$ is omitted and the two stator windings $\mathbf{k}_{ds}$ and $\mathbf{k}_{qs}$ are connected to the two-phase supply.

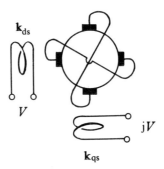

FIG. 17.1 A primitive machine with commutator brushgear short circuited

The rotor is short-circuited either (*a*) by short-circuiting the diametrically opposite tappings as would be done in the case of an actual induction wound-rotor machine or (*b*) by short-circuiting the commutator brushgear.

Either method will produce the same system of currents in the rotor conductors (as we have seen from Sec. 16.8.2) but the currents in the short-circuiting connections themselves will differ.

We shall assume that the commutator brushgear has been short-circuited so that we may use the primitive machine equations (16.29). These will enable us to determine the value of the axis currents which may be all that is necessary

If the actual rotor currents are required these can be obtained by applying. the transform to rotating axes (Eq. 16.69).

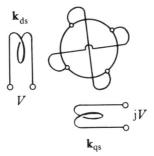

FIG. 17.2 **A primitive machine with slip rings short circuited**

17.2 Balanced voltage

17.2.1 Steady-state operation

Since the machine has a uniform air gap, the corresponding coefficients relating to the two axes will be equal to each other and also if we may assume sinusoidal distribution of the windings the G and M coefficients will be equal.

Thus
$$M_d = M_q = G_d = G_q = M$$
$$L_{ds} = L_{qs} = L_1$$
$$L_{dr} = L_{qr} = G_{dr} = G_{qr} = L_2 \tag{17.1}$$

The impedance matrix of the primitive machine (16.29) by omitting the winding k_f reduces to

$$[Z] = \begin{bmatrix} R_1 + L_1 p & M p & 0 & 0 \\ M p & R_2 + L_2 p & L_2(p\theta) & M(p\theta) \\ -M(p\theta) & -L_2(p\theta) & R_2 + L_2 p & M p \\ 0 & 0 & M p & R_1 + L_1 p \end{bmatrix} \tag{17.2}$$

We must return to this fundamental matrix whenever we are concerned with problems of transient response but under steady-state conditions further simplification is possible.

For steady-state operation at slip s the rotor speed $(p\theta)$ is constant and is equal to $(1-s)\omega$. If the voltage and currents are expressed as phasors, p can be replaced by $j\omega$. Finally with balanced primary voltage

$$\begin{bmatrix} V_{ds} \\ V_{dr} \\ V_{qr} \\ V_{qs} \end{bmatrix} = \begin{bmatrix} V \\ 0 \\ 0 \\ -jV \end{bmatrix} \tag{17.3}$$

Then the equations for the induction motor are

$$
\begin{bmatrix} V \\ 0 \\ 0 \\ -jV \end{bmatrix} = \begin{bmatrix} R_1+jX_1 & jX_m & 0 & 0 \\ jX_m & R_2+jX_2 & (1-s)X_2 & (1-s)X_m \\ -(1-s)X_m & -(1-s)X_2 & R_2+jX_2 & jX_m \\ 0 & 0 & jX_m & R_1+jX_1 \end{bmatrix} \cdot \begin{bmatrix} I_{ds} \\ I_{dr} \\ I_{qr} \\ I_{qs} \end{bmatrix}
$$

(17.4)

The values of the stator currents corresponding to a given value of slip s can be obtained by inversion I_{dr} and I_{qr} will also be obtained and from these rotor current will be determined by transform to rotating axes. The torque will be derived from the generalised torque equation (16.40) which for this machine is reduced to

$$
f = M(I_{dr}I_{qs} - I_{ds}I_{qr})
$$

(17.5)

17.2.2 Derivation of the equivalent circuit

A further simplification is possible which not only reduces the arithmetic involved in the previous section but helps us to understand more fully the relationships between the currents in the windings.

Under the balanced conditions we are considering and since the applied voltages are also balanced it is reasonable to expect that the stator and rotor currents will also be balanced, that is to say, the currents in the quadrature axes will be numerically equal to those in the direct axis but will lag by 90 degrees.

Consequently we may use the following transform which shows we are dealing with only two unknown currents I_1 and I_2, the other currents I_{ds}, I_{dr}, I_{qr} and I_{qs} being derived from them:

$$
\begin{bmatrix} I_{ds} \\ I_{dr} \\ I_{qr} \\ I_{qs} \end{bmatrix} = \begin{bmatrix} 1 & 0 \\ 0 & 1 \\ 0 & -j \\ -j & 0 \end{bmatrix} \cdot \begin{bmatrix} I_1 \\ I_2 \end{bmatrix}
$$

(17.6)

or
$$
[I] = [c][I']
$$
(17.7)

The four currents are thus transformed to two in a new frame of reference. Voltages are similarly transformed and the new impedance matrix becomes

$$
[Z'] = \tfrac{1}{2}[c_t{}^*][Z][c]
$$

(17.8)

according to the equation (4.33) the transform being complex (the fraction $\tfrac{1}{2}$ is introduced to maintain the invariance of power in the per-unit system, the number of circuits being reduced from 4 to 2). Thus

$$
\begin{bmatrix} V_1 \\ V_2 \end{bmatrix} = \tfrac{1}{2}\begin{bmatrix} 1 & 0 & 0 & j \\ 0 & 1 & j & 0 \end{bmatrix} \cdot \begin{bmatrix} R_1+jX_1 & jX_m & 0 & 0 \\ jX_m & R_2+jX_2 & (1-s)X_2 & (1-s)X_m \\ -(1-s)X_m & -(1-s)X_2 & R_2+jX_2 & jX_m \\ 0 & 0 & jX_m & R_1+jX_1 \end{bmatrix}
$$

$$\cdot \begin{bmatrix} 1 & 0 \\ 0 & 1 \\ 0 & -j \\ -j & 0 \end{bmatrix} \cdot \begin{bmatrix} I_1 \\ I_2 \end{bmatrix} \tag{17.9}$$

giving

$$\begin{bmatrix} V_1 \\ V_2 \end{bmatrix} = \begin{bmatrix} R_1 + jX_1 & jX_m \\ jsX_m & R_2 + jsX_2 \end{bmatrix} \cdot \begin{bmatrix} I_1 \\ I_2 \end{bmatrix} \tag{17.10}$$

For the usual case of the short-circuited rotor, $V_2 = 0$ the bottom row can be divided by s and the equation becomes

$$\begin{bmatrix} V_1 \\ 0 \end{bmatrix} = \begin{bmatrix} R_1 + jX_1 & jX_m \\ jX_m & R_2/s + jX_2 \end{bmatrix} \cdot \begin{bmatrix} I_1 \\ I_2 \end{bmatrix} \tag{17.11}$$

This matrix is now symmetrical and consequently can be represented by an equivalent circuit, which is shown in Fig. 17.3.

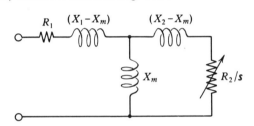

FIG. 17.3 **The equivalent circuit for an induction motor under balanced conditions**

17.3 Unbalanced applied voltages

The steady-state impedance matrix of the induction motor can be simplified by the use of another type of transform enabling a solution to be obtained when the applied voltages are unbalanced. This involves the use of two-phase symmetrical components.

17.3.1 Two-phase symmetrical components

If the currents in the two axis coils are I_{d1} and I_{q1} and they form a balanced two-phase system, they are equal in magnitude and I_{q1} lags behind I_{d1} by 90 degrees. That is to say

$$\begin{bmatrix} I_{d1} \\ I_{q1} \end{bmatrix} = a \begin{bmatrix} 1 \\ -j \end{bmatrix} I_f \tag{17.12}$$

It is well known that the resultant m.m.f. due to such current is constant in magnitude and rotates in the positive direction. But if

$$\begin{bmatrix} I_{d2} \\ I_{q2} \end{bmatrix} = a \begin{bmatrix} 1 \\ j \end{bmatrix} I_b \tag{17.13}$$

a rotating field in the negative direction is produced. I_f and I_b are phasors which determine the magnitude and phase relationship of these two fields and a is an arbitrary constant. Both these systems are balanced two-phase, but if we add them together the resulting currents I_d and I_q are unbalanced. Adding Eq. (17.12) and (17.13) and making a equal to $1/\sqrt{2}$, we obtain

$$\begin{bmatrix} I_d \\ I_q \end{bmatrix} = \frac{1}{\sqrt{2}} \begin{bmatrix} 1 & 1 \\ -j & j \end{bmatrix} \cdot \begin{bmatrix} I_f \\ I_b \end{bmatrix} \quad \text{or} \quad \begin{bmatrix} I_d \\ I_q \end{bmatrix} = [c] \begin{bmatrix} I_f \\ I_b \end{bmatrix} \tag{17.14}$$

and by inversion

$$\begin{bmatrix} I_f \\ I_b \end{bmatrix} = [c]^{-1} \begin{bmatrix} I_d \\ I_q \end{bmatrix} \quad \text{or} \quad \begin{bmatrix} I_f \\ I_b \end{bmatrix} = \frac{1}{\sqrt{2}} \begin{bmatrix} 1 & j \\ 1 & -j \end{bmatrix} \cdot \begin{bmatrix} I_d \\ I_q \end{bmatrix} \tag{17.15}$$

Currents I_d and I_q can thus be transformed to another system of co-ordinates, currents I_f and I_b being the components responsible for the separate fields rotating forward and backward respectively.

I_f and I_b may be termed the positive and negative symmetrical components corresponding to the unbalanced two phase system I_a and I_b.

The value of a is chosen as $1/\sqrt{2}$ to make $[c]^{-1}$ equal to $[c_t*]$. Thus the transform $[c]$ can also be used to transform voltages and the invariance of power is maintained (compare Sec. 16.8.1).

17.3.2 Components of stator and rotor currents

If we assume that both rotor and stator currents of the induction motor are unbalanced, then both systems should be converted to their symmetrical components

$$\begin{bmatrix} I_{ds} \\ I_{dr} \\ I_{qr} \\ I_{qs} \end{bmatrix} = [c] \cdot \begin{bmatrix} I_{1f} \\ I_{2f} \\ I_{2b} \\ I_{1b} \end{bmatrix} \tag{17.16}$$

where

$$[c] = \frac{1}{\sqrt{2}} \begin{bmatrix} 1 & 0 & 0 & 1 \\ 0 & 1 & 1 & 0 \\ 0 & -j & j & 0 \\ -j & 0 & 0 & j \end{bmatrix} \tag{17.17}$$

The impedance equations of the induction motor become

$$[V'] = [c_t*][Z][c][I'] \tag{17.18}$$

or

$$\begin{bmatrix} V_f \\ 0 \\ 0 \\ V_b \end{bmatrix} = [c_t*] \cdot \begin{bmatrix} R_1+jX_1 & jX_m & 0 & 0 \\ jX_m & R_2+jX_2 & (1-s)X_2 & (1-s)X_m \\ -(1-s)X_m & -(1-s)X_2 & R_2+jX_2 & jX_m \\ 0 & 0 & jX_m & R_2+jX_2 \end{bmatrix} \cdot [c] \cdot \begin{bmatrix} I_{1f} \\ I_{2f} \\ I_{2b} \\ I_{1b} \end{bmatrix} \tag{17.19}$$

which reduces to

$$
\begin{bmatrix} V_{\mathrm{f}} \\ 0 \\ 0 \\ V_{\mathrm{b}} \end{bmatrix}
=
\begin{bmatrix}
R_1+jX_1 & jX_m & 0 & 0 \\
jsX_m & R_2+jsX_2 & 0 & 0 \\
0 & 0 & R_2+j(2-s)X_2 & j(2-s)X_m \\
0 & 0 & jX_m & R_1+jX_1
\end{bmatrix}
\cdot
\begin{bmatrix} I_{1\mathrm{f}} \\ I_{2\mathrm{f}} \\ I_{2\mathrm{b}} \\ I_{1\mathrm{b}} \end{bmatrix}
$$

$$(17.20)$$

This new impedance matrix shows no mutual terms connecting forward and backward components and hence we may infer that the two components do not interact.

The corresponding equivalent circuit to Eq. (17.20) is shown in Fig. 17.4 where V_{f} is responsible for the forward components of current in the top part

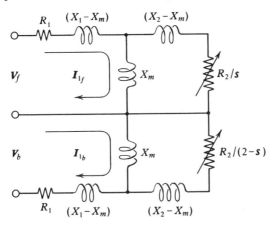

FIG. 17.4 The equivalent circuit for an induction motor with unbalanced voltage

of the diagram and V_{b} for the corresponding backward components in the lower part. We thus have the effect of two machines superimposed, one running with

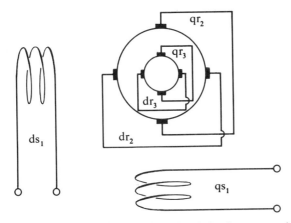

FIG. 17.5 Connections of a primitive machine as an induction motor with two sets of rotor windings

slip s and the other with slip $(2-s)$ both contributing torque but without mutual inductance between them.

17.3.3 Double-cage induction motor

A double-cage induction motor can be derived from a primitive machine which has two commutator windings on the rotor (Fig. 17.5).

The impedance matrix (17.2) for the single-cage rotor machine is increased to include two more lines and columns corresponding to the direct and quadrature circuits of the second rotor winding.

The new impedance matrix is:

$$[Z] =$$

$$\begin{array}{c} ds_1 \\ dr_2 \\ dr_3 \\ qr_3 \\ qr_2 \\ qs_1 \end{array}
\begin{bmatrix}
R_1+L_1p & M_{12}p & M_{13}p & 0 & 0 & 0 \\
M_{12}p & R_2+L_2p & R_4+M_{23}p & M_{23}(p\theta) & L_2(p\theta) & M_{12}(p\theta) \\
M_{13}p & R_4+M_{23}p & R_3+L_3p & L_3(p\theta) & M_{23}(p\theta) & M_{13}(p\theta) \\
-M_{13}(p\theta) & -M_{23}(p\theta) & -L_3(p\theta) & R_3+L_3p & R_4+M_{23}p & M_{13}p \\
-M_{12}(p\theta) & -L_2(p\theta) & -M_{23}(p\theta) & R_4+M_{23}p & R_2+L_2p & M_{12}p \\
0 & 0 & 0 & M_{13}p & M_{12}p & R_1+L_1p
\end{bmatrix}$$

$$(17.21)$$

There are now three coils on each axis with mutual inductances M_{12}, M_{13}, and M_{23} between them. If the end rings are common for the two rotor windings, this is represented by the mutual resistance R_4 added to the term $M_{23}p$.

To establish the steady state performance of this machine at slip s, the procedure is similar to that of Sec. 17.2.1.

(a) p and $(p\theta)$ are replaced by $j\omega$ and $(1-s)\omega$ respectively;
(b) assuming the currents are balanced, the six current phasors can be reduced to three by the following transformation:

$$\begin{bmatrix} I_{ds1} \\ I_{dr2} \\ I_{dr3} \\ I_{qr3} \\ I_{qr2} \\ I_{qs1} \end{bmatrix} = [c] \cdot \begin{bmatrix} I_1 \\ I_2 \\ I_3 \end{bmatrix}$$

$$(17.22)$$

where

$$[c]. = \begin{bmatrix} 1 & 0 & 0 \\ 0 & 1 & 0 \\ 0 & 0 & 1 \\ 0 & 0 & -j \\ 0 & -j & 0 \\ -j & 0 & 0 \end{bmatrix}$$

$$(17.23)$$

(c) the transformed impedance matrix is given by:

$$[Z'] = \tfrac{1}{2}[c_t{}^*][Z][c]$$

$$(17.24)$$

Thus, evaluating $[Z']$

$$\begin{bmatrix} V_1 \\ V_2 \\ V_3 \end{bmatrix} = \begin{bmatrix} R_1+jX_1 & jX_{12} & jX_{13} \\ jsX_{12} & R_2+jsX_2 & R_4+jsX_{23} \\ jsX_{13} & R_4+jsX_{23} & R_3+jsX_3 \end{bmatrix} \cdot \begin{bmatrix} I_1 \\ I_2 \\ I_3 \end{bmatrix} \quad (17.25)$$

and since $V_2 = 0$ and $V_3 = 0$ the second and third lines can be divided by s

$$\begin{bmatrix} V_1 \\ 0 \\ 0 \end{bmatrix} = \begin{bmatrix} R_1+jX_1 & jX_{12} & jX_{13} \\ jX_{12} & R_2/s+jX_2 & R_4/s+jX_{23} \\ jX_{13} & R_4/s+jX_{23} & R_3/s+jX_3 \end{bmatrix} \cdot \begin{bmatrix} I_1 \\ I_2 \\ I_3 \end{bmatrix} \quad (17.26)$$

Again we have a symmetrical matrix which can be represented by an equivalent circuit. In practice X_{12} and X_{13} are approximately equal and, under these circumstances, the circuit of Fig. 17.6 is suitable.

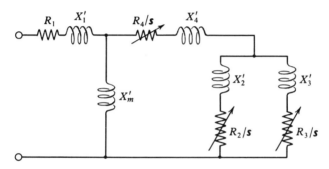

FIG. 17.6 The equivalent circuit for a double-cage-rotor induction motor

In the figure

$$X'_m = X_{13} = X_{12}$$
$$X'_1 + X'_m = X_1$$
$$X'_m + X'_4 = X_{23} \quad (17.27)$$
$$X_{23} + X'_2 = X_2$$
$$X_{23} + X'_3 = X_3$$

CHAPTER 18

The single-phase
induction motor

18.1 Introduction

An elementary single-phase induction motor has only one sinusoidally distributed stator winding and a cage rotor. A short-rated quadrature stator winding
is required for starting purposes, but this is open circuited under running
conditions.

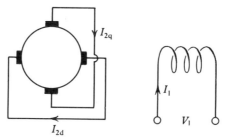

FIG. 18.1 Connections of a primitive machine as a single-phase induction motor

Figure 18.1 shows the connections of a primitive machine arranged as a
single-phase induction motor. The voltage V_1 is applied to the direct-axis stator
winding and the commutator brush gear is short-circuited.

18.2 The impedance matrix

The impedance matrix of the machine is obtained by omitting the last row and
column from the corresponding matrix for the two-phase machine.

The operation of the machine at slip s is defined by the following equations
derived from Eq. (17.4):

$$\begin{bmatrix} V_1 \\ 0 \\ 0 \end{bmatrix} = \begin{bmatrix} R_1+jX_1 & jX_m & 0 \\ jX_m & R_2+jX_2 & (1-s)X_2 \\ -(1-s)X_m & -(1-s)X_2 & R_2+jX_2 \end{bmatrix} \cdot \begin{bmatrix} I_1 \\ I_{2d} \\ I_{2q} \end{bmatrix} \quad (18.1)$$

Inversion of this matrix enables I_1, I_{2d} and I_{2q} to be determined and the torque is obtained from Eq. (16.38):

$$f = \text{Re}\{-I_{2q}{}^*(1-s)X_mI_1 - I_{2q}{}^*(1-s)X_2I_{2d} + I_{2d}{}^*(1-s)X_2I_{2q}\}$$
$$= \text{Re}\{-I_1I_{2q}{}^*(1-s)X_m\} \tag{18.2}$$

At this point it will be seen that when the slip is unity (i.e. the rotor is at standstill) I_{2q} will be zero. This is because when the rotor stands still neither transformer voltage nor rotational voltage will be induced in the quadrature axis.

In consequence (Eq. (18.2)), the starting torque is also zero.

18.3 Transform to symmetrical components

As with the two-phase induction motor, the impedance matrix will be simplified if a transform to symmetrical components is employed. This also will enable an equivalent circuit to be drawn.

Let
$$[c] = \frac{1}{\sqrt{2}}\begin{bmatrix} \sqrt{2} & 0 & 0 \\ 0 & 1 & 1 \\ 0 & -j & j \end{bmatrix} \tag{18.3}$$

so that
$$\begin{bmatrix} I_1 \\ I_{2d} \\ I_{2q} \end{bmatrix} = [c] \cdot \begin{bmatrix} I_1 \\ I_f \\ I_b \end{bmatrix} \tag{18.4}$$

I_1 is not affected in the transform which also satisfies the conditions for power invariance (see Sec. 4.4).

Thus
$$[Z'] = [c_t{}^*] \cdot \begin{bmatrix} R_1 + jX_1 & jX_m & 0 \\ jX_m & R_2 + jX_2 & (1-s)X_2 \\ -(1-s)X_m & -(1-s)X_2 & R_2 + jX_2 \end{bmatrix} \cdot [c] \tag{18.5}$$

and so
$$\begin{bmatrix} V \\ 0 \\ 0 \end{bmatrix} = \frac{1}{2}\begin{bmatrix} 2(R_1 + jX_1) & \sqrt{2}jX_m & \sqrt{2}jX_m \\ \sqrt{2}jsX_m & 2(R_2 + jsX_2) & 0 \\ \sqrt{2}j(2-s)X_m & 0 & 2(R_2 + j(2-s)X_2) \end{bmatrix} \cdot \begin{bmatrix} I_1 \\ I_f \\ I_b \end{bmatrix} \tag{18.6}$$

or
$$\begin{bmatrix} 2V \\ 0 \\ 0 \end{bmatrix} = \begin{bmatrix} 2(R_1 + jX_1) & jX_m & jX_m \\ jX_m & R_2/s + jX_2 & 0 \\ jX_m & 0 & R_2/(2-s) + jX_2 \end{bmatrix} \cdot \begin{bmatrix} I_1 \\ \sqrt{2}I_f \\ \sqrt{2}I_b \end{bmatrix} \tag{18.7}$$

Equation (18.7) can now be illustrated by the equivalent circuit of Fig. 18.2, where $X'_2 = (X_2 - X_m)$.

The torque (in synchronous power) is given by

$$f = I_t{}^2 R_2/s - I_b{}^2 R_2/(2-s) \qquad (18.8)$$

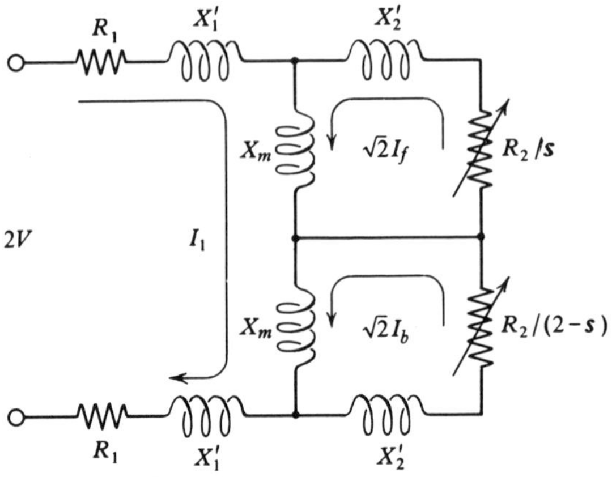

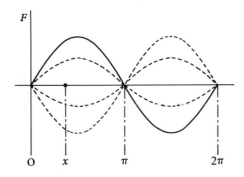

FIG. 18.2 The equivalent circuit of a single-phase induction motor

18.4 Alternative viewpoint

This equivalent circuit could have been derived by a more intuitive though less exact approach.

If we consider the flux distribution due to direct current I in a sinusoidally distributed stator winding, the m.m.f. at any point x on the circumference will be given by

$$F(x) = I(n/2) \sin x \qquad (18.9)$$

To obtain a satisfactory distribution there will have to be a large number of slots and the number of conductors per slot must be graded throughout the pole-pitch according to a cosine law. This distribution is shown by the continuous line in Fig. 18.3.

FIG. 18.3 Distribution of m.m.f. over two pole pitches

If alternating current given by

$$i = I \cos \omega t \qquad (18.10)$$

is applied to the winding, the m.m.f. distribution will be given by

$$F(x) = I(n/2) \cos \omega t \sin x \qquad (18.11)$$

This equation is also depicted in Fig. 18.3 by the three dotted lines, which correspond to three instants of time.

Equation (18.11) can be converted by trigonometrical transformations into the following expression

$$F(x) = \tfrac{1}{2}I(n/2)\{\sin(x - \omega t) + \sin(x + \omega t)\} \qquad (18.12)$$

Equation (18.11) depicts the m.m.f. as a sine wave whose amplitude varies with time, but the interpretation of Eq. (18.12) is that it can be represented by the sum of the two waves of fixed amplitude $\tfrac{1}{2}I(n/2)$ the first, $\sin(x - \omega t)$, moving forward with velocity ω and the second, $\sin(x + \omega t)$, moving backward. This is illustrated in Fig. 18.4 where the position of the waves at time t is given.

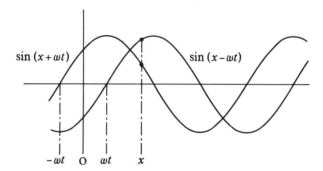

FIG. 18.4 Geometrical representation of Equation 18.12

It is well known that the m.m.f. due to the currents of a polyphase stator winding is a single forward rotating field and that this gives rise to the equivalent circuit of Fig. 18.5.*

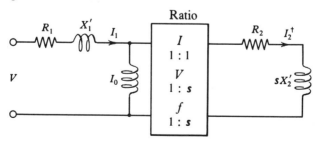

FIG. 18.5 The equivalent circuit of a polyphase induction motor

* See *Electrical Machines*, p. 122.

This equivalent circuit for the polyphase machine takes into account that

(*a*) the rotor frequency is equal to *s* times the stator frequency;
(*b*) the rotor voltage is *s* times the stator voltage;
(*c*) the rotor reactance varies with *s*;
(*d*) the m.m.f. of the magnetising current I_0 is the resultant of the stator and rotor current waves.

For the single-phase machine similar relationships exist for both the forward and backward rotating components that make up the stator m.m.f.

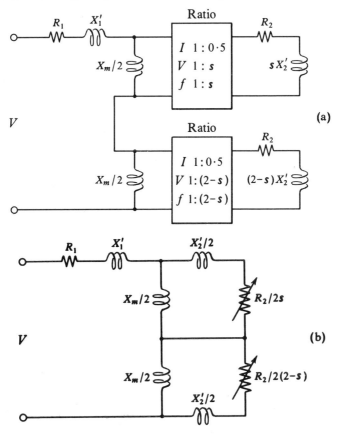

FIG. 18.6 (a) The equivalent circuit of a single-phase induction motor. (b) Alternative equivalent circuit diagram

At a given speed when the slip relative to the forward field is *s*, the corresponding slip relative to the backward rotating field is $(2-s)$.

Two equivalent transformer diagrams are thus drawn, one corresponding to each component of stator current and to ensure that the two currents are always equal in magnitude, the two diagrams are connected in series, Fig. 18.6(*a*).

Eliminating the transformer ratios, the equivalent circuit, Fig. 18.6(*b*) results.

Torque is obtained from the difference in the powers in the two resistances representing the secondary loads. It will be observed from the symmetry of the diagram that when s is equal to unity, the standstill torque is zero.

18.5 Starting

One of the important features of the single-phase induction motor is the absence of torque at standstill. To make a practically useful machine some modification must be made to ensure adequate starting torque. This can be done by adding a quadrature stator winding.

Although this winding is supplied in parallel with the main stator winding, a phase shift is introduced by connecting a capacitor in series or, alternatively, introducing excessive leakage reactance into the circuit. This converts the machine into a type of two-phase machine. The starting winding is usually short-time rated and is only used momentarily at starting. A centrifugal switch is arranged to cut out the starting winding when the machine reaches about three-quarters of its normal running speed.

The two types are known as capacitor-start and split-phase machines respectively.

Some motors are so designed that, after starting, the starting capacitor is re-connected in parallel with the running winding to improve the overall power factor.

If the capacitor machine is designed so that the quadrature winding is continuously rated and remains in circuit under running conditions, it is then known as a capacitor motor.

One of the main problems facing the designer of all these machines is to obtain adequate starting characteristics without impairing the running characteristics and, at the same time, ensuring a satisfactory efficiency at normal load.

18.6 The capacitor motor

18.6.1. Impedance matrix

The capacitor motor is derived from the primitive machine in the following way. The machine windings are not usually symmetrical, so let the quadrature winding have a times the turns of the direct axis winding and let the impedance of the capacitor be Z (Fig. 18.7).

To allow for the turns ratio a, the impedance matrix $[Z]$ in Eq. (17.4) is first transformed to $[Z']$ by

$$[Z'] = [c_t][Z][c] \tag{18.13}$$

where
$$[c] = \begin{bmatrix} 1 & 0 & 0 & 0 \\ 0 & 1 & 0 & 0 \\ 0 & 0 & 1 & 0 \\ 0 & 0 & 0 & a \end{bmatrix} \tag{18.14}$$

and the capacitor impedance $\mathbf{Z}$ is then added in the appropriate position. This gives

$$[\mathbf{Z}'] = \begin{bmatrix} R_1+jX_1 & jX_m & 0 & 0 \\ jX_m & R_2+jX_2 & (1-s)X_2 & (1-s)aX_m \\ -(1-s)X_m & -(1-s)X_2 & R_2+jX_2 & jaX_m \\ 0 & 0 & jaX_m & a^2(R_1+jX_1)+\mathbf{Z} \end{bmatrix} \quad (18.15)$$

and the equations to the capacitor motor are

$$\begin{bmatrix} V \\ 0 \\ 0 \\ 0 \end{bmatrix} = [\mathbf{Z}'] \cdot \begin{bmatrix} I_{1d} \\ I_{2d} \\ I_{2q} \\ I_{1q} \end{bmatrix} \quad (18.16)$$

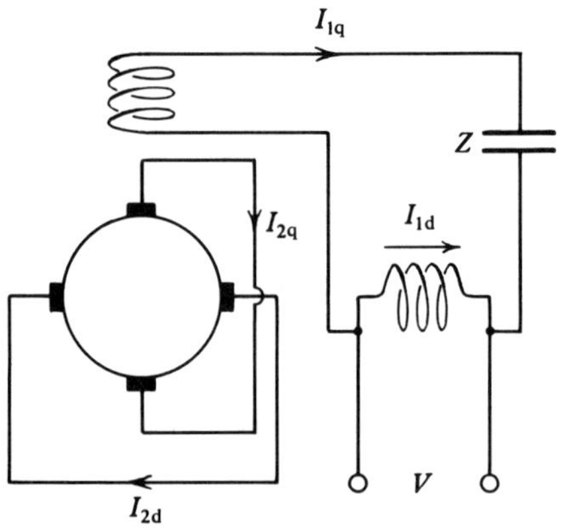

FIG. 18.7 Connections of a primitive machine as a capacitor motor

The total current

$$I_1 = I_{1d}+I_{1q} \quad (18.17)$$

can be obtained by inversion and the torque is

$$f = \mathrm{Re}\{(1-s)(I_{2d}*I_{1q}aX_m - I_{2q}*I_{1d}X_m)\} \quad (18.18)$$

To transform to symmetrical components

let
$$[c_1] = \frac{1}{\sqrt{2}} \begin{bmatrix} 1 & 0 & 0 & 1 \\ 0 & 1 & 1 & 0 \\ 0 & -j & j & 0 \\ -j/a & 0 & 0 & j/a \end{bmatrix} \quad (18.19)$$

which satisfies
$$[c_{1t}*][c_t][c][c_1] = [1] \quad (18.20)$$

for power invariance in the primitive machine.

Thus
$$
\begin{bmatrix} V_t \\ 0 \\ 0 \\ V_b \end{bmatrix} = [c_{1t}{}^*][Z'][c_1] \cdot \begin{bmatrix} I_{1f} \\ I_{2f} \\ I_{2b} \\ I_{1b} \end{bmatrix}
$$
(18.21)

$$
\begin{bmatrix} V_t \\ 0 \\ 0 \\ V_b \end{bmatrix} = [c_{1t}{}^*] \cdot \begin{bmatrix} V \\ 0 \\ 0 \\ V \end{bmatrix}
$$
(18.22)

$$
\frac{V}{\sqrt 2}\begin{bmatrix} 1+j/a \\ 0 \\ 0 \\ 1-j/a \end{bmatrix} = \begin{bmatrix} R_1+jX_1+Z/2a^2 & jX_m & 0 & -Z/2a^2 \\ jX_m & R_2/s+jX_2 & 0 & 0 \\ 0 & 0 & R_2(2-s)+jX_2 & jX_m \\ -Z/2a^2 & 0 & jX_m & R_1+jX_1+Z/2a^2 \end{bmatrix}
$$

$$
\begin{bmatrix} I_{1f} \\ I_{2f} \\ I_{2b} \\ I_{1b} \end{bmatrix}
$$
(18.23)

The equivalent circuit is therefore as shown in Fig. 18.8. It is similar to that of the two-phase machine with unbalanced voltages but the capacitor is included as a mutual impedance between loops **1** and **4**.

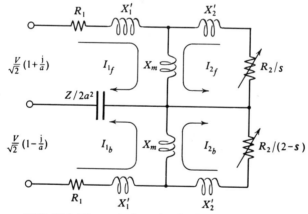

FIG. 18.8 The equivalent circuit of a capacitor motor

The direct and quadrature currents in the stator are obtained from

$$
\begin{bmatrix} I_{1d} \\ 0 \\ 0 \\ I_{1q} \end{bmatrix} = [c_1] \cdot \begin{bmatrix} I_{1f} \\ I_{2f} \\ I_{2b} \\ I_{1b} \end{bmatrix} = \frac{1}{\sqrt 2}\begin{bmatrix} I_{1f}+I_{1b} \\ 0 \\ 0 \\ -jI_{1f}/a+jI_{1b}/a \end{bmatrix}
$$
(18.24)

and the losses in R_2/s and $R_2/(2-s)$ are the positive and negative rotation torques (expressed as synchronous power) respectively.

CHAPTER 19

The amplidyne

The construction of this type of cross-field generator or rotating amplifier is described briefly elsewhere.* It is an obvious derivative of the primitive machine and, consequently, its impedance matrix can be obtained in the following manner.

19.1 The impedance matrix

Figure 19.1 shows the connection of the primitive machine when used as an amplidyne. Only one control field k_f is considered in the first instance and since

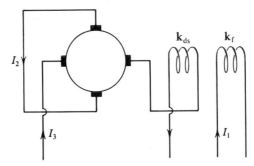

FIG. 19.1 Connections of a primitive machine to form an amplidyne

the quadrature stator winding k_{qs} is not used, the last line and column of the impedance matrix of the primitive machine are omitted. Thus

$$
\begin{bmatrix} V_f \\ V_{ds} \\ V_{dr} \\ V_{qr} \end{bmatrix} = \begin{bmatrix} R_f + L_f p & M_{fd}p & M_{fr}p & 0 \\ M_{fd}p & R_{ds} + L_{ds}p & M_d p & 0 \\ M_{fr}p & M_d p & R_r + L_{dr}p & G_{qr}(p\theta) \\ -G_f(p\theta) & -G_d(p\theta) & -G_{dr}(p\theta) & R_r + L_{qr}p \end{bmatrix} \cdot \begin{bmatrix} I_f \\ I_{ds} \\ I_{dr} \\ I_{qr} \end{bmatrix}
$$

$$(19.1)$$

* *Electrical Machines*, p. 243.

Although the amplidyne is a salient pole machine, the polar projections are usually symmetrical about both axes.

Hence $\qquad\qquad\qquad\qquad L_{dr} = L_{qr} = L_2$

and $\qquad\qquad\qquad\qquad G_{dr} = G_{qr}$ $\qquad\qquad$ (19.2)

The winding $\mathbf{k}_{ds}$ is the compensating winding and is connected in series opposition with $\mathbf{k}_{dr}$.

The quadrature brushes are short-circuited.

The transform matrix $[c]$ corresponding to this connection is, therefore, given by

$$\begin{bmatrix} I_f \\ I_{ds} \\ I_{dr} \\ I_{qr} \end{bmatrix} = \begin{bmatrix} 1 & 0 & 0 \\ 0 & -1 & 0 \\ 0 & 1 & 0 \\ 0 & 0 & 1 \end{bmatrix} \cdot \begin{bmatrix} I_f \\ I_d \\ I_q \end{bmatrix} \qquad (19.3)$$

Evaluating $\qquad\qquad\qquad\qquad [Z'] = [c_t][Z][c]$

we have the equations

$$\begin{bmatrix} V_f \\ V_d \\ V_q \end{bmatrix} = \begin{bmatrix} R_f + L_f p & (M_{fr} - M_{fd})p & 0 \\ (M_{fr} - M_{fd})p & (R_{ds} + R_r) + (L_{ds} - 2M_d + L_2)p & G_{qr}(p\theta) \\ -G_f(p\theta) & -(G_{dr} - G_d)(p\theta) & R_r + L_2 p \end{bmatrix} \begin{bmatrix} I_f \\ I_d \\ I_q \end{bmatrix} \qquad (19.4)$$

where $\qquad\qquad\qquad\qquad V_q = 0$

If the compensation of armature reaction in the direct axis is 100%, we may assume

$$\left.\begin{aligned} (M_{fr} - M_{fd}) &= 0 \\ (L_{ds} - 2M_d + L_2) &= L'_2 \simeq 0 \\ (G_{dr} - G_d) &= 0 \end{aligned}\right\} \qquad (19.5)$$

Therefore putting $R_2 = R_r$ and $R_3 = R_{ds} + R_r$

$$\begin{bmatrix} V_f \\ V_d \\ 0 \end{bmatrix} = \begin{bmatrix} R_f + L_f p & 0 & 0 \\ 0 & R_3 + L'_2 p & G_{qr}(p\theta) \\ -G_f(p\theta) & 0 & R_2 + L_2 p \end{bmatrix} \cdot \begin{bmatrix} I_f \\ I_d \\ I_q \end{bmatrix} \qquad (19.6)$$

Eliminating the last line and column by partitioning and putting $V_d = -I_d Z$ where Z is the load impedance

$$\begin{bmatrix} V_f \\ 0 \end{bmatrix} = \begin{bmatrix} R_f + L_f p & 0 \\ G_f G_{qr}(p\theta)^2/(R_2 + L_2 p) & R_3 + L'_2 p + Z \end{bmatrix} \cdot \begin{bmatrix} I_f \\ I_d \end{bmatrix} \qquad (19.7)$$

Under normal working conditions of constant speed this simplifies to

$$\begin{bmatrix} V_f \\ 0 \end{bmatrix} = \begin{bmatrix} R_1 + L_1 p & 0 \\ k R_2 & (R_2 + L_2 p)(R_3 + L_3 p) \end{bmatrix} \cdot \begin{bmatrix} I_f \\ I_d \end{bmatrix} \qquad (19.8)$$

where
$$(R_1+pL_1) = (R_f+pL_f)$$
$$(R_2+pL_2) = (R_r+pL_{qr})$$
$$(R_3+pL_3) = (R_{ds}+R_r+L'_{2p}+Z)$$
$$kR_2 = G_fG_{qr}(p\theta)^2 \qquad (19.9)$$

and*
$$E_0 = kI_f \qquad (19.10)$$

The coefficient k is known as the 'volts per amp' ratio. It it the open-circuit voltage produced by unit control field current and is an easily measured parameter of the machine.

Thus from Eq. (19.8)
$$I_f = V_f/(R_1+pL_1) \qquad (19.12)$$

and the output current
$$I_d\dagger = -I_d = V_fkR_2/(R_1+pL_1)(R_2+pL_2)(R_3+pL_3) \qquad (19.13)$$

or in the form usually needed in regulating system problems
$$I_d\dagger = -I_d = \frac{V_fk}{R_1R_3} \cdot \frac{1}{1+pT_1} \cdot \frac{1}{1+pT_2} \cdot \frac{1}{1+pT_3} \qquad (19.14)$$

where
$$T_1 = L_1/R_1 \quad T_2 = L_2/R_2 \quad T_3 = L_3/R_3$$

The amplidyne is thus represented by the block diagram of Fig. 19.2.

Equivalent gain   Field circuit   Quadrature circuit   Load circuit
                          lag   lag   lag

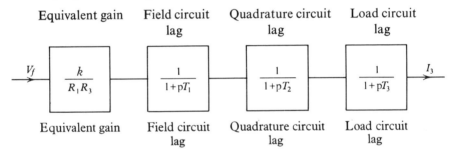

Equivalent gain   Field circuit   Quadrature circuit   Load circuit
                          lag   lag   lag

FIG. 19.2 Block diagram of a simple amplidyne

19.2 Multiple fields

When the amplidyne is used as an element in a control system, and this is its principal function, it is usually wound with several control fields. This enables

* From the second line of Eq. (19.6) after elimination of the third line and column
$$V_d = \frac{G_fG_{qr}(p\theta)^2I_f}{R_2+L_{2p}}+(R_3+L'_{2p})I_d \qquad (19.11)$$

The open-circuit voltage E_0 at constant speed and corresponding to constant control field current is therefore obtained by putting $I_d = 0$ and $L_{2p} = 0$

Hence
$$E_0 = \frac{G_fG_{qr}(p\theta)^2I_f}{R_2}$$
$$= kI_f \qquad (19.10)$$

a number of different voltages from reference sources, feed-back and stabilising circuits, to be applied simultaneously.

Under these circumstances, the time constants of the circuits are modified by mutual induction between the control fields, and allowance has to be made in the following way.

A diagram of an amplidyne with three control fields is shown in Fig. 19.3.

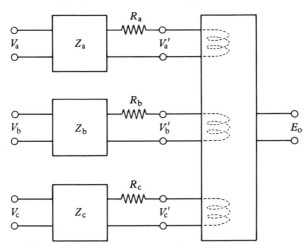

FIG. 19.3 Block diagram of the control fields

Each field is assumed to be preceded by some device, a potential divider, transformer, or phase modifying network.

In the diagram each field circuit is represented by a two-port device which includes the field resistance but not the inductance.

V_a is the voltage applied to the first field circuit and V'_a is the voltage across the inductance. If the short-circuit transfer impedance from port **1** to port **2** is Z_{1a} and the short-circuit driving-point impedance at port **2** is Z_{2a}, the current I_a can be obtained by applying the principle of superposition.

Hence $$I_a = V_a/Z_{1a} - V'_a/Z_{2a} \tag{19.15}$$

and so

$$\begin{bmatrix} I_a \\ I_b \\ I_c \end{bmatrix} = \begin{bmatrix} 1/Z_{1a} & \cdot & \cdot \\ \cdot & 1/Z_{1b} & \cdot \\ \cdot & \cdot & 1/Z_{1c} \end{bmatrix} \cdot \begin{bmatrix} V_a \\ V_b \\ V_c \end{bmatrix} - \begin{bmatrix} 1/Z_{2a} & \cdot & \cdot \\ \cdot & 1/Z_{2b} & \cdot \\ \cdot & \cdot & 1/Z_{2c} \end{bmatrix} \begin{bmatrix} V'_a \\ V'_b \\ V'_c \end{bmatrix} \tag{19.16}$$

If all the coils are single turn coils each having inductance L and with a coupling factor of unity (i.e. $M = L$), the volt drop across the inductances are given by

$$\begin{bmatrix} V'_a \\ V'_b \\ V'_c \end{bmatrix} = pL \begin{bmatrix} 1 & 1 & 1 \\ 1 & 1 & 1 \\ 1 & 1 & 1 \end{bmatrix} \cdot \begin{bmatrix} I_a \\ I_b \\ I_c \end{bmatrix} \tag{19.17}$$

The open-circuit output voltage is now the sum of the effects of all the currents and so Eq. (19.10) becomes

$$E_0 = [k][I_f]$$

$$= k \begin{bmatrix} 1 & 1 & 1 \end{bmatrix} \cdot \begin{bmatrix} I_a \\ I_b \\ I_c \end{bmatrix} \tag{19.18}$$

Converting to the general case when the field turns are n_a, n_b and n_c respectively, Eq. (19.18) becomes

$$E_0 = k \begin{bmatrix} n_a & n_b & n_c \end{bmatrix} \begin{bmatrix} I_a \\ I_b \\ I_c \end{bmatrix} \tag{19.19}$$

and the connection matrix

$$[c] = \begin{bmatrix} n_a & . & . \\ . & n_b & . \\ . & . & n_c \end{bmatrix} \tag{19.20}$$

must be used to transform the impedance matrix (19.17).
Thus

$$\begin{bmatrix} V'_a \\ V'_b \\ V'_c \end{bmatrix} = pL \begin{bmatrix} n_a & 0 & 0 \\ 0 & n_b & 0 \\ 0 & 0 & n_c \end{bmatrix} \cdot \begin{bmatrix} 1 & 1 & 1 \\ 1 & 1 & 1 \\ 1 & 1 & 1 \end{bmatrix} \cdot \begin{bmatrix} n_a & 0 & 0 \\ 0 & n_b & 0 \\ 0 & 0 & n_c \end{bmatrix} \cdot \begin{bmatrix} I_a \\ I_b \\ I_c \end{bmatrix} \tag{19.21}$$

$$= pL \begin{bmatrix} n_a & 0 & 0 \\ 0 & n_b & 0 \\ 0 & 0 & n_c \end{bmatrix} \cdot \begin{bmatrix} n_a & n_b. & n_c \\ n_a & n_b & n_c \\ n_a & n_b & n_c \end{bmatrix} \cdot \begin{bmatrix} I_a \\ I_b \\ I_c \end{bmatrix} \tag{19.22}$$

$$= pL \begin{bmatrix} n_a \\ n_b \\ n_c \end{bmatrix} E_0/k \tag{19.23}$$

From Eqs. (19.16) and (19.19)

$$E_0 = k \begin{bmatrix} n_a & n_b & n_c \end{bmatrix} \cdot \begin{bmatrix} I_a \\ I_b \\ I_c \end{bmatrix}$$

$$= k \begin{bmatrix} n_a & n_b & n_c \end{bmatrix} \left\{ \begin{bmatrix} V_a/Z_{1a} \\ V_b/Z_{1b} \\ V_c/Z_{1c} \end{bmatrix} - pLE_0/k \begin{bmatrix} n_a/Z_{2a} \\ n_b/Z_{2b} \\ n_c/Z_{2c} \end{bmatrix} \right\} \tag{19.24}$$

or $\qquad E_0 \left\{ 1 + p \sum \dfrac{L_a}{Z_{2a}} \right\} = \sum V_a \dfrac{k_a}{Z_{1a}} \tag{19.25}$

where

$$\begin{bmatrix} L_a \\ L_b \\ L_c \end{bmatrix} = L \begin{bmatrix} n^2{}_a \\ n^2{}_b \\ n^2{}_c \end{bmatrix} \tag{19.26}$$

and

$$\begin{bmatrix} k_a \\ k_b \\ k_c \end{bmatrix} = k \begin{bmatrix} n_a \\ n_b \\ n_c \end{bmatrix} \tag{19.27}$$

Equations (19.26) and (19.27) represent the measured inductances and the 'volts per amp.' coefficients of the individual windings.

19.3 Block diagram of amplidyne

Equation (19.25) is depicted diagrammatically by the block diagrams of Fig. 19.4. This shows that the input voltages must be operated upon by the transfer impedances of the input circuits before summating.

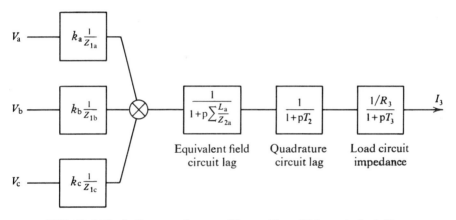

FIG. 19.4 **Block diagram of an amplidyne with multiple control windings**

The equivalent lag introduced by the field circuit is obtained by summing the ratios L_a/Z_{2a}. If Z_{2a} is resistive, this summation is a simple addition of the time constants of the individual field circuits.

CHAPTER 20

The synchronous machine

20.1 Introduction

The unified approach to electrical machine theory has particular significance with respect to synchronous machines.

In the first place both salient pole and uniform air gap machines are manufactured for a wide range of ratings. It is useful to have a comprehensive treatment which enables the difference in performance of the two types to be noted when and where this is significant. Secondly, the transient performances of these machines differ considerably from their steady-state operation. The parameters affecting transient performance tend to be overlooked when the treatment is only concerned with steady state.

The synchronous machine is derived from the primitive machine in the following manner.

In Fig. 20.1, the winding k_f represents the field and the current I_f is the d.c. excitation current.

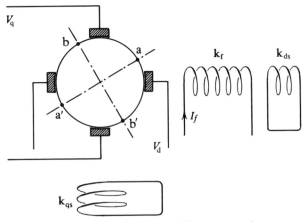

FIG. 20.1 Connections of a primitive machine as a synchronous machine

The main a.c. windings are the rotor windings accessible by the slip-rings **aa'** and **bb'** forming a two-phase system as in Sec. 17.3. The two-phase currents I_a and I_b can be transformed to equivalent axis currents I_d and I_q for the purpose of analysis, as described in that section. The stator windings k_{ds} and k_{qs} are short-circuited and represent the damper windings.

20.2 Steady-state analysis

First we consider a salient-pole machine under steady-state conditions. If speed and load angle are constant and the line-current is steady, there will be no current in the damper windings and so these can be ignored.

20.2.1 Steady-state impedance matrix for the salient-pole machine

The impedance matrix for the primitive machine is reduced to that of Eq. (20.1) by omitting the second and fifth rows and columns. Thus

$$\begin{bmatrix} v_f \\ v_{dr} \\ v_{qr} \end{bmatrix} = \begin{bmatrix} R_f + L_f p & M_{fr} p & 0 \\ M_{fr} p & R_r + L_{dr} p & G_{qr}(p\theta) \\ -G_f(p\theta) & -G_{dr}(p\theta) & R_r + L_{qr} p \end{bmatrix} \cdot \begin{bmatrix} i_f \\ i_{dr} \\ i_{qr} \end{bmatrix} \quad (20.1)$$

Lower-case letters will be used for instantaneous values of time-varying quantities in order to distinguish between them and the corresponding maximum values of alternating quantities.

Simplifying the subscripts and assuming sinusoidal m.m.f. distribution

$$\left. \begin{aligned} R_f &= R_1 \\ R_2 &= R_2 \\ G_{dr} &= L_{dr} = L_{2d} \\ G_{qr} &= L_{qr} = L_{2q} \\ G_f &= M_{fd} = M_d \end{aligned} \right\} \quad (20.2)$$

$$\begin{bmatrix} v_f \\ v_d \\ v_q \end{bmatrix} = \begin{bmatrix} R_1 + L_1 p & M_d p & 0 \\ M_d p & R_2 + L_{2d} p & L_{2q}(p\theta) \\ -M_d(p\theta) & -L_{2d}(p\theta) & R_2 + L_{2q} p \end{bmatrix} \cdot \begin{bmatrix} i_f \\ i_d \\ i_q \end{bmatrix} \quad (20.3)$$

Currents i_d and i_q are the axis currents in the primitive machine and these are related to the actual currents in the rotor by the transform to rotating axes developed in Sec. 16.8.2.

20.2.2 Transform to rotating axes

From Eq. (16.68)

$$\begin{bmatrix} i_d \\ i_q \end{bmatrix} = \begin{bmatrix} \cos\theta & -\sin\theta \\ \sin\theta & \cos\theta \end{bmatrix} \cdot \begin{bmatrix} i_a \\ i_b \end{bmatrix} \quad (16.68)$$

we see how currents in the slip rings **a–a′** and **b–b′** (Fig. 20.1) are related to the axis currents, when the coils are inclined at an angle θ to the main axes.

Under balanced steady-state conditions the rotor is rotating at synchronous speed and the current in the **a** winding is represented by the phasor I. This is usually resolved into two components at right angles* I_d and I_q such that

$$I = I_d + I_q$$
$$= I_d + jI_q \tag{20.4}$$

and

$$\theta = \omega t \tag{20.5}$$

These components are not the same as the currents i_d and i_q of Eq. (16.68), but they are related to them in the following manner.

For balanced two-phase conditions I_b can be derived from I_a, hence

$$\begin{bmatrix} I_a \\ I_b \end{bmatrix} = \begin{bmatrix} 1 \\ j \end{bmatrix} I \tag{20.6}$$

The instantaneous values of the rotor currents are thus

$$\begin{bmatrix} i_a \\ i_b \end{bmatrix} = \mathrm{Re} \begin{bmatrix} 1 \\ j \end{bmatrix} (I_d + jI_q) \exp j\omega t$$

$$= \mathrm{Re} \begin{bmatrix} I_d + jI_q \\ jI_d - I_q \end{bmatrix} (\cos \omega t + j \sin \omega t)$$

$$= \begin{bmatrix} \cos \omega t & -\sin \omega t \\ -\sin \omega t & -\cos \omega t \end{bmatrix} \cdot \begin{bmatrix} I_d \\ I_q \end{bmatrix} \tag{20.7}$$

Combining Eqs. (16.68) and (20.7)

$$\begin{bmatrix} i_d \\ i_q \end{bmatrix} = \begin{bmatrix} \cos \theta & -\sin \theta \\ \sin \theta & \cos \theta \end{bmatrix} \cdot \begin{bmatrix} \cos \omega t & -\sin \omega t \\ -\sin \omega t & -\cos \omega t \end{bmatrix} \cdot \begin{bmatrix} I_d \\ I_q \end{bmatrix} \tag{20.8}$$

and putting $\theta = \omega t$ when the speed is synchronous

$$\begin{bmatrix} i_d \\ i_q \end{bmatrix} = \begin{bmatrix} 1 & 0 \\ 0 & -1 \end{bmatrix} \cdot \begin{bmatrix} I_d \\ I_q \end{bmatrix} \tag{20.9}$$

and similarly for voltage

$$\begin{bmatrix} v_d \\ v_q \end{bmatrix} = \begin{bmatrix} 1 & 0 \\ 0 & -1 \end{bmatrix} \cdot \begin{bmatrix} V_d \\ V_q \end{bmatrix} \tag{20.10}$$

Equation (20.9) shows that time has been eliminated from the transform matrix and consequently from the axis currents i_d and i_q. This is the mathe-

---

* *Electrical Machines*, sec. 7.4.1.

matical interpretation of the well-known fact that the armature m.m.f. of a synchronous machine rotates with the poles but its phase depends on the ratio of the inphase and quadrature components of the armature current.

20.2.3 Derivation of the phasor equations

We can now apply the transform equation (20.9) to the impedance matrix of the synchronous machine (Eq. (20.3)) and make allowance for the other special conditions relating to steady state.

Since speed is constant and synchronous, we first put $(p\theta)$ equal to ω. Furthermore, since we have established that i_t, i_d and i_q are all constant, pi_t, pi_d and pi_q are all zero. Hence Eq. (20.3) reduces to

$$\begin{bmatrix} v_f \\ v_d \\ v_q \end{bmatrix} = \begin{bmatrix} R_1 & 0 & 0 \\ 0 & R_2 & X_{2q} \\ -X_m & -X_{2d} & R_2 \end{bmatrix} \cdot \begin{bmatrix} i_f \\ i_d \\ i_q \end{bmatrix} \qquad (20.11)$$

where

$$\left. \begin{aligned} X_m &= \omega M_d \\ X_{2d} &= \omega L_{2d} \\ X_{2q} &= \omega L_{2q} \end{aligned} \right\} \qquad (20.12)$$

At this stage it is worth comparing Eqs. (20.3) and (20.11) and noting how many terms of (20.3) are eliminated.

This shows the many factors which influence the transient behaviour of the machine but play no part in its steady-state operation. Applying the transform to phasors.

$$\begin{bmatrix} V_f \\ V_d \\ V_q \end{bmatrix} = \begin{bmatrix} 1 & 0 & 0 \\ 0 & 1 & 0 \\ 0 & 0 & -1 \end{bmatrix} \cdot \begin{bmatrix} R_1 & 0 & 0 \\ 0 & R_2 & X_{2q} \\ -X_m & -X_{2d} & R_2 \end{bmatrix}$$

$$\begin{bmatrix} 1 & 0 & 0 \\ 0 & 1 & 0 \\ 0 & 0 & -1 \end{bmatrix} \cdot \begin{bmatrix} I_f \\ I_d \\ I_q \end{bmatrix} \qquad (20.13)$$

$$\begin{bmatrix} V_f \\ V_d \\ V_q \end{bmatrix} = \begin{bmatrix} R_1 I_f \\ R_2 I_d - X_{2q} I_q \\ X_m I_f + X_{2d} I_d + R_2 I_q \end{bmatrix} \qquad (20.14)$$

and thus

$$V_f = R_1 I_f \qquad (20.15)$$

$$\begin{aligned} V &= V_d + jV_q \\ &= jX_m I_f + R_2(I_d + jI_q) + jX_{2d}I_d + jX_{2q}(jI_q) \\ &= E_0 + IR_2 + I_d jX_{2d} + I_q jX_{2q} \end{aligned} \qquad (20.16)$$

Equation (20.16) is substantially the same as that developed by two axis theory using phasor methods.*

It should be remembered that the present equations use the general impedance matrix conventions whereas those in *Electrical Machines* are virtually two-port network conventions.

Since the output current from a network

$$I_2\dagger = -I_2$$

in terms of output currents, Eq. (20.16) becomes

$$E_0 = V + I\dagger R_2 + I_d\dagger j X_{2d} + I_q\dagger j X_{2q} \qquad (20.17)$$

Further consideration will be given to the steady-state condition in Sec. 22.1.

20.3 Symmetrical short-circuit of an alternator

20.3.1 Waveforms of line current

When an alternator is suddenly short-circuited with oscillograph recorders suitably connected so as to measure the line currents immediately subsequent to the short circuit, the current waveform is usually found to be similar to that shown in Fig. 20.2.

Such oscillographs when analysed are observed to have the following prominent features:

1. The waveform is initially asymmetrical only becoming symmetrical after several hundred cycles. In other words the alternating current has a superimposed direct-current transient.
2. The alternating component commences with large amplitude reducing exponentially with time to a final steady-state value.
3. The rate at which the alternating component is reduced is much greater over the first few cycles than it is subsequently. This suggests the presence of two transient alternating components in addition to the final steady-state value, one having a much smaller time constant than the other.

These features are demonstrated if we examine the waveform shown in Fig. 20.2. In the diagram, the envelopes of the curve have been drawn and by bisecting the ordinates between these envelopes, the d.c. transient is clearly shown.

In Fig. 20.3, the d.c. transient has been subtracted leaving only the a.c. components. By drawing in the envelopes of this waveform the manner in which the amplitude changes can be seen. The initial rapid fall soon changes to a recognisable exponential shape and, projecting the latter curve backwards by the dotted lines, the additional transient which occurs in the first few cycles is separated.

Three maximum current values are important: $\sqrt{2}I''$ is the initial peak value; $\sqrt{2}I$ is the peak value of the final steady-state current and $\sqrt{2}I'$ is initial peak value neglecting the small highly damped component.

* *Electrical Machines*, p. 174, Eq. (7.41).

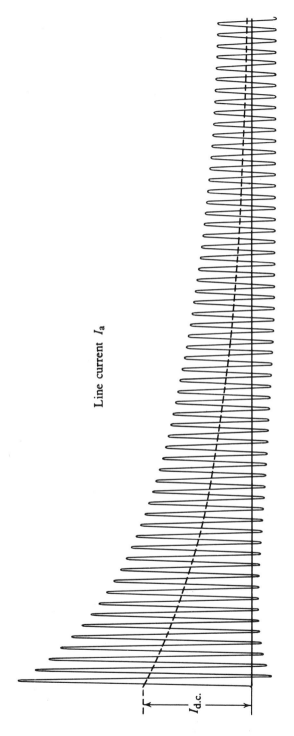

FIG. 20.2 Waveform of the line current of an alternator suddenly shortcircuited

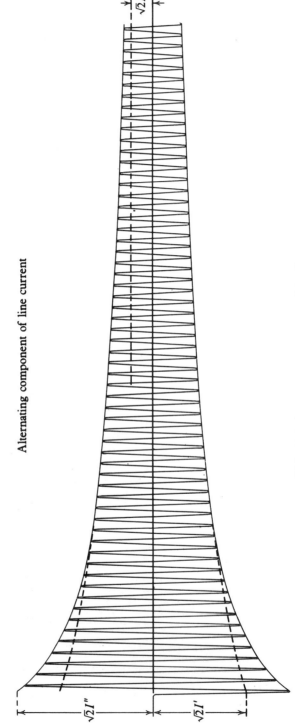

Alternating component of line current

FIG. 20.3 Alternating component of the line current

The values of the a.c. components depend solely on the machine parameters and on the excitation current but the magnitude of the d.c. transient also depends on the instant in the cycle when the switching takes place (see example 7.8.7).

Since the three phases are necessarily at different points in their cycles at the instant of switching, the magnitudes and directions of the d.c. transient in each of the phases usually differ considerably but their time constants are similar.

The expression for the current waveform is thus expected to have the following form.

$$I_{dc} \exp(-t/T_0) + \sqrt{2}\{I + (I' - I) \exp(-t/T') + (I'' - I') \exp(-t/T'')\} \sin(\omega t + \phi)$$
(20.18)

where the first term is the d.c. transient with time constant T_0, the second term is the steady-state value, the third term is known as the main transient with time constant T' and the fourth term is termed the sub-transient with time constant T''. It will be shown that other small components exist including a second harmonic term but these are not immediately apparent from an examination of the waveform and are of secondary importance.

The magnitude of the current components are almost independent of the resistance parameters of the machine, hence it is customary to assume that each current is equal to the open-circuit voltage divided by a reactance value

$$I = V_0/x_d \tag{20.19}$$
$$I' = V_0/x'_d \tag{20.20}$$
$$I'' = V_0/x''_d \tag{20.21}$$

The reactances defined by these equations are given the following terms:

x_d is the direct-axis synchronous reactance.
x'_d is the direct-axis transient reactance.
x''_d is the direct-axis sub-transient reactance.

20.3.2 Field current waveform

Oscillograms taken of the current in the field windings during the transient period following the short circuit also show the addition of transient components. Figure 20.4 is typical. It shows that the normal steady-state field current is suddenly increased by a large d.c. transient whose time constant is similar to that of the main a.c. transient in the phase windings, the current only returning to normal when this has become negligible.

Also we see that this transient is reduced by a smaller d.c. transient with the time constant corresponding to T'' and finally an a.c. transient of fundamental frequency is present with an exponential decay similar to that of the d.c. transient in the main windings.

The expression for the field current is, therefore, of the form

$$i_f = I_{f0} + I'_f \exp(-t/T') - I''_f \exp(-t/T'') + I'''_f \exp(-t/T_0) \sin(\omega t + d) \tag{20.22}$$

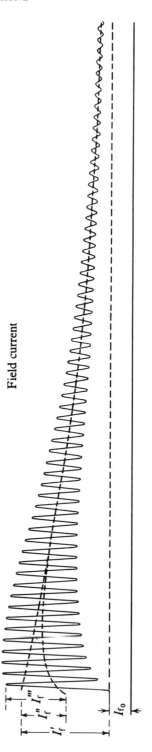

FIG. 20.4 Waveform of the field current

Before we attempt to relate these various transients and their time constants to the machine parameters as given in the impedance matrix of the primitive machine, it is useful to discuss the cause of these components in terms of the chamges of m.m.f. and flux in the machine.

20.3.3 Flux changes due to the short circuit

On open circuit the main flux is produced solely by the field m.m.f. acting along the direct axis and consequently the terminal voltage is equal to the generated voltage due to the movement of the a.c. winding conductors through this flux.

With a sustained short circuit of the polyphase armature windings it is well known* that the armature m.m.f. acts along the direct axis in opposition to the field m.m.f. and the resultant flux is much smaller than the corresponding value on open circuit.

When the machine is on open-circuit and the short-circuit is suddenly applied, the transient period is the time taken for this flux to reduce from the one value to the other.

Although the armature m.m.f. acts immediately the short-circuit is applied, the flux cannot change instantaneously and, consequently, the initial value of the a.c. component of short-circuit current is very nearly equal to the open circuit voltage applied to the leakage reactance of the armature. In other words the sub-transient reactance and the leakage reactance are nearly equal.

The armature m.m.f. is opposed by d.c. transients induced by the rate of change of flux in all circuits linking with the main flux. It is only as these transients diminish that the flux is reduced and when they become negligible the short-circuit current reaches its steady-state value.

The most important circuit linking the flux is, of course, the main-field circuit and the time constant of this circuit determines the decay of both its own d.c. transient and also the main a.c. transient in the a.c. winding.

In addition, however, d.c. transients are induced in all other circuits linking the flux, principally in pole-face damper windings if these are fitted, but also in solid pole cores and shoes where eddy currents linking the flux have possible paths. These paths have a lower reactance-to-resistance ratio and hence a smaller time constant than that of the main-field circuit. These currents are the first to die away and are responsible for the sub-transient condition.

A d.c. transient is produced in all inductive circuits when alternating voltage is suddenly applied, unless the switch is closed at a voltage maximum (Sec. 7.8.7). Such a coincidence cannot occur for more than one phase winding.

D.C. transients induced in the armature winding decay with a time constant, to be determined later, which is mainly dependent on the resistance/reactance ratio of the armature itself.

The resultant m.m.f. due to these d.c. components rotates with the armature relative to the field at fundamental frequency and so induces an a.c. component in the field winding, both components having the same time constant.

---

* *Electrical Machines*, p. 168.

It will be seen that the effect of saliency is only of secondary importance and for a salient-pole machine the direct-axis parameters dominate the situation.

Sudden changes of load produce similar transients though of smaller amplitude and the sudden application of unbalanced load introduces further complications beyond the scope of this book.

In the next sections it is hoped to demonstrate how the transient reactances and their time constants can be derived from the impedance parameters of the primitive machine. The analysis is presented in considerable detail in order to demonstrate the method and show why certain simplifying assumptions are justified.

20.3.4 The impedance matrix

Commencing once again with the impedance matrix of the primitive machine, Eq. (20.23) is obtained by rearranging the order of the rows and columns:

$$
\begin{bmatrix} v_\mathrm{f} \\ v_\mathrm{dr} \\ v_\mathrm{qr} \\ v_\mathrm{ds} \\ v_\mathrm{qs} \end{bmatrix} = \begin{bmatrix} R_\mathrm{f}+L_\mathrm{f}p & M_\mathrm{fr}p & 0 & M_\mathrm{fd}p & 0 \\ M_\mathrm{fr}p & R_\mathrm{r}+L_\mathrm{dr}p & G_\mathrm{qr}(p\theta) & M_\mathrm{d}p & G_\mathrm{q}(p\theta) \\ -G_\mathrm{f}(p\theta) & -G_\mathrm{dr}(p\theta) & R_\mathrm{r}+L_\mathrm{qr}p & -G_\mathrm{d}(p\theta) & M_\mathrm{q}p \\ M_\mathrm{fd}p & M_\mathrm{d}p & 0 & R_\mathrm{ds}+L_\mathrm{ds}p & 0 \\ 0 & 0 & M_\mathrm{q}p & 0 & R_\mathrm{qs}+L_\mathrm{qs}p \end{bmatrix} \cdot \begin{bmatrix} i_\mathrm{f} \\ i_\mathrm{dr} \\ i_\mathrm{qr} \\ i_\mathrm{ds} \\ i_\mathrm{qs} \end{bmatrix}
$$
(20.23)

Simplifying the nomenclature we obtain:

$$
\begin{bmatrix} v_\mathrm{f} \\ v_\mathrm{d} \\ v_\mathrm{q} \\ 0 \\ 0 \end{bmatrix} = \begin{bmatrix} R_1+L_1p & M_\mathrm{d}p & 0 & M_\mathrm{d}p & 0 \\ M_\mathrm{d}p & R_2+L_{2\mathrm{d}}p & L_{2\mathrm{q}}\omega & M_\mathrm{d}p & M_\mathrm{q}\omega \\ -M_\mathrm{d}\omega & -L_{2\mathrm{d}}\omega & R_2+L_{2\mathrm{q}}p & -M_\mathrm{d}\omega & M_\mathrm{q}p \\ M_\mathrm{d}p & M_\mathrm{d}p & 0 & R_{3\mathrm{d}}+L_{3\mathrm{d}}p & 0 \\ 0 & 0 & M_\mathrm{q}p & 0 & R_{3\mathrm{q}}+L_{3\mathrm{q}}p \end{bmatrix} \cdot \begin{bmatrix} i_\mathrm{f} \\ i_{2\mathrm{d}} \\ i_{2\mathrm{q}} \\ i_{3\mathrm{d}} \\ i_{3\mathrm{q}} \end{bmatrix}
$$
(20.24)

where

$$\omega = (p\theta)$$
$$R_1 = R_\mathrm{f}$$
$$R_2 = R_\mathrm{r}$$
$$R_{3\mathrm{d}} = R_\mathrm{ds}$$
$$R_{3\mathrm{q}} = R_\mathrm{qs}$$

and

$$L_{2\mathrm{d}} = L_\mathrm{dr} = G_\mathrm{dr}$$
$$L_{2\mathrm{q}} = L_\mathrm{qr} = G_\mathrm{qr}$$
$$M_\mathrm{d} = M_\mathrm{fd} = G_\mathrm{d} - G_\mathrm{f}$$
$$M_\mathrm{q} = G_\mathrm{q}$$
$$L_{3\mathrm{d}} = L_\mathrm{ds}$$
$$L_{3\mathrm{q}} = L_\mathrm{qs}$$

(20.25)

Sinusoidal distribution of m.m.f. in each axis is assumed, hence the corresponding G and M parameters are equal, and it is also assumed that the mutual inductances between coils on the same axis are alike.

The damper windings are short-circuited hence v_{3d} and v_{3q} are zero.

The first step in reducing the matrix is to eliminate the line and column representing the field winding using the method of partitioning (see (3.8)).

Since $$[V'] = [V_1] - [Z_2][Z_4]^{-1}[V_2] \tag{3.47}$$

and $$[Z'] = [Z_1] - [Z_2][Z_4]^{-1}[Z_3] \tag{3.48}$$

the reduced matrix becomes

$$
\begin{bmatrix}
V_d - \dfrac{M_d p V_f}{R_1 + L_1 p} \\[2mm]
V_q + \dfrac{M_d \omega V_f}{R_1 + L_1 p} \\[2mm]
- \dfrac{M_d p V_f}{R_1 + L_1 p} \\[2mm]
0
\end{bmatrix}
=
$$

$$
\begin{bmatrix}
R_2 + L_{2d} p - \dfrac{M_d^2 p^2}{R_1 + L_1 p} & L_{2q} \omega & M_d p - \dfrac{M_d^2 p^2}{R_1 + L_1 p} & M_q \omega \\[3mm]
-L_{2d} \omega + \dfrac{M_d \omega p}{R_1 + L_1 p} & R_2 + L_{2q} p & -M_d \omega + \dfrac{M_d^2 \omega p}{R_1 + L_1 p} & M_q p \\[3mm]
M_d p - \dfrac{M_d^2 p^2}{R_1 + L_1 p} & 0 & R_{3d} + L_{3d} p - \dfrac{M_d^2 p^2}{R_1 + L_1 p} & 0 \\[3mm]
0 & M_q p & 0 & R_{3q} + L_{3q} p
\end{bmatrix}
\cdot
\begin{bmatrix}
i_{2d} \\[2mm]
i_{2q} \\[2mm]
i_{3d} \\[2mm]
i_{3q}
\end{bmatrix}
$$

$$\tag{20.26}$$

This can be simplified since V_f the exciter voltage is assumed to be constant, thus $p V_f$ is zero and $V_f/(R_1 + p L_1)$ is equal to V_f/R_1. When the armature windings are short-circuited V_d and V_q are also zero. Thus

$$
\begin{bmatrix}
0 \\
V_{q0} \\
0 \\
0
\end{bmatrix}
=
\begin{bmatrix}
R_2 + L'_{2d}(p)p & \omega L_{2q} & M'_d(p)p & \omega M_q \\
-\omega L'_{2d}(p) & R_2 + L_{2q} p & -\omega M'_d(p) & M_q p \\
M'_d(p)p & 0 & R_{3d} + L'_{3d}(p)p & 0 \\
0 & M_q p & 0 & R_{3q} + L_{3q} p
\end{bmatrix}
\cdot
\begin{bmatrix}
i_{2d} \\
i_{2q} \\
i_{3d} \\
i_{3q}
\end{bmatrix}
$$

$$\tag{20.27}$$

where*

$$
\left.
\begin{aligned}
L'_{2d}(p) &= L_{2d} - M_d^2 p/(R_1 + L_1 p) \\
L'_{3d}(p) &= L_{3d} - M_d^2 p/(R_1 + L_1 p) \\
M'_d(p) &= M_d - M_d^2 p/(R_1 + L_1 p) \\
V_{q0} &= \omega M_d V_f/R_1
\end{aligned}
\right\}
\tag{20.28}
$$

* The expression $L'_{2d}(p)$ is used to show that this is a function of p and not a simple inductance.

If the machine has no damper windings and the effects of eddy currents in the pole faces are ignored, Eq. (20.27) can be simplified by omitting the last two lines and columns. Thus without damper windings

$$\begin{bmatrix} 0 \\ V_{q0} \end{bmatrix} = \begin{bmatrix} R_2 + L'_{2d}(p)p & \omega L_{2q} \\ -\omega L'_{2d}(p) & R_2 + L_{2q}p \end{bmatrix} \cdot \begin{bmatrix} i_d \\ i_q \end{bmatrix} \qquad (20.29)$$

Otherwise, to eliminate the lines and columns corresponding to permanently short-circuited damper windings the partitioning procedure must be applied once more resulting in

$$\begin{bmatrix} 0 \\ V_{q0} \end{bmatrix} = \begin{bmatrix} R_2 + L''_{2d}(p)p & \omega L''_{2q}(p) \\ -\omega L''_{2d}(p) & R_2 + L''_{2q}(p)p \end{bmatrix} \cdot \begin{bmatrix} i_d \\ i_q \end{bmatrix} \qquad (20.30)$$

where

$$L''_{2d}(p) = L'_{2d}(p) - M'^2_d(p) \, p/[R_{3d} + L'_{3d}(p) \, p] \qquad (20.31)$$

and

$$L''_{2q}(p) = L_{2q} - M_q^2 \, p/(R_{3q} + L_{3q} \, p) \qquad (20.32)$$

The machine equations are thus reduced to two. If these are solved under the conditions imposed by the suddenly applied short-circuit i_d and i_q can be determined. Before this is attempted, we will give further thought to the significance of $L'_{2d}(p)$ and $L''_{2d}(p)$.

20.3.5 The operators $L''_{2d}(p)$, $L'_{2d}(p)$ and $L''_{2q}(p)$

Whereas L_{2d}, L_{2q} and M_d are straightforward inductance parameters of the machine, $L''_{2d}(p)$, $L'_{2d}(p)$ and $L''_{2q}(p)$ are functions of p as their symbols indicate. Their presence in Eq. (20.30) increases the difficulty of finding a solution, since high orders of p are encountered.

The expressions for these quantities ((20.28), (20.31), (20.32)) will first be simplified in terms of the time constants and the coupling factors of the circuits to which they relate.
Thus

$$L'_{2d}(p) = L_{2d} - M_d^2 p/(R_1 + L_1 p) \qquad (20.28)$$

$$= \frac{L_{2d}(R_1 + L_1 p) - M_d^2 p}{(R_1 + L_1 p)}$$

$$= L_{2d} \frac{[1 + pL_1(1 - M_d^2/L_1 L_{2d})/R_1]}{1 + pL_1/R_1}$$

$$= L_{2d} \frac{1 + p(1 - k_{21}^2)T_1}{1 + pT_1}$$

$$= L_{2d} \frac{1 + pc_{21}T_1}{1 + pT_1} \qquad (20.33)$$

where

$$T_1 = L_1/R_1 \qquad (20.34)$$

the time constant of the field circuit, and

$$k_{21}{}^2 = M_d{}^2/L_1 L_{2d} \qquad (20.35)$$

k_{21} is the coupling factor between the inductances L_1 and L_{2d}, and c_{21} is defined by

$$c_{21} = (1 - k_{21}{}^2) * \qquad (20.36)$$

$L'_{2d}(p)$ is thus to be seen as L_{2d} multiplied by the ratio of two operational impedances, the effective time constant $c_{21}T_1$ being much less than T_1.

Similarly the other two equations of (20.28) can be reduced to give

$$L'_{3d}(p) = L_{3d} \frac{1 + pc_{31}T_1}{1 + pT_1} \qquad (20.37)$$

where

$$c_{31} = (1 - M_d{}^2/L_{3d}L_1) \qquad (20.38)$$

and

$$M'_d(p) = M_d \frac{1 + pc_{M1}T_1}{1 + pT_1} \qquad (20.39)$$

where

$$c_{M1} = (1 - M_d/L_1) \qquad (20.40)$$

Applying the same treatment to Eq. (20.32),

$$L''_{2q}(p) = L_{2q} \frac{1 + pc_{23q}T_3}{1 + pT_3} \qquad (20.41)$$

* If a coil whose inductance is L_1 is coupled to a secondary coil of inductance L_2 the mutual inductance being M, and resistance is neglected,

$$\begin{bmatrix} V_1 \\ V_2 \end{bmatrix} = p \begin{bmatrix} L_1 & M \\ M & L_2 \end{bmatrix} \cdot \begin{bmatrix} I_1 \\ I_2 \end{bmatrix}$$

If the secondary coil is open-circuited

$$\begin{bmatrix} V_1 \\ V_2 \end{bmatrix} = p \begin{bmatrix} L_1 & M \\ M & L_2 \end{bmatrix} \cdot \begin{bmatrix} I_1 \\ 0 \end{bmatrix}$$

and

$$V_{10} = pL_1 I_1$$

If the secondary coil is short-circuited,

$$\begin{bmatrix} V_1 \\ 0 \end{bmatrix} = p \begin{bmatrix} L_1 & M \\ M & L_2 \end{bmatrix} \cdot \begin{bmatrix} I_1 \\ I_2 \end{bmatrix}$$

and eliminating the second line and column,

$$\begin{aligned} V_{1x} &= p(L_1 - M^2/L_2)I_1 \\ &= p(1 - k^2)L_1 I_1 \\ &= pcL_1 I_1 \end{aligned}$$

The result of the short-circuit is thus to reduce the effective inductance of the primary coil in the ratio c where

$$c = (1 - k^2)$$

and k is the coupling factor.

where

$$c_{23q} = (1 - M_q^2/L_{2q}L_{3q}) \qquad (20.42)$$

The reduction of $L''_{2d}(p)$ is more complicated:

$$L''_{2d}(p) = L'_{2d}(p) - M'_d{}^2(p)p/(R_{3d} + L'_{3d}(p)p) \qquad (20.31)$$

$$= L'_{2d}(p) \cdot \frac{R_{3d} + p\left(1 - \dfrac{M'_d{}^2(p)}{L'_{2d}(p)L'_{3d}(p)}\right)L'_{3d}(p)}{R_{3d} + pL'_{3d}(p)}$$

$$= L'_{2d}(p) \cdot \frac{R_{3d} + p\left(1 - \dfrac{M_d{}^2(1 + pc_{M1}T_1)^2}{L_{2d}L_{3d}(1 + pc_{21}T_1)(1 + pc_{31}T_1)}\right)L'_{3d}(p)}{R_{3d} + pL'_{3d}(p)}$$

Now the time constant T_1 of the field winding is much greater than that of the damper winding T_3. Hence if R_1 is neglected where it occurs inside the brackets* then

$$\frac{M_d{}^2}{L_{2d}L_{3d}} \cdot \frac{(1 + pc_{M1}T_1)^2}{(1 + pc_{21}T_1)(1 + pc_{31}T_1)} \simeq \frac{k_{23}{}^2 c_{M1}{}^2}{c_{21}c_{31}}$$

and

$$L''_{2d}(p) = L'_{2d}(p) \cdot \frac{R_{3d} + p\left(1 - \dfrac{k_{23}{}^2 c_{M1}{}^2}{c_{21}c_{31}}\right)L'_{3d}(p)}{R_{3d} + pL'_{3d}(p)}$$

Again, it is sufficiently accurate if R_1 is neglected and hence

$$L'_{3d} \simeq c_{31}L_{3d}$$

We therefore find, putting $T_3 = L_{3d}/R_3$,

$$L''_{2d}(p) \simeq L'_{2d}(p) \cdot \frac{1 + p\left(1 - \dfrac{k_{23}{}^2 c_{M1}{}^2}{c_{21}c_{31}}\right)c_{31}T_3}{1 + pc_{31}T_3}$$

or

$$\simeq L_{2d} \cdot \frac{1 + pc_{21}T_1}{1 + pT_1} \cdot \frac{1 + p(c_{31} - k_{23}{}^2 c_{M1}{}^2/c_{21})T_3}{1 + pc_{31}T_3}$$

or

$$\simeq L_{2d} \cdot \frac{1 + pc_{21}T_1}{1 + pT_1} \cdot \frac{1 + pc'_{31}T_3}{1 + pc_{31}T_3} \qquad (20.43)$$

where

$$c'_{31} = c_{31} - k_{23}{}^2 c_{M1}{}^2/c_{21} \qquad (20.44)$$

It it were possible to neglect all the resistances (this is satisfactory for the determination of the initial values of current but there would be no subsequent damping), Eqs. (20.33), (20.41), (20.37), (20.39) and (20.43) reduce to

$$L'_{2d} = c_{21}L_{2d} \qquad (20.45)$$

* See example 7.8.6.

$$L''_{2q} = c_{23q}L_{2q} \tag{20.46}$$

$$L'_{3d} = c_{31}L_{3d} \tag{20.47}$$

$$M'_d = c_{M1}M_d \tag{20.48}$$

$$L''_{2d} = (c_{21} - c_{M1}{}^2 k_{23}{}^2 / c_{31})L_{2d} \tag{20.49}$$

These values are true effective inductances and are not functions of p.

20.3.6 Determination of initial values of short-circuit current

Before attempting the complete solution of Eq. (20.30), we will observe the effect of neglecting all resistances. Under these conditions the equations simplify to

$$\begin{bmatrix} 0 \\ V_{q0} \end{bmatrix} = \begin{bmatrix} L''_{2d}p & \omega L''_{2q} \\ -\omega L''_{2d} & L''_{2q}p \end{bmatrix} \cdot \begin{bmatrix} i_d(t) \\ i_q(t) \end{bmatrix} \tag{20.50}$$

The effect of the short-circuit is as though V_q were applied to the circuit as a step function at time zero, so transforming to the s domain

$$\begin{bmatrix} 0 \\ V_{q0}/s \end{bmatrix} = \begin{bmatrix} L''_{2d}s & \omega L''_{2q} \\ -\omega L''_{2d} & L''_{2q}s \end{bmatrix} \cdot \begin{bmatrix} i_d(s) \\ i_q(s) \end{bmatrix} \tag{20.51}$$

and inverting

$$\begin{bmatrix} i_d(s) \\ i_q(s) \end{bmatrix} = \frac{1}{L''_{2d}L''_{2q}(s^2+\omega^2)} \begin{bmatrix} L''_{2q}s & -\omega L''_{2q} \\ \omega L''_{2d} & L''_{2d}s \end{bmatrix} \cdot \begin{bmatrix} 0 \\ V_{q0}/s \end{bmatrix} \tag{20.52}$$

hence

$$i_d(s) = \frac{V_{q0}}{\omega L''_{2d}} \frac{-\omega^2}{s(s^2+\omega^2)} \tag{20.53}$$

and

$$i_q(s) = \frac{V_{q0}}{\omega L''_{2q}} \frac{\omega}{(s^2+\omega^2)} \tag{20.54}$$

From the table of transform pairs lines 18 and 16 we have

$$i_d(t) = \frac{V_{q0}}{\omega L''_{2d}} (1 - \cos \omega t) \tag{20.55}$$

and

$$i_q(t) = \frac{V_{q0}}{\omega L''_{2q}} \sin \omega t \tag{20.56}$$

The next step is to transform i_d and i_q to rotating axes and so to determine the winding currents. It is sufficient if i_a alone is found.

The transform is

$$i_a = i_d \cos \theta + i_q \sin \theta \tag{16.69}$$

where

$$\theta = (\omega t + \lambda) \tag{20.57}$$

the angle λ being determined by the instant in the cycle when switching takes place. Thus

$$
\begin{aligned}
i_a &= i_d \cos (\omega t + \lambda) + i_q \sin (\omega t + \lambda) \\
&= -V_{q0}[(1/\omega L''_{2d}) \cos (\omega t + \lambda) - (1/\omega L''_{2d}) \cos \omega t \cos (\omega t + \lambda) \\
&\qquad\qquad\qquad\qquad\qquad - (1/\omega L''_{2q}) \sin \omega t \sin (\omega t + \lambda)]
\end{aligned} \tag{20.58}
$$

$$
\begin{aligned}
i_a\dagger = -i_a = V_{q0} \Bigg\{ &\frac{1}{\omega L''_{2d}} \cos (\omega t + \lambda) - \frac{1}{2} \left(\frac{1}{\omega L''_{2d}} + \frac{1}{\omega L''_{2q}} \right) \cos \lambda \\
&- \frac{1}{2} \left(\frac{1}{\omega L''_{2d}} - \frac{1}{\omega L''_{2q}} \right) \cos (2\omega t + \lambda) \Bigg\}
\end{aligned} \tag{20.59}
$$

The sign of $i_a\dagger$ is that usually accepted for a generator output current.

As expected this expression does not include any damping since resistances have been neglected. The first term is the most important—the fundamental frequency term. Its magnitude

$$
I'' = V_{q0}/\omega L''_{2d} \tag{20.60}
$$

corresponds with the initial value defined by Eq. (20.21). Thus the direct axis sub-transient reactance

$$
x''_d = \omega L''_{2d} \tag{20.61}
$$

If the machine had no damper windings L''_{2d} would revert to L'_{2d} and I'' would become I'.

Hence the direct-axis transient reactance

$$
x'_d = \omega L'_{2d} \tag{20.62}
$$

Similarly $\omega L''_{2q}$ is known as x''_q, the quadrature-axis sub-transient reactance.

The second term in Eq. (20.59) is a d.c. component dependent on $\cos \lambda$ and the mean of $1/x''_d$ and $1/x''_q$. In the complete solution it appears as the d.c. transient. Its time constant will be determined in the next section. Its presence and magnitude clearly depends on the instant when the short circuit takes place.

The final term is a second harmonic introduced by the saliency and is zero for a uniform air gap machine where L''_{2d} and L''_{2q} are equal.

20.3.7 Complete solution for the line currents

We are now in a position to attempt a solution of Eq. (20.30). If

$$
\begin{bmatrix} 0 \\ V_{q0} \end{bmatrix} = \begin{bmatrix} R_2 + L''_{2d}(p)p & \omega L''_{2q}(p) \\ -\omega L''_{2d}(p) & R_2 + L''_{2q}(p)p \end{bmatrix} \cdot \begin{bmatrix} i_d(t) \\ i_q(t) \end{bmatrix} \tag{20.30}
$$

we follow the procedure of the previous section, transforming to the s domain, putting $V_{q0}(s)$ equal to V_{q0}/s, and inverting

$$
\begin{bmatrix} i_d(s) \\ i_q(s) \end{bmatrix} = \frac{V_{q0}}{s} \frac{1}{\Delta} \begin{bmatrix} -\omega L''_{2q}(s) \\ R_2 + L''_{2d}(s)s \end{bmatrix} \tag{20.63}
$$

where

$$\Delta = R_2{}^2 + sR_2[L''_{2d}(s) + L''_{2q}(s)] + L''_{2d}(s)L''_{2q}(s)(s^2 + \omega^2)$$

and neglecting $R_2{}^2$

$$\frac{1}{\Delta} = \frac{1/L''_{2d}(s)L''_{2q}(s)}{s^2 + 2\alpha s + \omega^2} \qquad (20.64)$$

where

$$2\alpha = R_2[\{1/L''_{2d}(s)\} + \{1/L''_{2q}(s)\}]$$

Strictly, α is a function of (s), but it is sufficiently accurate to use the numerical values of L''_{2d} and L''_{2q} within the bracket, so that

$$2\alpha = R_2[(1/L''_{2d}) + (1/L''_{2q})] \qquad (20.65)$$

and the denominator of equation (20.64) is a quadratic in s.*

Substituting for Δ in (20.63) and again neglecting R_2

$$\begin{bmatrix} i_d(s) \\ i_q(s) \end{bmatrix} = \frac{V_{q0}}{s} \frac{1}{s^2 + 2\alpha s + \omega^2} \begin{bmatrix} -\omega/L''_{2d}(s) \\ s/L''_{2q}(s) \end{bmatrix} \qquad (20.66)$$

These expressions must now be converted to partial fractions in order that they may be recognised as transform pairs.

Beginning with the equation of $i_d(s)$ and substituting for $L''_{2d}(s)$

$$i_d(s) = \frac{-V_{q0}}{\omega L_{2d}} \cdot \frac{1}{s} \cdot \frac{\omega^2}{s^2 + 2\alpha s + \omega^2} \cdot \frac{1 + T_1 s}{1 + c_{21} T_1 s} \cdot \frac{1 + c_{31} T_3 s}{1 + c'_{31} T_3 s} \qquad (20.67)$$

Turning into partial fractions by the 'cover-up' rule,

$$i_d(s) = \frac{-V_{q0}}{\omega L_{2d}} \left(\frac{A}{s} + \frac{B}{1 + c_{21} T_1 s} + \frac{C}{1 + c'_{31} T_3 s} + \frac{D}{s + a_1} + \frac{E}{s + a_2} \right) \qquad (20.68)$$

where a_1 and a_2 are the complex conjugate roots of the binomial factor $s^2 + 2\alpha s + \omega^2$ and

$$A = \omega^2/a_1 a_2 = 1$$

$$B = \frac{-\omega^2(1 - 1/c_{21})(1 - c_{31}T_3/c_{21}T_1)}{(1/c_{21}T_1)(1 - c'_{31}T_3/c_{21}T_1)(1/c_{21}{}^2T_1{}^2 - 2\alpha/c_{21}T_1 + \omega^2)}$$

$$\simeq c_{21}T_1(1 - 1/c_{21})$$

$$C = \frac{-\omega^2(1 - T_1/c'_{31}T_3)(1 - c_{31}/c'_{31})}{(1/c'_{31}T_3)(1 - c_{21}T_1/c'_{31}T_3)(1/c'_{31}{}^2T_3{}^2 - 2\alpha/c'_{31}T_3 + \omega^2)}$$

$$\simeq c'_{31}T_3(1/c_{21} - c_{31}/c_{21}c'_{31})$$

$$D = \frac{-\omega^2(1 - T_1 a_1)(1 - c_{31}T_3 a_1)}{a_1(1 - c_{21}T_1 a_1)(1 - c'_{31}T_3 a_1)(a_2 - a_1)}$$

$$\simeq \frac{c_{31}}{c_{21}c'_{31}} \cdot \frac{\omega^2}{a_1(a_1 - a_2)}$$

* Retaining $L''_{2d}(s)$ and $L''_{2q}(s)$ in the expression for a would introduce additional factors in the denominator but these correspond to highly damped transients of small magnitude and can be ignored.

$$E = \frac{-\omega^2(1-T_1a_2)(1-c_{31}T_3a_2)}{a_2(1-c_{21}T_1a_2)(1-c'_{31}T_3a_2)(a_1-a_2)}$$

$$\simeq \frac{c_{31}}{c_{21}c'_{31}} \cdot \frac{-\omega^2}{a_2(a_1-a_2)}$$

These approximations are justified since

$$\omega \gg 1$$

and also

$$\omega \gg \alpha$$

and

$$T_1/T_3 \gg 1$$

thus

$$a_1 = \alpha + j\omega$$

and

$$a_2 = \alpha - j\omega$$

Combining the last two fractions

$$\frac{D}{s+a_1} + \frac{E}{s+a_2} \simeq \frac{c_{31}}{c_{21}c'_{31}} \cdot \frac{-s}{s^2+2\alpha s+\omega^2}$$

Thus

$$i_d(s) = -\frac{V_{q0}}{\omega L_{2d}}\left(\frac{1}{s} - (1-1/c_{21})\frac{c_{21}T_1}{1+c_{21}T_1s} - (1/c_{21}-c_{31}/c_{21}c'_{31})\frac{c'_{31}T_3}{1+c'_{31}T_3s}\right)$$

$$-\frac{c_{31}}{c_{21}c'_{31}}\frac{s}{s^2+2\alpha s+\omega^2} \qquad (20.69)$$

and from the table of Laplace transform pairs

$$i_d(t) = \frac{-V_{q0}}{\omega L_{2d}}[1-(1-1/c_{21})\exp(-t/c_{21}T_1)$$

$$-(1/c_{21}-c_{31}/c_{21}c'_{31})\exp(-t/c'_{31}T_3)$$

$$-(c_{31}/c_{21}c'_{31})\exp(-\alpha t)\cos \omega t]$$

$$= -V_{q0}\left\{\frac{1}{x_d} + \left(\frac{1}{x'_d} - \frac{1}{x_d}\right)\exp(-t/T') + \left(\frac{1}{x''_d} - \frac{1}{x'_d}\right)\exp(-t/T'')\right.$$

$$\left. -\frac{1}{x''_d}\exp(-t/T_a)\cos \omega t\right\} \qquad (20.70)$$

where

$$T' = c_{21}T_1$$

$$T'' = c'_{31}T_3$$

$$T_a = 1/\alpha = \frac{2L''_{2d}L''_{2q}}{R_2(L''_{2d}+L''_{2q})} \qquad \text{from} \quad (20.65)$$

$$x_d = \omega L_{2d}$$

$$x'_d = \omega c_{21} L_{2d}$$

and

$$x''_d = \omega \frac{c_{21} c'_{31}}{c_{31}} L_{2d}$$

The expression for i_q is obtained in a similar manner. From Eq. (20.66)

$$i_q(s) = \frac{V_{q0}}{\omega L''_{2q}} \frac{\omega s}{s^2 + 2\alpha s + \omega^2} \frac{1 + c_{32q} T_{3q}}{1 + T_{3q}} \tag{20.71}$$

giving after resolving into partial fractions and transforming

$$i_q(t) = \frac{V_{q0}}{x''_q} \exp\left(-t/T_a\right) \sin \omega t \tag{20.72}$$

Finally, as in the previous section, i_d and i_q are transformed to rotating axes to determine the line current i_a and the corresponding generated current $i_a\dagger$. Hence

$$i_a = i_d \cos(\omega t + \lambda) + i_q \sin(\omega t + \lambda) \tag{17.26}$$

$$i_a\dagger = -i_a$$

$$= \frac{V_{q0}}{x_d} \cos(\omega t + \lambda) + V_{q0}\left(\frac{1}{x'_d} - \frac{1}{x_d}\right) \exp\left(-t/T'\right) \cos(\omega t + \lambda)$$

$$+ V_{q0}\left(\frac{1}{x''_d} - \frac{1}{x'_d}\right) \exp\left(-t/T''\right) \cos(\omega t + \lambda)$$

$$- \frac{V_{q0}}{2}\left(\frac{1}{x''_d} + \frac{1}{x''_q}\right) \exp\left(-t/T_a\right) \cos \lambda$$

$$- \frac{V_{q0}}{2}\left(\frac{1}{x''_d} - \frac{1}{x''_q}\right) \exp\left(-t/T_a\right) \cos(2\omega t - \lambda) \tag{20.73}$$

This equation should be compared with the forecast Eq. (20.18).

The following table gives a summary of the terms of which the line current is composed.

COMPONENT	INITIAL VALUE	TIME CONSTANT
Steady-state fundamental frequency	V_{q0}/x_d	—
Main transient fundamental frequency	$V_{q0}(1/x'_d - 1/x_d)$	T'
Sub transient fundamental frequency	$V_{q0}(1/x''_d - 1/x'_d)$	T''
D.C. transient	$\frac{V_{q0}}{2}(1/x''_d + 1/x''_q) \cos \lambda$	T_a
Second harmonic	$\frac{V_{q0}}{2}(1/x''_d - 1/x''_q)$	T_a

Table continued overleaf

where $T' = c_{21}T_1$
$T'' = c'_{31}T_3$
$T_a = 2L''_{2d}L''_{2q}/R_2(L''_{2d}+L''_{2q})$
$x_d = \omega L_{2d}$
$x'_d = \omega L'_{2d} = \omega L_{2d}c_{21}$
$x''_d = \omega L''_{2d} = \omega L_{2d}(c_{21}-c_{M1}{}^2k_{23}{}^2/c_{31})$
$x''_q = \omega L''_{2q} = \omega L_{2q}c_{23q}$
$V_{q0} = \omega M_d V_t/R_1$

$c_{21} = (1-M_d{}^2/L_{2d}\,L_1)$
$c_{31} = (1-M_d{}^2/L_{3d}\,L_1)$
$c_{M1} = (1-M_d/L_1)$
$c_{23q} = (1-M_q{}^2/L_{2q}\,L_{3q})$
$k_{23}{}^2 = M_d{}^2/L_{2d}\,L_{3d}$

20.3.8 Solution for the field current

To obtain the transient components of the field current we return to the reduced impedance matrix (Eq. (20.27)). From the third line

$$0 = M'_d(p)pi_d + [R_{3d}+L'_{3d}(p)p]i_{3d}$$

thus

$$i_{3d} = i_d[-M'_d(p)p]/[R_{3d}+L'_{3d}(p)p] \qquad (20.74)$$

Substituting in the first line of the complete impedance matrix, Eq. (20.24)

$$V_f = (R_1+L_1p)i_f + M_dp(i_d+i_{3d}) \qquad (20.75)$$

and so

$$i_f(R_1+L_1p) = V_f - M_dp[1 - M'_d(p)p/(R_{3d}+L'_{3d}(p)p]i_d$$

$$= V_f - M_dp\,\frac{R_{3d}+(L_{3d}-M_d)p}{R_{3d}+L'_{3d}(p)p}\,i_d \qquad (20.76)$$

since from (20.28)

$$L'_{3d}(p) - M'_d(p) = L_{3d} - M_d. \qquad (20.77)$$

If R_1 is ignored in $L'_{3d}(p)$ we have

$$i_f(R_1+L_1p) = V_f - M_dp\,\frac{1+T_{3dl}p}{1+c_{31}T_{3d}p}\,i_d \qquad (20.78)$$

where

$$T_{3dl} = (L_{3d}-M_d)/R_{3d} \qquad (20.79)$$

(T_{3dl} is the time constant of the damper winding due to its resistance and leakage inductance only).

Under the conditions of the short-circuit, transforming to the s domain and substituting for $i_d(s)$ in Eq. (20.78) using (20.67)

$i_f(s)(R_1+L_1s)$

$$= V_{f0}+M_ds\,\frac{1+T_{3dl}s}{1+c_{31}T_{3d}s}\cdot\frac{\omega M_d V_{f0}}{R_1\omega L_{2d}}\cdot\frac{1}{s}\cdot\frac{\omega^2}{s^2+2\alpha s+\omega^2}\cdot\frac{1+T_1s}{1+T's}\cdot\frac{1+c_{31}T_{3d}s}{1+T''s} \qquad (20.80)$$

giving

$$i_f(s) = \frac{V_{f0}}{R_1}+\frac{V_{f0}\omega M_d{}^2}{R_1{}^2\omega L_{2d}}\cdot\frac{1+T_{3dl}s}{(1+T's)(1+T''s)}\cdot\frac{\omega^2}{s^2+2\alpha s+\omega^2} \qquad (20.81)$$

Converting to partial fractions, making the same assumptions as for i_d and substituting for $M_d{}^2/L^2_d$,

$$i_f(s) = \frac{V_{f0}}{R_1} + \frac{V_{f0}(1-c_{21})T'}{R_1} \frac{}{c_{21}} \left(\frac{A}{1+T's} + \frac{B}{1+T''s} + \frac{C}{s+a_1} + \frac{D}{s+a_2} \right) \qquad (20.82)$$

where

$$A = \frac{(1-T_{3dl}/T')\omega^2}{(1-T''/T')(1/T'^2 - 2\alpha/T' + \omega^2)} \simeq \frac{1-T_{3dl}/T'}{1-T''/T'}$$

$$B = \frac{(1-T_{3dl}/T'')\omega^2}{(1-T'/T'')(1/T''^2 - 2\alpha/T'' + \omega^2)} \simeq -\frac{T''}{T'} \frac{1-T_{3dl}/T''}{1-T''/T'}$$

$$C = \frac{(1-a_1 T_{3dl})\omega^2}{(1-a_1 T')(1-a_1 T'')(a_2-a_1)} \simeq \frac{T_{3dl}}{T'T''} \frac{\omega^2}{a_1(a_1-a_2)}$$

$$D = \frac{(1-a_2 T_{3dl})\omega^2}{(1-a_2 T')(1-a_2 T'')(a_1-a_2)} \simeq \frac{T_{3dl}}{T'T''} \frac{-\omega^2}{a_2(a_1-a_2)}$$

and combining the last two fractions

$$\frac{C}{s+a_1} + \frac{D}{s+a_2} = \frac{T_{3dl}}{T'T''} \frac{-s}{s^2 + 2\alpha s + \omega^2}$$

Thus

$$i_f(s) = \frac{V_{f0}}{R_1} + \frac{V_{f0}}{R_1} \frac{1-c_{21}}{c_{21}} \cdot$$

$$\left(\frac{1-T_{3dl}/T'}{1-T''/T'} \frac{T'}{1+T's} - \frac{1-T_{3dl}/T''}{1-T''/T'} \frac{T''}{1+T''s} - \frac{T_{3dl}}{T''} \frac{s}{s^2 + 2\alpha s + \omega^2} \right) \qquad (20.83)$$

which further simplifies to

$$i_f(s) = \frac{V_{f0}}{R_1} + \frac{V_{f0}}{R_1} \frac{1-c_{21}}{c_{21}} \cdot$$

$$\left(\frac{T'}{1+T's} - (1-T_{3dl}/T'') \frac{T''}{1+T''s} - \frac{T_{3dl}}{T''} \frac{s}{s^2 + 2\alpha s + \omega^2} \right)$$

if T' is much greater than T'' and T_{3dl}.

Transforming to the time domain, we obtain the final expression for the field current

$$i_f(t) = \frac{V_{f0}}{R_1} + \frac{V_{f0}}{R_1} \frac{1-c_{21}}{c_{21}} \cdot$$

$$(\exp(-t/T') - (1-T_{3dl}/T'') \exp(-t/T'') - (T_{3dl}/T'') \exp(-t/T_a) \cos \omega t)$$

The various components of the field current are summarised in the following table which may be seen to correspond to the forecast made in Sec. 20.3.2.

COMPONENT	INITIAL VALUE	TIME CONSTANT
Steady-state d.c. value	V_f/R_1	—
D.C. main transient	$(V_f/R_1)[(1/c_{21})-1]$	T'
D.C. sub-transient	$(V_f/R_1[(1/c_{21})-1](1-T_{3dl}/T'')$	T''
A.C. transient	$(V_f/R_1)[(1/c_{21})-1](T_{3dl}/T'')$	T_a

CHAPTER 21

Magnetic circuits and flux plots

21.1 Introduction

A study of electrical circuits such as we have undertaken in the previous chapters cannot be regarded as sufficient until more time has been devoted to the relationship between the lumped parameters R, L and C and the physical size and configuration of the quantities they represent. If one is not sufficiently perceptive it can be easily assumed that proficiency in the manipulation of the many equations of electrical circuit theory is all that is required of the electrical engineer. The very elegance of circuit analysis tends to obscure the fact that the parameters used in the various formulae are derived from actual physical systems and these factors are related to the size, shape and physical dimensions of real practical structures.

Mention has been made elsewhere of the analogies between conduction, magnetic and electric fields.* We now look into these analogues more deeply and determine their connection with circuits. We shall see that simple electric circuit theory is derived from conduction field theory with the arbitrary addition of lumped elements representing L and C. In a similar manner 'circuits' can be set up as approximations to complicated magnetic and dielectric field systems. Particularly the representation of magnetic fields in this way has important applications and these applications are worth more than a passing reference. Finally it will be shown that a composite circuit can be derived to determine the overall performance of a complex magnetic field and its energising electrical system.

21.2 Laplace's equation

Faraday was the first engineer to point out the immense significance of the similarity which exists between electric and magnetic field patterns and the

---

* *Electrical Machines*, Sec. 15.2.8.

corresponding diagrams of streamline flow in incompressible liquids. He developed these analogues and was the first to use the terms electric and magnetic flux. The reason why these analogues are possible is because all three systems have a common mathematical basis in Laplace's equation.

The fundamental condition for electric current flow in a conductor is expressed by the statement that at any point in the conductor the current density is proportional to the voltage gradient.

$$J = E/\rho \qquad (21.1)$$

or
$$J = \sigma E \qquad (21.2)$$

where
$$J = \text{current density}$$
$$E = \text{voltage gradient}$$
$$\rho = \text{resistivity}$$
$$\sigma = \text{conductivity}$$

Considering first the flow at a point **P** in a wire where the current density is known to be uniform and thinking of a small rectangular cell at the point **P**, all the current entering the front face of the prism leaves at the rear face

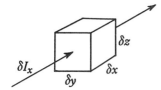

The dimensions of the cell are δx, δy and δz and the current direction is assumed to be along the x axis.

FIG. 21.1 Elementary cell at point P in a conductor with flow in the x axis only

The current density

$$J_x = \frac{\delta I_x}{\delta y \delta z} = \frac{\delta V/\delta x}{\rho} \qquad (21.3)$$

and since J is known to be constant

$$\delta V/\delta x \text{ must also be constant}$$

hence
$$\delta^2 V/\delta x^2 = 0 \qquad (21.4)$$

21.2.1 General case

In an irregularly shaped conductor the current density will not be constant at all points through the conductor. Again considering an elementary cell at a point **P** within the conductor (Fig. 21.2), if current flows through the elementary block obliquely then the components of current entering the faces normally will be δI_x, δI_y and δI_z. If there is a change of current density the current leaving will be $\delta' I_x$, $\delta' I_y$ and $\delta' I_z$. The change in current density will be

$$\delta J_x = (\delta' I_x - \delta I_x)/\delta y \delta z$$
$$\delta J_y = (\delta' I_y - \delta I_y)/\delta z \delta x$$
$$\delta J_z = (\delta' I_z - \delta I_z)/\delta x \delta y$$

and the space rates of change of flux density

$$\delta J_x/\delta x = (\delta'I_x - \delta I_x)/\delta x \delta y \delta z$$
$$\delta J_y/\delta y = (\delta'I_y - \delta I_y)/\delta x \delta y \delta z$$
$$\delta J_z/\delta z = (\delta'I_z - \delta I_z)/\delta x \delta y \delta z$$

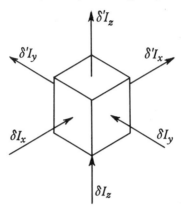

FIG. 21.2 Elementary cell at point P where flow is inclined and divergent

Now unless current is generated within the block it is obvious that

$$\delta I_x + \delta I_y + \delta I_z = \delta'I_x + \delta'I_y + \delta'I_z$$

or

$$(\delta'I_x - \delta I_x) + (\delta'I_y - \delta I_y) + (\delta'I_y - \delta I_y) = 0$$

and hence

$$\frac{\delta J_x}{\delta x} + \frac{\delta J_y}{\delta y} + \frac{\delta J_z}{\delta z} = 0 \qquad (21.5)$$

But

$$\rho J_x = \delta V_x/\delta x$$

hence

$$\rho \frac{\delta J_x}{\delta x} = \frac{\delta^2 V_x}{\delta x^2}$$

and similarly

$$\rho \frac{\delta J_y}{\delta y} = \frac{\delta^2 V_y}{\delta y^2}$$

$$\rho \frac{\delta J_z}{\delta z} = \frac{\delta^2 V_z}{\delta z^2}$$

Substituting in Eq. (22.5) we have

$$\frac{\delta^2 V_x}{\delta x^2} + \frac{\delta^2 V_y}{\delta y^2} + \frac{\delta^2 V_z}{\delta z^2} = 0 \qquad (21.6)$$

which is known as Laplace's Equation

21.2.2 Flow in sheets

When there is no divergence in the z axis, that is when the thickness of the conductor is uniform, there can be no flow in the z axis and

$$\delta^2 V_z/\delta z^2 = 0$$

Laplace's equation for sheet material is reduced to

$$\frac{\delta^2 V}{\delta x^2} + \frac{\delta^2 V}{\delta y^2} = 0 \tag{21.7}$$

The solution of this equation shows that (a) flow lines and equipotential lines drawn on the surface are always at right angles and (b) if flow lines and equipotentials are drawn at uniform intervals, that is if the same value of current is enclosed between all pairs of flow lines and the same numerical value also applies to the potential difference between adjacent equipotential lines, the resulting pattern will consist of a mosaic of 'curvilinear squares'.*

21.3 Principles of flux plotting

If the electrodes and the non-conducting boundaries of a sheet conductor are defined, trial and error methods of sketching the flow lines and equipotentials can be applied since curvilinear squares can easily be recognised.

Principles

1. Lines of symmetry are always equipotential or flow lines.

2. Lines always enter a boundary at right angles.

3. Curvilinear squares can always be subdivided into four smaller squares each of which is more nearly a true square.

4. A curvilinear square is recognised by dividing it into four such smaller squares by drawing the mid-flow and mid-potential lines. For a true square (a) these two mid-dimensions will be of the same length and (b) all the angles will be right angles.

5. The total resistance of sheet conductor is given by

$$R = (\text{Resistance of one square}) \frac{\text{number in series}}{\text{number in parallel}} \tag{21.8}$$

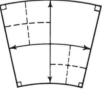

FIG. 21.3 A curvilinear square showing how it can be divided into smaller squares

The resistance of one square $= \rho/t$

where ρ = resistivity, t = thickness of the sheet.

By the application of these rules, together with some slight sketching ability and a little practice, approximate flux plots can be drawn which enable calcu-

* See also p. 327 *et seq.*, *Electrical Machines*.

lations of conductance or resistance to be made with sufficient accuracy for use in subsequent design calculations.

21.4 Electric and magnetic fields

The basic laws of all types of Laplacian field, conduction, electric, magnetic and thermal, depend on the constant relationship between flow density and

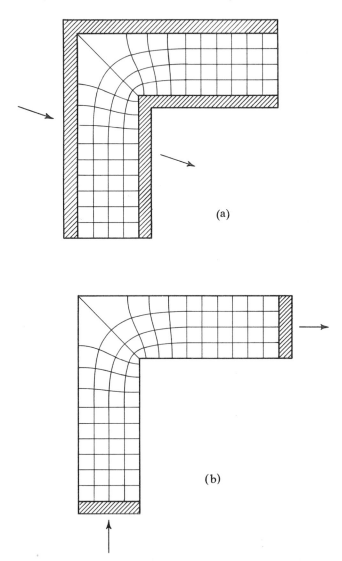

FIG. 21.4 A two-dimensional field system and its inverse

potential gradient (Eq. (2.1)) for the conduction field).

For a conduction field	$J = \sigma E$	(21.9)
For an electric field	$D = \mathrm{d}Q/\mathrm{d}A = \epsilon\epsilon_0 E$	(21.10)
For a magnetic field	$B = \mathrm{d}\phi/\mathrm{d}A = \mu\mu_0 H$	(21.11)

These basic equations are similar in form (flux density proportional to potential gradient), hence the Laplace equation applies to all the three systems and each system can be used as an analogue for the others. One significant difference between electric and magnetic fields remains to be considered in the next section.

21.4.1 Inversion

It will be noted that if the conducting and non-conducting boundaries of a model are interchanged, the curvilinear pattern of the plot remains the same but the flow lines in the first system become the equipotentials of the second and vice-versa. The second system is said to be the inverse of the first. Figure 21.4 shows two such dual fields. In (a) the flow is from left to right in the diagram and in (b) the flow is upwards.

21.5 Conduction, electric and magnetic fields compared

21.5.1 Conduction field

Figure 21.5(a) shows the flux plot of the flow from a central conductor to an outer sheath. It corresponds to the leakage current in the insulation of a single-core cable. The flow lines are radial and the equipotentials are circular. For the plot shown (assuming unit depth)

Resistance
$$R = \frac{V}{I} = \rho\frac{4}{16}$$

or Conductance
$$G = \frac{I}{V} = \sigma\frac{16}{4}$$

where I is the current flowing between the electrodes and V is the potential difference between them.

21.5.2 Electric field

Faraday suggested that the same pattern could be used for the electric field in the dielectric between two concentric conductors. He used the term electric flux or lines of force for the flow lines and retained the term equipotential for the orthogonal lines.

In Fig. 21.5(b), $+Q$ is the charge on the inner conductor (per unit depth) and $-Q$ the charge on the outer one (that is to say the charge Q has been transferred from outer to inner to create the field) when V is the voltage between electrodes.

The parameter C, the capacitance of the system, corresponds to the conductance G in the system (a) and

$$C = \frac{Q}{V} = \epsilon\epsilon_0 \frac{16}{4}$$

21.5.3 Magnetic field—direct analogue

Faraday also suggested that magnetic field could be represented in terms of lines of force or magnetic flux which he suggested 'flowed' from **N** to **S** of a magnet system.

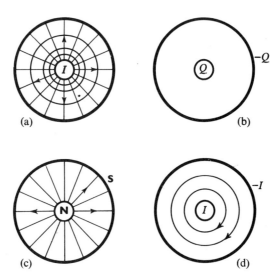

(a)

(b)

(c)

(d)

FIG. 21.5 Flux plots of a cylindrical coaxial system (a) leakage current between core and sheath, (b) capacitance between core and sheath, (c) magnetic field in a radial gap, (d) magnetic field due to coaxial conductors

Thus the same flow diagram we have been considering (Fig. 21.5(c)), corresponds to the radial field in the gap of the magnetic system for a moving-coil speaker with a central **N** pole and concentric **S** pole. Faraday did not at first attach importance to the equipotentials.

In this system the 'conductance' parameter is Λ the permeance of the field and

$$\Lambda = \frac{\phi}{\int H . dx} = \frac{\phi}{I} = \mu\mu_0 \frac{\text{squares in parallel}}{\text{squares in series}} \tag{21.12}$$

$= \mu\mu_0 \, 16/4$ for the example.

When the direct analogue is used for a magnetic field:

Current flow lines   correspond to magnetic flux lines

Conduction field equipotentials   correspond to magnetic equipotentials

Conductance	corresponds to magnetic permeance
Resistance	corresponds to magnetic reluctance
Figure 21.5(a)	corresponds to Fig. 21.5(c)
A conducting boundary	corresponds to an infinitely permeable iron boundary.

21.5.4 Magnetic field—inverted analogue

The same pattern of field (Fig. 21.5(d)) can be used to represent the magnetic field of a concentric cable system in which the current in the central conductor is I returning in the outer sheath, but in this case we know that the magnetic flux lines are the concentric lines encircling the central conductor.

Thus if we are to assume this analogue, we are representing the magnetic flux lines by the equipotentials of the original conduction field and in the same way we are taking the original flow lines to represent the new magnetic equipotentials. In other words we are using for our analogue the inverse of the previous conduction field.

The permeance of this magnetic field (Fig. 21.5(d)) will be the reciprocal of that of the previous magnetic field (Fig. 21.5(c)) and for the example chosen is therefore $\mu\mu_0$ 4/16.

Although the direct magnetic field analogue has its uses—indeed it is the reason why magnetic flux is so called—it is perhaps unfortunate that too much attention is given to it. Exactly the same information can be obtained from the inverted analogue and later sections will show that this analogue enables magnetic fields to be more readily mapped, particularly those fields due to current carrying conductors.

The direct analogue is possibly more suitable for fields due to permanent magnets.

When the inverted (or dual) magnetic analogue is used:

Conduction field equipotentials	correspond to magnetic flux lines
Current flow lines	correspond to magnetic equipotentials
Resistance	corresponds to magnetic permeance
Conductance	corresponds to magnetic reluctance
Fig. 21.5(a)	corresponds to Fig. 21.5(d)
A non-conducting boundary	corresponds to an infinitely permeable iron boundary.

21.5.5 Examples

It will be useful to discuss a number of important examples at this stage and develop formulae for calculating the field parameters for these practical cases.

21.5.5.1 *The insulation resistance of a cable.* Figure 21.5(a) can be used to show the leakage current in the homogeneous dielectric of a coaxial cable.

Let the diameter of the conductor be d_1 and that of the sheath d_2. We will assume 1 to be the leakage current per unit length of the conductor. Examination of the flow pattern of the leakage current shows that the current density decreases uniformly with radius, hence at radius r the current density is given by

$$J = \frac{i}{2\pi r \times 1} \tag{21.13}$$

E (the voltage gradient at r) $= \rho . J = \rho(1/2\pi r)$ where ρ is the resistivity of the insulation.

The voltage between conductors

$$V = \int_{\frac{1}{2}d_1}^{\frac{1}{2}d_2} E . dr = \frac{\rho i}{2\pi} \int_{\frac{1}{2}d_1}^{\frac{1}{2}d_2} dr/r$$

$$= \frac{\rho i}{2\pi} \log h \frac{d_2}{d_1} \tag{21.14}$$

The insulation resistance is thus

$$R]_{\text{leakage per unit length}} = \frac{\rho}{2\pi} \log h \frac{d_2}{d_1} \tag{21.15}$$

21.5.5.2 *The capacitance of a coaxial cable.* We can use the results of the previous section to determine the capacitance of the same system.

Since by considering this flow system as a conduction field we have proved

$$R = \frac{\rho}{2\pi} \log h \, (d_2/d_1) \tag{21.16}$$

the corresponding conductance equation will be given by

$$G = 2\pi\sigma \frac{1}{\log h \, (d_2/d_1)} \tag{21.17}$$

Using the capacitance analogue (Sec. 21.5.2).

$$C]_{\text{per unit length}} = \frac{2\pi\epsilon\epsilon_0}{\log h \, (d_2/d_1)} \tag{21.18}$$

21.5.5.3 *The inductance of a coaxial cable.* Since a coaxial cable is a single turn system, the inductance of the loop is equal to the permeance of the magnetic field bounded by the conductor.

From the inverted analogue described in Sec. 21.5.4 (Fig. 21.5(d)) we infer that there is similarity between the permeance of a magnetic system and the resistance of the inverted flow pattern.

Thus by analogy with Eq. (21.15) we are able to infer

$$L]_{\text{per unit length}} = \Lambda = \frac{\mu\mu_0}{2\pi} . \log h \, (d_2/d_1) \tag{21.19}$$

This equation relates only to the magnetic flux in the space between the conductors. To this we must add the effect of internal flux within the conductor itself.

If we assume that the current density within the conductor is uniform we see that within the conductor itself the centre part of the conductor has more magnetic flux linking with it than has the conductor as a whole.

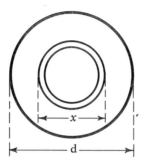

FIG. 21.6 An elementary ring of diameter x within the conductor

Examining an elementary ring within the conductor of diameter x (where $x < d_1$) and thickness dx the permeance of this ring will be

$$d\Lambda = \mu\mu_0 \frac{dx}{2\pi x} \tag{21.20}$$

The current within this ring, which is the cause of flux in the ring and with which this flux links, is a fraction $(x^2/d_1{}^2)$ of the total conductor current. This fraction takes the place of N in Eq. (21.51). N usually stands for 'number of turns' and generally is greater than one. It is a term signifying summation of current and there is no mathematical reason why it should not have a fractional value.

Hence $$dL = N^2 \cdot d\Lambda \tag{21.21}$$

$$= (x^2/d_1{}^2)^2 \mu\mu_0 \frac{dx}{2\pi x}$$

Integrating

$$L = \frac{\mu\mu_0}{2\pi} \int_0^{d_1} \frac{x^3\, dx}{d_1{}^4}$$

$$= \frac{\mu\mu_0}{2\pi} (\tfrac{1}{4}) \tag{21.22}$$

It should be noted that Eq. (21.22) is independent of d.

The total inductance of the cable is derived from Eqs. (21.19) and (21.22).

$$L]_{\text{per unit length}} = \frac{\mu\mu_0}{2\pi} (\tfrac{1}{4} + \text{logh}\,(d_2/d_1)) \tag{21.23}$$

21.5.5.4 *Inductance and capacitance of parallel conductors.* The equations for parallel conductors can be derived from those of the previous section by considering these as particular examples of a two-dimensional flow problem between two small electrodes.

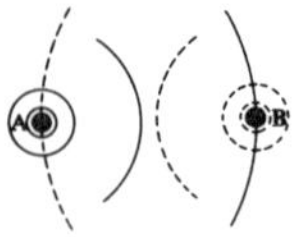

FIG. 21.7 Cross-section of two parallel conductors A and B

We commence by thinking of the flow between two small cylinders **A** and **B** of unit depth in a flux plotting tank, each conductor having a diameter d and with spacing D. If current flow I takes place between **A** and a circumferential electrode of large radius, the equipotential pattern will be a series of concentric circles about **A**. (Fig. 21.7).

The potential difference between **A** and the equipotential passing through **B** will be

$$V_{AB}]_a = I \cdot R = I \cdot \frac{\rho}{2\pi} \log h \frac{D}{d/2} \qquad (21.24)$$

from Eq. (21.15).

Similarly if the flow takes place from the infinite electrode toward **B** the voltage between **A** and **B** will also be

$$V_{AB}]_b = I \cdot \frac{\rho}{2\pi} \log h \frac{D}{d/2} \qquad (21.25)$$

In this way we are led to the observation that if flow takes place between **A** and **B** in a sufficiently large tank, these two conditions occur simultaneously.*

Applying the principle of superposition we obtain

$$V_{AB} = V_{AB}]_a + V_{AB}]_b = I \cdot \frac{\rho}{\pi} \log h \frac{D}{d/2} \qquad (21.26)$$

$$R]_{\text{per unit depth}} = \frac{\rho}{\pi} \log h \frac{D}{d/2} \qquad (21.27)$$

* Strictly, this only applies if the diameter of the conductor is small compared with the spacing. If a flux plot in an electrolytic tank is made for an example in which d is not very small compared with D it will be observed that the equipotentials close to the conductors, though still circular, are no longer concentric. The proximity of the other conductor causes the current density in the tank at the surface of the conductors to vary, being a maximum at the point where they are nearest and a minimum on the other side. It is not correct therefore to assume that the superposition of the two systems of flow to infinity can always be made without any alteration to the boundary condtiions. Moreover, under these conditions the electric and magnetic fields are not true analogues of each other.

The capacitance field follows the conduction analogue and the charge density at the conductor surface follows the current density in the analogue. But in the magnetic field, since the conductor current density in the conductor is constant, the magnetic flux density at the conductor surface must also be constant. Equation (21.28) can therefore be applied for higher values of the ratio d/D than is possible with equation (21.29).

To construct a resistance analogue to correspond exactly with the magnetic field it is necessary to subdivide the conductors into segments and to ensure that the same value of current enters or leaves the bath through each segment by connecting a high resistance between each and the supply.

In sections 21.5.5.5 and 21.5.5.6 as with 21.5.5.4 it must be assumed that the conductor diameter is small compared with the spacing.

Completing the analogues for inductance and capacitance

$$L]_{\text{per unit length}} = \frac{\mu\mu_0}{\pi}\left[\tfrac{1}{4}+\log_h\frac{D}{d/2}\right] \tag{21.28}$$

$$C]_{\text{per unit length}} = \frac{\pi\epsilon\epsilon_0}{\log_h\dfrac{D}{d/2}} \tag{21.29}$$

21.5.5.5 *Inductance and capacitance between a line and an earthed-plane conductor.* The flow pattern between two electrodes discussed in the previous section can be seen to be symmetrical about a plane midway between them. Flow taking place between one of the electrodes and a linear conductor placed on the line of symmetry will give an identical flow pattern and this tank model represents the

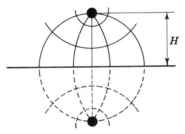

FIG. 21.8 Cross-section of a single conductor and earth plane showing the image conductor

two-dimensional field between an overhead line and the earth plane shown in Fig. 21.8. The resistance of this half section is obviously half that given by Eq. (21.27).

Consequently in the present case, inductance will therefore be half as much and capacitance twice the value for the corresponding parallel lines. If H is the height of the conductor above the earth plane.

$$L]_{\text{per unit length}} = \frac{\mu\mu_0}{2\pi}\left[\tfrac{1}{4}+\log_h\frac{4H}{d}\right] \tag{21.30}$$

$$C]_{\text{per unit length}} = \frac{2\pi\epsilon\epsilon_0}{\log_h\dfrac{4H}{d}} \tag{21.31}$$

21.5.5.6 *Inductance and capacitance of a three-phase overhead-line circuit.* The three lines of an overhead circuit are usually hung with slightly unequal spacing as shown in Fig. 21.9.

Assuming a conductor diameter d and spacings D_{AB}, D_{BC} and D_{CA} respectively, to determine the equivalent inductance of the system we assess the flux links with each individual conductor due to currents, i_A, i_B and i_C in the lines up to a considerable distance R using the method of Sec. 22.5.5.4.

This gives rise to the following equations expressed in matrix form by

$$
\begin{bmatrix} \text{Flux links} \\ \text{with} \\ \text{conductors} \\ \text{up to} \\ \text{distance} \\ R \end{bmatrix}
\begin{bmatrix} \phi_A \\ \phi_B \\ \phi_C \end{bmatrix}
= \frac{\mu\mu_0}{2\pi}
\begin{bmatrix}
\tfrac{1}{4}+\operatorname{logh}\dfrac{R}{d/2} & \operatorname{logh}\dfrac{R}{D_{AB}} & \operatorname{logh}\dfrac{R}{D_{AC}} \\[2mm]
\operatorname{logh}\dfrac{R}{D_{AB}} & \tfrac{1}{4}+\operatorname{logh}\dfrac{R}{d/2} & \operatorname{logh}\dfrac{R}{D_{BC}} \\[2mm]
\operatorname{logh}\dfrac{R}{D_{AC}} & \operatorname{logh}\dfrac{R}{D_{BC}} & \tfrac{1}{4}+\operatorname{logh}\dfrac{R}{d/2}
\end{bmatrix}
\cdot
\begin{bmatrix} i_A \\ i_B \\ i_C \end{bmatrix}
\tag{21.32}
$$

$$
= \frac{\mu\mu_0}{2\pi}\operatorname{logh} R
\begin{bmatrix} i_A+i_B+i_C \\ i_A+i_B+i_C \\ i_A+i_B+i_C \end{bmatrix}
$$

$$
+ \frac{\mu\mu_0}{2\pi}
\begin{bmatrix}
\tfrac{1}{4}+\operatorname{logh}\dfrac{1}{d/2} & \operatorname{logh}\dfrac{1}{D_{AB}} & \operatorname{logh}\dfrac{1}{D_{AC}} \\[2mm]
\operatorname{logh}\dfrac{1}{D_{AB}} & \tfrac{1}{4}+\operatorname{logh}\dfrac{1}{d/2} & \operatorname{logh}\dfrac{1}{D_{BC}} \\[2mm]
\operatorname{logh}\dfrac{1}{D_{AC}} & \operatorname{logh}\dfrac{1}{D_{BC}} & \tfrac{1}{4}+\operatorname{logh}\dfrac{1}{d/2}
\end{bmatrix}
\cdot
\begin{bmatrix} i_A \\ i_B \\ i_C \end{bmatrix}
\tag{21.33}
$$

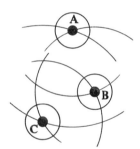

FIG. 21.9 Cross-section of three conductors A, B and C with unequal spacing

Assuming that no zero sequence current is flowing (we have seen in Chap. 13 that zero-sequence currents take a different path)

$$
i_A+i_B+i_C = 0 \tag{21.34}
$$

so that the first matrix in Eq. (21.33) vanishes. The arbitrary value assumed by R is seen to have no significance.

Rate of change of the flux ϕ_A ϕ_B ϕ_C is responsible for the reactance voltages in the lines and so from Eq. (21.33)

$$
\begin{bmatrix} v_{AA'} \\ v_{BB'} \\ v_{CC'} \end{bmatrix}
= p\begin{bmatrix} \phi_A \\ \phi_B \\ \phi_C \end{bmatrix}
= \begin{bmatrix} L_A & M_{AB} & M_{AC} \\ M_{AB} & L_B & M_{BC} \\ M_{AC} & M_{BC} & L_C \end{bmatrix}
p\begin{bmatrix} i_a \\ i_b \\ i_c \end{bmatrix}
\tag{21.35}
$$

where
$$
L_A = L_B = L_C = \frac{\mu\mu_0}{2\pi}\left(\tfrac{1}{4}+\operatorname{logh}\dfrac{1}{d/2}\right)
$$

$$M_{AB} = \frac{\mu\mu_0}{2\pi} \operatorname{logh} \frac{1}{D_{AB}}$$

$$M_{BC} = \frac{\mu\mu_0}{2\pi} \operatorname{logh} \frac{1}{D_{BC}}$$

$$M_{AC} = \frac{\mu\mu_0}{2\pi} \operatorname{logh} \frac{1}{D_{AC}} \tag{21.36}$$

This equation shows the results of the unequal spacings of the conductors and that some unbalance in the system must be expected.

However if we can assume that the lines are transposed at regular intervals the mutual inductance coefficients can be given average values.

If this is the case the effective mutual inductance M is given by

$$M = \tfrac{1}{3}(M_{AB} + M_{BC} + M_{AC})$$

$$= \frac{\mu\mu_0}{2\pi} \cdot \tfrac{1}{3} \left(\operatorname{logh} \frac{1}{D_{AB}} + \operatorname{logh} \frac{1}{D_{BC}} + \operatorname{logh} \frac{1}{D_{AC}} \right)$$

$$= \frac{\mu\mu_0}{2\pi} \operatorname{logh} \frac{1}{\sqrt[3]{(D_{AB} D_{BC} D_{AC})}} \tag{21.37}$$

Finally we remember from Sec. 13.7.2. that if we are confining our attention to positive or negative sequence currents alone (as given in Sec. 13.7.2) the system of self and mutual inductances can be reduced to one containing only equivalent inductances.

$$L_{\text{equivalent}} = L - M \tag{21.38}$$

The equivalent line-to-neutral inductance of the line (or in other words the positive-sequence or negative-sequence inductance of the line) is equal to

$$L]_{\text{per unit length}} = \frac{\mu\mu_0}{2\pi} \left[\tfrac{1}{4} + \operatorname{logh} \frac{\sqrt[3]{(D_{AB} D_{BC} D_{AC})}}{d/2} \right] \tag{21.38}$$

By pursuing the analogue for electric fields we can also write down the value for the equivalent positive-sequence (or negative-sequence) capacitance of the system.

$$C]_{\text{per unit length}} = \frac{2\pi\epsilon\epsilon_0}{\operatorname{logh} \dfrac{\sqrt[3]{(D_{AB} D_{BC} D_{AC})}}{d/2}} \tag{21.39}$$

The term μ has been included in all inductance formulae for completeness but of course for air and most normal dielectric materials

$$\mu = 1$$

21.6 The concept of a magnetic circuit

Pursuing the direct analogue between magnetic flux and current flow, if the elements of a magnetic field are lumped, the field will be reduced to a magnetic circuit. Such circuits can then be analysed using mesh and nodal analysis methods developed for equivalent electric circuits.

If the method is to be satisfactory it can only be applied in conditions where
(a) the elements of the circuits, magneto-motive forces, and reluctances can be lumped without excessive error;
(b) linear relationships (constant reluctances) can be assumed.

Since for a simple magnetic loop

$$F = S\Phi \qquad (21.40)$$

where
$$F = \text{m.m.f.}$$
$$S = \text{reluctance of the loop}$$
$$\Lambda = \text{permeance} = 1/S$$
$$\Phi = \text{flux}$$

if it can be assumed that lumping of m.m.f. and reluctance will not introduce errors, the circuit can be represented by Fig. 21.10 showing an m.m.f. source F connected to a reluctance S. The 'current' in this circuit is the flux Φ.

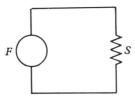

FIG. 21.10 A simple magnetic circuit

In a more complicated system we have a number of m.m.f. sources and reluctance elements forming meshes for which the following mesh equations exist

$$[F] = [S] \, . \, [\Phi] \qquad (21.41)$$

where
$[F]$ are the loop m.m.f.'s
$[\Phi]$ are the mesh fluxes
$[S]$ is a matrix of self and mutual reluctances.

Equation (21.41) is the analogue of the mesh equations of an electrical network

$$[E] = [Z] \, . \, [I] \qquad (21.42)$$

Such a network is known as the direct electrical analogue of the magnetic network. A more complicated magnetic circuit and its direct electrical analogue is shown in Fig. 21.11.

Such simple analogues are helpful in understanding magnetic problems and are used in many text-books for illustrating principles.*

In the direct analogue, e.m.f. corresponds to m.m.f. current to flux and resistance to reluctance.

* *Electrical Machines* (Sec. 5.1).

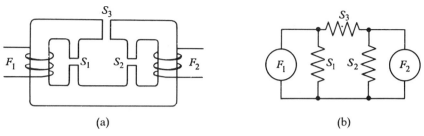

FIG. 21.11 (a) Diagram of a simple magnetic circuit comprising an iron core with three
air gaps and two magnetising coils. (b) The corresponding magnetic circuit

21.6.1 The principle of duality

This principle arises from the topology of networks. It is really geometrical in
character and refers to grids that can be drawn on a closed surface (e.g. a sphere).

If such a network has M loops and N nodes, then a second grid can be drawn
interlaced with the first having N loops and M nodes. This is done by placing a
node in the centre of each loop of the original and interconnecting these nodes
by branches which cross all the corresponding elements of the original network.

The new network is said to be the dual of the original.

In Fig. 21.12 we see how the the loops **A**, **B** and **C** of the original network
(Fig. 21.12(a) and (b)) become the nodes of the dual network **A**, **B** and **C** in
Fig. 21.12(c).

Properties of dual networks were discussed in Chap. 8.

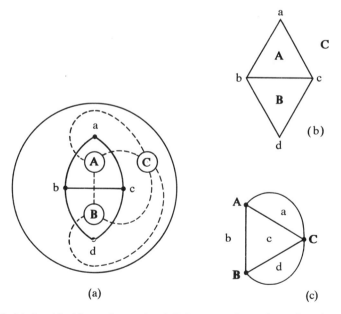

FIG. 21.12 (a) A grid with two loops A and B drawn on the surface of a sphere. (b) The
original network of two loops. (c) The dual network with loops and nodes interchanged

21.6.2 The dual electrical analogue

The dual form of Eq. (21.42) is given by Eq. (21.44).

$$[E] = [Z][I] \tag{21.42}$$

$$[I] = [Y][V] \tag{21.44}$$

These dual equations are related to a dual electrical circuit derived from the original network by the methods of Chap. 8.

The dual circuit corresponding to the original electrical circuit in Fig. 21.11(*b*) is given in Fig. 21.13.

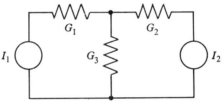

FIG. 21.13 The dual magnetic circuit corresponding to Fig. 21.11b

$$
\begin{array}{rcl}
I_1 & \text{corresponds to} & F_1 \\
I_2 & \text{corresponds to} & F_2 \\
G_1 G_2 G_3 & \text{correspond to} & S_1 S_2 S_3 \\
V_1 V_2 & \text{correspond to} & \Phi_1 \Phi_2
\end{array}
$$

In this circuit the magnitudes of the current sources correspond to the m.m.f.s of the original magnetic circuit, conductances correspond to the reluctances and voltages represent fluxes.

Reconsidering Eq. (21.41)

$$[F] = [S][\Phi] \tag{21.41}$$

where each loop m.m.f. is due to a single current (the magnetising current) flowing in a given number of turns N

$$F = NI \tag{21.45}$$

Also if all the flux Φ links with each of the N turns then

$$V = p\Phi N \tag{21.46}$$

or

$$\Phi = (1/p)(V/N) \tag{21.47}$$

and Eq. (21.41) becomes

$$[IN] = [S][(1/p)(V/N)] \tag{21.48}$$

or

$$[(1/p)(V/N)] = [S]^{-1} \cdot [IN]$$

$$= [\Lambda][IN] \tag{21.49}$$

hence

$$[V] = [L_{pq}]p[I] \tag{21.50}$$

where

$$[L_{pq}] = [\Lambda_{pq} N_p N_q] \tag{21.51}$$

These equations relate to the voltages and currents in the actual magnetising coils of the original magnetic circuit.

Consequently if we take the dual electrical circuit of Fig. 21.13 a stage further we arrive at Fig. 21.14.

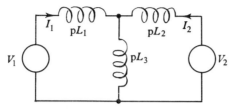

FIG. 21.14 The dual electrical circuit corresponding to Figs. 21.11b and 21.13

In Fig. 21.14 we have an actual electrical circuit of the same configuration as Fig. 21.13 but now we have multiplied the 'resistances' by the appropriate pN_pN_q.

These elements are thus reactances pL and the circuit corresponds to Eqs. (21.50) and not (21.41).

This circuit is therefore an equivalent circuit showing the relationships between the electrical parameters of the magnetising coils and shows how these relationships are affected by the magnetic circuit elements L_1, L_2 and L_3 which are derived from S_1, S_2 and S_3.

The dual electrical analogue is a more powerful tool than the direct analogue since it is an equivalent electrical network which can be directly linked with any external network to which the magnetising coils may be connected.

21.7 Examples of magnetic circuits

In the following examples it is convenient to distinguish between flux paths in air and flux paths in iron.

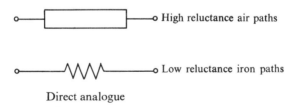

Direct analogue

FIG. 21.15 Symbols used for magnetic branches when the direct analogue is used

The symbols for the high reluctance air paths and the low reluctance iron paths for use in the direct analogy are shown in Fig. 21.15.

It will be remembered that the reluctances of the iron paths are really non-linear in practice, but they are usually small in value relative to the air path reluctances.

For the dual analogue the symbols for the inductances are given in Fig. 21.16.

In these circuits the iron paths are represented by high value non-linear inductances and the air paths by low value constant inductances.

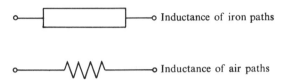

FIG. 21.16 Symbols used for magnetic branches when the dual electric analogue is used

For many purposes the inductances of the iron paths, which form high impedance parallel paths in the dual circuit, can safely be ignored or assumed to be linear.

21.7.1 The 'theoretical' single-phase transformer

A transformer which has its primary and secondary windings wound on different legs is shown in Fig. 21.17(a). This is not a satisfactory way of arranging the windings in practice but many elementary text books show this arrangement since it is easy to recognise the paths of main and leakage flux. Typical flux paths for main, primary leakage and secondary leakage flux are shown as Φ_m, Φ_1 and Φ_2.

The corresponding direct electric analogy is shown in Fig. 21.17(b) and this follows the geometry of the pictorial diagram reasonably closely. The dual circuit, drawn using the principles described in Sec. 21.6.1 is given in Fig. 21.17(c). The current sources represent the windings, the resistors **AB** and **BC** are the result of leakage reactance and the reluctances of the yokes and core contribute the parallel high resistances.

21.7.2 The core-type single-phase transformer

In a core-type transformer the path of the leakage flux is the duct between the windings.* The direct analogue Fig. 21.18(b) shows the result of lumping the elements. Loops **A** and **B** enclose the ampere turns of the primary and the secondary windings respectively. Loop **A** includes the reluctances of the core and loop **B** the yoke, both loops including the reluctance of the air leakage path.

The dual analogue shows the equivalent current sources with a delta system of resistances representing the inductances of the system.

It is interesting to note that the dual system shown in Fig. 21.17(a) for the transformer with the windings on two legs can be converted to that of Fig. 21.18(c) by a star-mesh transformation.

* *Electrical Machines*, Sec. 5.3, p. 110.

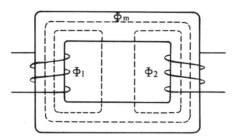

(a) Magnetic circuit

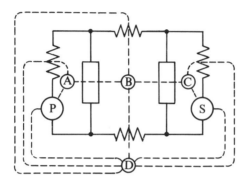

(b) Lumped magnetic circuit

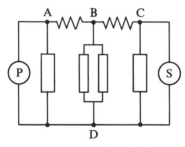

(c) Dual electrical circuit

FIG. 21.17 The magnetic circuit of a simple transformer (a) appearance of the core and coils, (b) lumped magnetic circuit, (c) dual electrical analogue

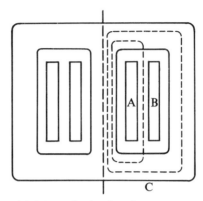

(a) Magnetic circuits of core-type
single-phase transformers

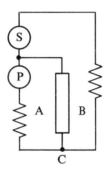

(b) Direct analogue

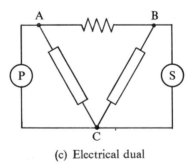

(c) Electrical dual

**FIG. 21.18 The magnetic circuit of a single-phase core-type transformer (a) appearance
of the core and coils, (b) lumped magnetic circuit, (c) dual electrical analogue**

21.7.3 The three-limb core-type transformer

It is assumed that the core-type three-phase transformer can be assembled by taking three separate cores and adding the yokes at the top and the bottom of the diagram.

In the direct analogy, Fig. 21.19(*b*), we have three diagrams similar to Fig. 21.18(*b*) with mesh-connected systems of resistance representing the flux path in the yokes connected at the top and bottom of the diagram.

The dual electric circuit Fig. 21.19(*c*) shows the effective inductances of the system. The diagram is nearly symmetrical and consequently with normal balanced voltages applied to the primary windings, the magnetising currents will be nearly symmetrical. It should be noted that there are three groups of resistors representing inductances p, q and r. The three resistances in each group are nearly alike. The resistors p represent the inductances of the leakage paths between the windings (the normal leakage inductance of a core-type transformer). The resistors q represent inductances corresponding to leakage across from yoke to yoke. Under balanced conditions it will be noted that no current flows in these resistors which form a delta circuit in Fig. 21.19(*c*).

The resistors r correspond to the magnetising inductances of the cores and d to the magnetising inductances of the separate flux paths in the yokes. The magnitudes of these resistances are such that normally

$$p > q$$
$$r \gg p$$
$$d \simeq r \qquad\qquad (21.52)$$

21.7.4 The five-limb three-phase core-type transformer

The addition of the two extra limbs to the core of a three-phase transformer introduce two extra resistors S and S' in the dual equivalent circuit of Fig. 21.20(*c*) which are not present in Fig. 21.19(*c*).

For this circuit Fig. 21.20(*c*)

$$S \simeq d \simeq r \gg p > q \qquad\qquad (21.53)$$

The high value resistors S appear in series with the low value resistors q.

A study of the equivalent circuits shown in Figs. 21.19(*c*) and 21.20(*c*) will enable us to understand the factors on which magnetising current depends for both three-limb and five-limb transformers. We will first consider the effect of positive-sequence voltage (or negative-sequence) at the primary winding terminals. The total value of the voltage sources acting around the delta circuit C_{12}, C_{23}, C_{31} will be zero and consequently only a very small current will flow through the resistors q due to the presence of the leakage reactances d.

As to be expected the current in the primary windings for both types of transformer depends almost entirely on the value of the resistances p representing core reactance, the high impedances S having negligible effect.

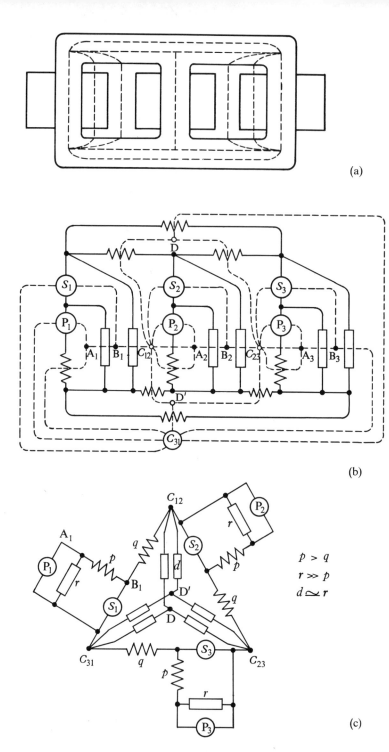

FIG. 21.19 The three-limb core-type transformer (a) appearance of the core and coils, (b) lumped magnetic circuit, (c) dual electrical analogue

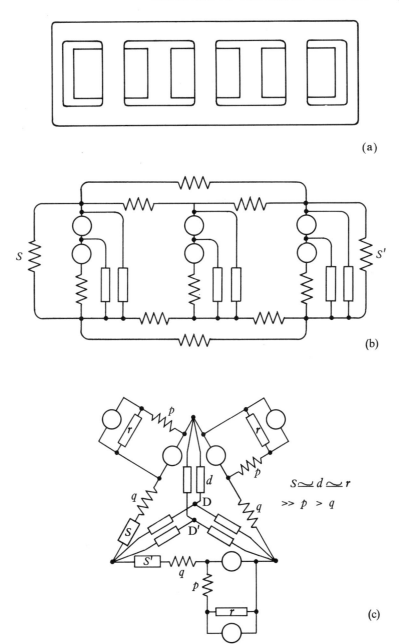

FIG. 21.20 The five-limb core-type transformer (a) appearance of core and coils,
(b) lumped magnetic circuit, (c) dual electrical analogue

The condition differs for zero-sequence voltages. For the three-limb trans-former, when the threee primary voltage sources are equal, there will be no potential difference between points C_{12}, C_{23} and C_{31}, and the resistors q are effectively connected across their respective sources. These yoke leakage reactances are effectively paralleled with the core reactances and being of a much lower magnitude the value of the zero-sequence magnetising current will be increased considerably. For the five-limb transformer the presence of the high impedance S limits the current in the delta path to a much lower figure and so the zero-sequence magnetising current will only be slightly greater than the positive sequence magnetising current.

21.8 Relationship between duality and inversion

The foregoing sections show that there is limited application for the direct analogue between electric and magnetic circuits which is based on the assumption that magnetic flux is the analogue of current flow. The dual equivalent electric circuit, which shows actual currents and voltages in magnetising windings is much more satisfactory.

If we examine the grid of a Laplacian field plot, shown by unbroken lines in Fig. 21.21(a) and think of this diagram as a network of electrical conductors and if we draw the corresponding dual network using the technique established in Sec. 21.6.1 we obtain a second pattern of curvilinear squares shown by broken lines interlaced with the first. This dual network has entirely the same configuration as the original and either (or both) represent the field. It is important to notice, however, that in the process of drawing the dual network the loops of the former become the nodes of the latter and so the equipotential lines and flow lines have been interchanged. This will be seen by examining the portion $aBbA$ of Fig. 21.21(a) shown again in Fig. 21.21(b). In other words the dual network is the inverse of the original plot.

It has been shown (Sec. 21.5.3) that a magnetic field plot corresponds to the inverse of a flow chart and this is the reason why the dual circuit is the true analogue of a magnetic circuit problem.

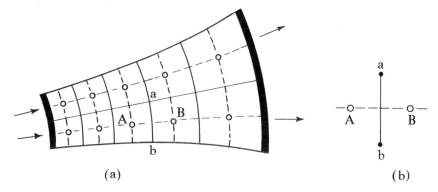

(a)   (b)

FIG. 21.21 (a) A Laplacian field plot. (b) Detail between meshes A and B

21.9 To determine the inductance of a conductor from a flux plot of the associated magnetic field

We have seen in Sec. 21.5 that a flux plot obtained from the flow of current between two electrodes in an electric bath can be used as an analogue of a magnetic field system. It is a useful method of representation but the current in the bath has no direct relationship to the current in the conductor system which is the cause of the magnetic field under investigation except in the following terms.

Consider Fig. 21.22 which is a plot determined in a tank as a result of flow between two circular electrodes some distance apart.

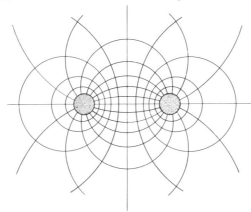

FIG. 21.22 Tank plot of flow between two circular electrodes

If this represents an actual current flow system, the resistance between the electrodes is given by

$$\text{resistance of one cell (of given depth)} \times \frac{\text{number of squares in series}}{\text{number of squares in parallel}} \quad (21.54)$$

$$= R_C \times \frac{10}{16} \text{ for the plot shown in Fig. 21.22.}$$

If on the other hand the system represented is the magnetic field between two parallel conductors we can say that the inductance of the loop

$$= \frac{\text{permeance of one cell}}{\text{(of depth equal to that of the loop)}} \times \frac{\text{number of squares in parallel}}{\text{number of squares in series}} \quad (21.55)$$

$$= L_C \times \frac{10}{16}$$

Note that the fraction is the same in the two cases since magnetic flux lines of the second case are the equipotentials of the first case.

In other words the magnetic analogue uses the inverse of the current flow-diagram with the result that inductance in the magnetic analogue is represented by resistance in the inverted flow-diagram.

It will usually be found straightforward to represent magnetic problems in this way rather than attempting to form a direct analogue in which magnetic flow-lines correspond to actual current flow-lines.

To set up such inverted analogues the following points must be observed.

1. Iron surfaces (assumed to be infinitely permeable) are represented by non-conducting tank walls.
2. A cross section of a magnetising coil is represented by a number of electrodes representing the turns, each one connected to a current source and introducing the same current into the bath.

21.10 Mutual inductances

An example of how a flux plot can be used to determine the value of a mutual inductance is shown in Fig. 21.23. This represents a stator coil **P** and a rotor coil **Q** of an induction motor each consisting of two parallel conductors embedded in slots and with an air gap between them. We shall consider the effect of only the parallel part of the conductors and neglect the end connections. The iron is assumed to be infinitely permeable and each coil to have only one turn.

A tank model to represent the flux in the air gap is shown in Fig. 21.23(b). The flux plot corresponding to equal currents entering the tank at electrodes **P** and **Q** and leaving at a common electrode on the axis of symmetry is also shown in the figure.

The significance of the equipotential line **xx** should be noted. This line divides the plot into three parts, a resistance R_A between conductor **P** and **xx**, a resistance R_B between conductor **Q** and **xx** and a resistance R_C between **xx** and the negative electrode. In terms of A, B and C the corresponding values of the ratios (number of squares in series)/(number of squares in parallel)

$$R_A = A \text{ (resistance of one square)}$$
$$R_B = B \text{ (resistance of one square)}$$
$$R_C = C \text{ (resistance of one square)} \qquad (21.56)$$

For the plot shown in Fig. 21.23(b) $A = 4/2$, $B = 0.5/2$ $C = 7.5/4$ (21.23)

The equivalent circuit of the actual flow in the tank is given in Fig. 21.23(c) where the current I_1 and I_2 are the currents entering the tank at **P** and **Q** respectively. The voltage V_1 and V_2 are the potentials of these electrodes.

Mesh analysis for this circuit shows that

$$\begin{bmatrix} V_1 \\ V_2 \end{bmatrix} = \begin{bmatrix} R_A + R_C & R_C \\ R_C & R_B + R_C \end{bmatrix} \cdot \begin{bmatrix} I_1 \\ I_2 \end{bmatrix} \qquad (21.57)$$

$$= R \begin{bmatrix} A + C & C \\ C & B + C \end{bmatrix} \cdot \begin{bmatrix} I_1 \\ I_2 \end{bmatrix} \qquad (21.58)$$

where R is the actual resistance of the water in the tank for one square.

The corresponding permeance equation for a magnetic field represented by the tank plot (per unit depth in air)

$$\begin{bmatrix} \phi_1 \\ \phi_2 \end{bmatrix} = \mu_0 \begin{bmatrix} A+C & C \\ C & B+C \end{bmatrix} \cdot \begin{bmatrix} I_1 \\ I_2 \end{bmatrix} \qquad (21.50)$$

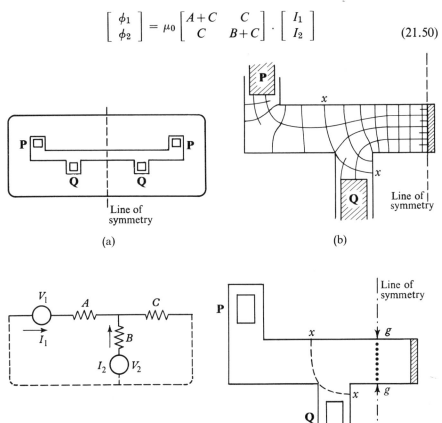

FIG. 21.23 (a) Tank model to represent a stator coil and a rotor coil and part of the air gap of a machine. (b) Flow pattern due to equal currents entering at P and Q. (c) Equivalent circuit. (d) Extension of the model for calibration purposes

and in terms of voltage generated in the conductors (per unit depth) when current changes

$$\begin{bmatrix} V_1 \\ V_2 \end{bmatrix} = \mu_0 \begin{bmatrix} A+C & C \\ C & B+C \end{bmatrix} \cdot p \begin{bmatrix} I_1 \\ I_2 \end{bmatrix} \qquad (21.60)$$

more usually expressed as

$$\begin{bmatrix} V_1 \\ V_2 \end{bmatrix} = \begin{bmatrix} L_1 & M \\ M & L_2 \end{bmatrix} \cdot p \begin{bmatrix} I_1 \\ I_2 \end{bmatrix} \qquad (21.61)$$

where

$$L_1 = \mu_0(A+C)$$
$$L_2 = \mu_0(B+C)$$
$$M = \mu_0 C \qquad (21.62)$$

21.10.1 To determine the value of R

The value of R, the resistance of one square, is required if Eq. (21.58) is used to determine the values of A, B, and C and it is useful if this can be measured experimentally.

This can be arranged if the tank is extended beyond the original negative electrode by one square and the original electrode at the line of symmetry is replaced by a conducting grid.

The current I_1 flowing from electrode **P** to the negative electrode **N** and current I_2 from **Q** to **N** must be adjusted to the same value I and the voltages V_{PN}, V_{QN}, V_{xN} and V_{gN} should be measured. V_{xN} is the potential difference between the equipotential **xx** and electrode **N**, **xx** touching the corner of the slot as shown in Fig. 21.23(d).

The resistance of one square (the calibrating square)

$$R = V_{gn}/2I \tag{21.63}$$

similarly
$$R_C = (V_{xN} - V_{gN})/2I \tag{21.64}$$
$$R_A = (V_{PN} - V_{xN})/I \tag{21.65}$$
$$R_B = (V_{QN} - V_{xN})/I \tag{21.66}$$

Hence
$$A = R_A/R = 2(V_{PN} - V_{xN})/V_{gn} \tag{21.67}$$
$$B = R_B/R = 2(V_{QN} - V_{xN})/V_{gn} \tag{21.68}$$
$$C = R_C/R = (V_{xN} - V_{gn})/V_{gn} \tag{21.69}$$

and L_1, L_2 and M are obtained from Eq. (21.62).

21.11 The direct-axis magnetic circuit of a synchronous machine

21.11.1 Single-slot windings

Assuming in the first instance that the stator and pole-face winding are both concentrated in single slots, and that all windings consist of single turns the direct-axis magnetic circuit of a salient pole machine can be represented by the simplified flux plot shown in Fig. 21.24(a).*

The plot consists of the total air gap between the direct axis and quadrature axis of a machine with three electrodes **x**, **y** and **z** representing the stator winding, the pole-face winding and the field winding respectively. The direct axis forms the negative electrode and the quadrature axis is a non-conducting boundary. The equivalent circuit is shown in Fig. 21.24(b) and forms a ladder network of resistances. An accurate plot will enable the values of the resistances to be found and from these the equivalent inductances self and mutual of the windings.

a_1 and a_2 are the values of total resistance measured from point **O** to the tapping points. These will be mutual resistances in the three loops and if b_1, b_2

---

* As in the previous example Sec 21.10, the calculations are concerned with slot and gap flux only. No account is taken of flux linkages with the end windings.

and d are the resistances measured from **O** to the electrodes these will be contour resistances in the loops.

a_1, a_2, b_1, b_2 and d must be expressed in per-unit values (by dividing actual resistances by the resistance of one square).

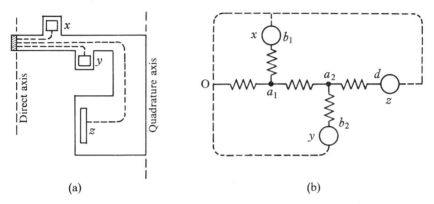

$$\text{(a)} \qquad\qquad\qquad\qquad \text{(b)}$$

FIG. 21.24 (a) Tank model of a salient-pole machine. (b) Equivalent circuit

By mesh analysis of the equivalent circuit and applying equation 21.50,

$$\begin{bmatrix} \dot{V}_x \\ V_y \\ V_z \end{bmatrix} = \mu_0 \begin{bmatrix} b_1 & a_1 & a_1 \\ a_1 & b_2 & a_2 \\ a_1 & a_2 & d \end{bmatrix} \cdot p \begin{bmatrix} I_x \\ I_y \\ I_z \end{bmatrix} \tag{21.70}$$

giving the relationships between the voltages per unit depth of conductor in the actual machine and the currents in them.

If the tank has a calibrating section (as described in Sec. 21.10) the values of the coefficients will be found by applying the currents to the bath through each electrode in turn in each test, measuring all electrode voltages including the grid and expressing the main electrode voltages in per-unit values.

Test 1. I_x only

$$\begin{bmatrix} V_x \\ V_y \\ V_z \end{bmatrix}_{pu} = \begin{bmatrix} b_1 & a_1 & a_1 \\ a_1 & b_2 & a_2 \\ a_1 & a_2 & d \end{bmatrix} \cdot \begin{bmatrix} 1 \\ 0 \\ 0 \end{bmatrix} = \begin{bmatrix} b_1 \\ a_1 \\ a_1 \end{bmatrix} \tag{21.71}$$

Test 2. I_y only

$$\begin{bmatrix} V_x \\ V_y \\ V_z \end{bmatrix}_{pu} = \begin{bmatrix} b_1 & a_1 & a_1 \\ a_1 & b_2 & a_2 \\ a_1 & a_2 & d \end{bmatrix} \cdot \begin{bmatrix} 0 \\ 1 \\ 0 \end{bmatrix} = \begin{bmatrix} a_1 \\ b_2 \\ a_2 \end{bmatrix} \tag{21.72}$$

Test 3. I_z only

$$\begin{bmatrix} V_x \\ V_y \\ V_z \end{bmatrix}_{pu} = \begin{bmatrix} b_1 & a_1 & a_1 \\ a_1 & b_2 & a_2 \\ a_1 & a_2 & d \end{bmatrix} \cdot \begin{bmatrix} 0 \\ 0 \\ 1 \end{bmatrix} = \begin{bmatrix} a_1 \\ a_2 \\ d \end{bmatrix} \tag{21.73}$$

21.11.2 Distributed windings

If the stator winding and the pole-face winding each consist of distributed windings with coils connected in series, having different numbers of turns in each coil so as to obtain sinusoidal or other specified waveforms, we shall have to take account of a number of electrodes in a tank plot as shown in Fig. 21.25(a).

The equivalent circuit is a ladder network shown in Fig. 21.25(b). The contour resistances to the tapping points will be $[b_1 - - - b_n]$ the mutual resistances will be $[a_1 - - - a_n]$ and the contour resistance to the field electrode will be d. It will be necessary to define for each electrode (a) to which winding the coil it represents is connected and (b) the number of turns of the coil.

For this purpose it is necessary to set up a connection matrix $[c]$ showing the relationship between slot ampere-conductors and the three currents I_x, I_y and I_z.

If the slot ampere conductors are given by the column matrix $[x]$ and the winding currents by I_x, I_y and I_z as before

$$[x] = [c] \cdot [I] \tag{21.74}$$

$$
\begin{bmatrix}
x_1 \\ x_2 \\ x_3 \\ | \\ | \\ | \\ x_n \\ x_f
\end{bmatrix}
=
\left[
\begin{array}{ccc}
& | & | & \\
& | & | & \\
c_x & | & c_y & | & 0 \\
& | & | & \\
& | & | & \\
& | & | & \\
& | & | & \\
---&|&---&|&--- \\
0 & | & 0 & | & g
\end{array}
\right]
\begin{bmatrix}
I_x \\ I_y \\ I_z
\end{bmatrix}
\tag{21.75}
$$

The elements of c_x and c_y are formed by the numbers of turns in the slots, each appearing in one column only according to whether the slot is in the stator or in the pole face. g is the number of turns in the field winding. Each column should have zeros where figures exist in the other column.

The impedance matrix of the ladder network of Fig. 21.25(b) has an interesting symmetrical arrangement

$$
[Z] =
\begin{bmatrix}
b_1 & a_1 & a_1 & - & - & - & a_1 & a_1 \\
a_1 & b_2 & a_2 & - & - & - & a_2 & a_2 \\
a_1 & a_2 & b_3 & - & - & - & a_3 & a_3 \\
| & | & | & & & & | & | \\
| & | & | & & & & | & | \\
| & | & | & & & & | & | \\
a_1 & a_2 & a_3 & - & - & - & b_n & a_n \\
a_1 & a_2 & a_3 & - & - & - & a_n & d
\end{bmatrix}
\tag{21.76}
$$

which can be partitioned to give

$$[Z] = \begin{bmatrix} [Z'] & [a] \\ [a_t] & d \end{bmatrix} \tag{21.77}$$

where

$$[Z'] = \begin{bmatrix} b_1 & a_1 & a_1 & - & - & - & a_1 \\ a_1 & b_2 & a_2 & - & - & - & a_2 \\ a_1 & a_2 & b_3 & - & - & - & a_3 \\ | & | & | & & & & \\ | & | & | & & & & \\ | & | & | & & & & \\ a_1 & a_2 & a_3 & - & - & - & b_n \end{bmatrix} \tag{21.78}$$

and

$$[a] = \begin{bmatrix} a_1 \\ a_2 \\ | \\ | \\ | \\ a_n \end{bmatrix} \tag{21.79}$$

The impedance matrix for the connected slots can now be set up

$$[Z''] = [c_t].[Z].[c] = \begin{bmatrix} c_{xt} & 0 \\ c_{yt} & 0 \\ 0 & g \end{bmatrix} \begin{bmatrix} Z' & a \\ a_t & d \end{bmatrix} \begin{bmatrix} c_x & c_y & 0 \\ 0 & 0 & g \end{bmatrix} \tag{21.80}$$

$$= \begin{bmatrix} c_{xt}Z'c_x & c_{xt}Z'c_y & gc_{xt}a \\ c_{yt}Z'c_x & c_{yt}Z'c_y & gc_{yt}a \\ ga_tc_x & ga_tc_y & g^2d \end{bmatrix} \tag{21.81}$$

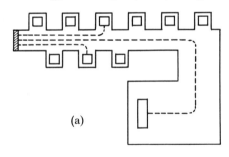

(a)

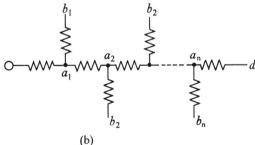

(b)

FIG. 21.25 (a) Tank model with several slots. (b) Equivalent circuit

Each of the terms in Eq. (21.81) reduces to a simple number although the elements c_x, c_y and Z comprising them are matrices. $[Z'']$ is thus a normal impedance matrix similar to Eq. (21.70).

To enable the value of $[Z'']$ to be obtained in a given case a computer program is required to assemble the data in the following steps.

21.11.3 Computer program to determine the impedance matrix

1. Assemble the input data in the form of column matrices
   $a = [a_1 \ a_2 \ a_3 \quad a_n]$ mutual resistances from the flux plot
   $b = [b_1 \ b_2 \ b_3 \quad b_n]$ contour resistances from the flux plot
   n = number of terms in these matrices
   c_x = connection matrix in terms of the stator slot turns and connections
   c_y = connection matrix in terms of the pole slot turns and connections
   d = contour resistance to field winding in the flux plot
   g = number of field winding turns

2. A general program can be arranged to generate $[Z']$ in terms of $[a]$, $[b]$ and n. Actual values of $[a]$ and $[b]$ must be substituted and the matrix $[Z']$ obtained.

3. The following matrix operations are then carried out in a matrix interpretive scheme.

$$P = Z'c_x \qquad Q = Z'c_y$$
$$A = c_{xt}P \qquad C = ga_t c_x$$
$$B = c_{xt}Q \qquad E = ga_t c_y$$
$$D = c_{yt}Q \qquad F = g^2 d \tag{21.82}$$

4. The final matrix for Z can now be assembled

$$[Z] = \begin{bmatrix} A & B & C \\ B & D & E \\ C & E & F \end{bmatrix} \tag{21.83}$$

21.11.4 The effect of separating the windings into three-phase groups

In the above analysis (Sec. 21.11.2) it is assumed that all the conductors of the stator winding in the half-pole-pitch under consideration are connected in series and similarly for the pole-face winding.

If, on the other hand windings are arranged for three-phase operation, as is the more usual condition it means that windings x and y are each divided into three

$$[x_A \ x_B \ x_C] \qquad \text{and} \qquad [y_A \ y_B \ y_C]$$

This results in the connection matrices $[c_x]$ and $[c_y]$ being extended from simple column matrices to rectangular $3 \times n$ matrices. These matrices can be manipulated in exactly the same manner as in the previous section and the final impedance matrix becomes

$$[Z] = \begin{bmatrix} A & B & C \\ B & D & E \\ C & E & F \end{bmatrix} = \left[\begin{array}{ccc|ccc|c} A_A & A_{AB} & A_{AC} & B_A & B_{AB} & B_{AC} & C_A \\ A_{BA} & A_B & A_{BC} & B_{BA} & B_B & B_{BC} & C_B \\ A_{CA} & A_{CB} & A_C & B_{CA} & B_{CB} & B_C & C_C \\ \hline B_A & B_{AC} & B_{AC} & D_A & D_{AB} & D_{AC} & E_A \\ B_{BA} & B_B & B_{BC} & D_{BA} & D_B & D_{BC} & E_B \\ B_{CA} & B_{CB} & B_C & D_{CA} & D_{CB} & D_C & E_C \\ \hline C_A & C_B & C_C & E_A & E_B & E_C & F \end{array} \right]$$

(21.84)

The sub-matrix

$$[A] = \begin{bmatrix} A_A & A_{AB} & A_{AC} \\ A_{BA} & A_B & A_{BC} \\ A_{CA} & A_{CB} & A_C \end{bmatrix}$$

(21.85)

represents the impedance of the armature and consists of self and mutual terms for the three phases.

If the magnetic circuits were symmetrical for the phases there would only be two terms A'_A and A'_{AB} and so to reduce the variation and approximate to symmetry we take the average of the self and mutual terms.

$$A'_A = \tfrac{1}{3}(A_A + A_B + A_C)$$

(21.86)

$$A'_{AB} = \tfrac{1}{6}(A_{AA} + A_{BC} + A_{CA} + A_{BA} + A_{CB} + A_{AC})$$

(21.87)

Finally if we are to assume balanced current only* this matrix is reduced to

$$[A] = \begin{bmatrix} A & 0 & 0 \\ 0 & A & 0 \\ 0 & 0 & A \end{bmatrix}$$

(21.88)

where

$$A = (A_A - A_{AB})$$

(21.89)

Thus A is the equivalent positive-sequence inductance for one phase of the winding, making due allowance for the mutual inductance between phases.

The equivalent values of B, C, D and E are obtained in a similar manner.

21.11.5 The effect of a double-layer winding

In a modern double-layer winding there are two coil sides per slot and if the coils are short pitched the two coil sides in a given slot may be connected to different phases.

In the analogue of a system the conductors should be represented by electrodes placed at one-third the depth of the conductor to introduce a factor corresponding to the effect of internal magnetic flux.†

* Section 13.7.2.
† *Electrical Machines*, p. 111, Eq. (5.34).

When there are two conductors per slot the tank has electrodes situated as shown in Fig. 21.26(*b*) and consequently the ladder network is extended in each limb beyond **b** to **c**.

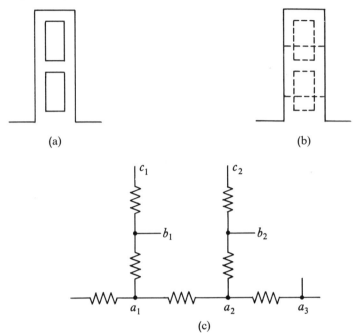

(a)   (b)

(c)

FIG. 21.26 (a) Slot with two conductors. (b) Tank model with grid electrodes. (c) Equivalent circuit

This extension can be dealt with if every *b* term in the previous section is replaced by the matrix

$$\begin{bmatrix} b & b \\ b & c \end{bmatrix}$$
    (21.90)

The form of this sub-matrix arises since the contour impedance of *b* is also the mutual impedance with *c* for all such pairs of terms.

The matrix for *Z* will be increased fourfold in size.

$$[Z] = \begin{bmatrix} b_1 & b_1 & a_1 & 0 \\ b_1 & c_1 & 0 & a_1 \\ \hline a_1 & 0 & b_2 & b_2 \\ 0 & a_1 & b_2 & c_2 \end{bmatrix}$$
    (21.91)

Both the connection matrices c_x and c_y will have twice as many elements in each column and take the *b* and *c* connections alternately.

With these modifications the program developed in Sec. 21.11.3 can be extended to determine the connected impedance matrix.

21.12 The quadrature axis magnetic circuit of a synchronous machine

The tank plot is now arranged with the common electrode on the quadrature-axis boundary and the insulating boundary on the direct axis.

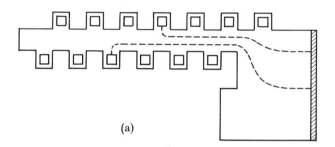

(a)

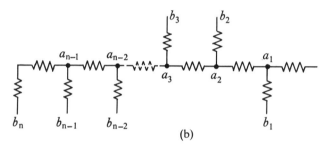

(b)

FIG. 21.27 (a) Tank model of a salient-pole machine for determining quadrature-axis parameters. (b) Equivalent circuit

The field windings are now neglected since there is no deliberate excitation about the quadrature axis.

Again we have a ladder network of contour and mutual resistances b and a and these should now be numbered from the opposite side, the quadrature axis. The corresponding impedance matrix for this ladder network is

$$[Z] = \begin{bmatrix} b_1 & a_1 & a_1 & a_1 & - & - & a_1 \\ a_1 & b_2 & a_2 & a_2 & - & - & a_2 \\ a_1 & a_2 & b_3 & a_3 & - & - & a_3 \\ a_1 & a_2 & a_3 & b_4 & - & - & a_4 \\ | & | & | & | & & & | \\ | & | & | & | & & & a_{n-1} \\ a_1 & a_2 & a_3 & a_4 & - & a_{n-1} & b_n \end{bmatrix} \qquad (21.92)$$

Connection matrices c_x and c_y must be established for the two windings to show how the slots are connected and to multiply by the correct number of turns.

Thus
$$[Z''] = \begin{bmatrix} c_{xt} \\ -- \\ c_{yt} \end{bmatrix} [Z][c_x \mid c_y]$$

$$= \begin{bmatrix} c_{xt}Zc_x & c_{xt}Zc_y \\ c_{yt}Zc_x & c_{yt}Zc_y \end{bmatrix} \tag{21.93}$$

In a similar manner to Sec. 21.11.3, computer programs can be written to generate Z, $c_{xt}Zc_x$, $c_{yt}Zc_y$, $c_{xt}Zc_y$ using a matrix interpretive scheme and using as input the values of

$$n, [a_1 \quad a_2 --- a_{n-1}], [b_1 \quad b_2 \quad b_3 --- b_n], c_x \text{ and } c_y.$$

CHAPTER 22

Appendix

22.1 The synchronous machine in steady-state operation

22.1.1 Phasor diagram

The phasor equation for the synchronous machine in steady-state operation was derived in Sec. 20.2.3, Eq. (20.16):

$$V = E_0 + IR_2 + I_\mathrm{d}jX_{2\mathrm{d}} + I_\mathrm{q}jX_{2\mathrm{q}} \qquad (20.16)$$

where

$$E_0 = jX_\mathrm{m}I_\mathrm{f}$$

and

$$I = I_\mathrm{d} + I_\mathrm{q} = I_\mathrm{d} + jI_\mathrm{q} \qquad (20.4)$$

If we assume positive values of I_d and I_q, the phasor diagram will be as shown in Fig. 22.1. This corresponds to the under-excited condition of the machine, acting as a motor connected to busbars whose voltage is V, and since E_0 is less than V the power-factor is seen to be lagging.*

If I_d is negative, we have Fig. 22.2 corresponding to a motor operating with leading power factor. The condition, due to negative values of both I_d and I_q, is shown in Fig. 22.3. In this diagram, the phasors of the components of output current I_d† and I_q† are also shown and, consequently, it will be seen that this diagram corresponds to that of a generator delivering lagging current.

It is useful to compare these diagrams with those derived in Chap. 9 of *Electrical Machines*. Those in *Electrical Machines* have been built up in a different way, commencing with the V phasor directed along the real axis, whereas the present phasor diagrams are related to the direct and quadrature axes of the machine. In other respects, the diagrams are identical.

---

* It should be clear to the reader that we are implicitly assuming motoring conditions, since Eq. (20.16) is derived from general impedance theory where the power sources are external to the network and the network includes the sinks.

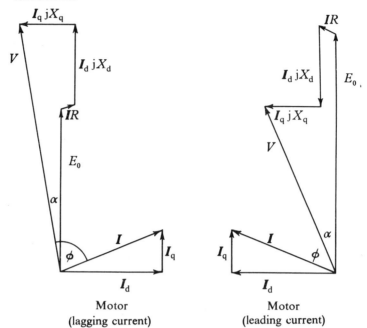

Motor
(lagging current)

FIG. 22.1 Phasor diagram of a synchronous motor under-excited

Motor
(leading current)

FIG. 22.2 Phasor diagram of a synchronous motor over-excited

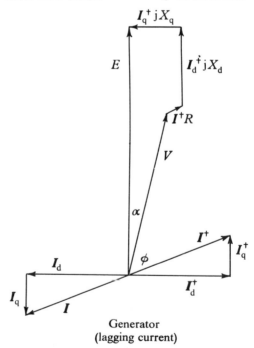

Generator
(lagging current)

FIG. 22.3 Phasor diagram of a synchronous generator delivering lagging current

22.1.2 Torque

The steady-state equations of the synchronous machine are given by—

$$\begin{bmatrix} v_f \\ v_d \\ v_q \end{bmatrix} = \begin{bmatrix} R_1 & 0 & 0 \\ 0 & R_2 & X_{2q} \\ -X_m & -X_{2d} & R_2 \end{bmatrix} \cdot \begin{bmatrix} i_f \\ i_d \\ i_q \end{bmatrix} \qquad (20.11)$$

and, consequently, the torque can be obtained by substituting these terms in the torque equation of the primitive machine (Eq. (16.36)). Thus the torque (in synchronous power units)—

$$f_\omega = [i_d \ \ i_q] \cdot \begin{bmatrix} 0 & 0 & X_{2q} \\ -X_m & -X_{2d} & 0 \end{bmatrix} \cdot \begin{bmatrix} i_f \\ i_d \\ i_q \end{bmatrix} \qquad (22.1)$$

$$= -X_m i_f i_q - i_d i_q (X_{2d} - X_{2q}) \qquad (22.2)$$

Remembering that

$$E_0 = X_m i_f$$

and

$$I_d = i_d \qquad (20.9)$$

$$I_q = -i_q$$

where I_d and I_q are the components of the current phasor—

$$I = I_d + jI_q$$

we obtain by substitution in (22.2)

$$f_\omega = E_0 I_q + I_d I_q (X_{2d} - X_{2q}) \qquad (22.3)$$

which is the torque equation of the synchronous machine in a suitable form if the components I_d and I_q are known. The accuracy of Eq. (22.3) can be checked from the phasor diagram by multiplying the components of the voltage V by the corresponding in phase components of I. This is another way of determining the power.

22.1.3 Power/load-angle relationships

In Secs. 9.3 and 9.5, *Electrical Machines*, the very useful formulae were derived relating power and load angle for a machine connected to constant-voltage, constant-frequency busbars. An alternative derivation is as follows:
    From Eq. (20.11)

$$\begin{bmatrix} v_f \\ v_d \\ v_q \end{bmatrix} = \begin{bmatrix} R_1 & 0 & 0 \\ 0 & R_2 & X_{2q} \\ -X_m & -X_{2d} & R_2 \end{bmatrix} \cdot \begin{bmatrix} i_f \\ i_d \\ i_q \end{bmatrix}$$

and transforming to phasor components, as in Sec. 20.2.3

$$
\begin{bmatrix} V_t \\ V_d \\ V_q \end{bmatrix} = \begin{bmatrix} R_1 & 0 & 0 \\ 0 & R_2 & -X_{2q} \\ X_m & X_{2d} & R_2 \end{bmatrix} \cdot \begin{bmatrix} I_t \\ I_d \\ I_q \end{bmatrix} \tag{20.14}
$$

Eliminating the first line and column, using Eqs. (3.47) and (3.48)

$$
\begin{bmatrix} V_d \\ V_q - V_t X_m/R_1 \end{bmatrix} = \begin{bmatrix} V_d \\ V_q - E_0 \end{bmatrix} = \begin{bmatrix} R_2 & -X_{2q} \\ X_{2d} & R_2 \end{bmatrix} \cdot \begin{bmatrix} I_d \\ I_q \end{bmatrix} \tag{22.4}
$$

Inverting

$$
\begin{bmatrix} I_d \\ I_q \end{bmatrix} = \frac{1}{R^2 + X_{2d}X_{2q}} \begin{bmatrix} R & X_{2q} \\ -X_{2d} & R_2 \end{bmatrix} \cdot \begin{bmatrix} V_d \\ V_q - E_0 \end{bmatrix} \tag{22.5}
$$

In terms of the load angle α we see from Fig. 22.1

$$
\begin{bmatrix} V_d \\ V_q \end{bmatrix} = \begin{bmatrix} -\sin \alpha \\ \cos \alpha \end{bmatrix} V \tag{22.6}
$$

hence if R_2 can be neglected

$$
\begin{bmatrix} I_d \\ I_q \end{bmatrix} = \frac{1}{X_{2d}X_{2q}} \begin{bmatrix} 0 & X_{2q} \\ -X_{2d} & 0 \end{bmatrix} \cdot \begin{bmatrix} -V \sin \alpha \\ V \cos \alpha - E_0 \end{bmatrix} = \begin{bmatrix} (V \cos \alpha - E_0)/X_{2d} \\ (V \sin \alpha)/X_{2q} \end{bmatrix} \tag{22.7}
$$

Since $P = E_0 I_q + I_d I_q (X_{2d} - X_{2q})$ \hfill (22.3)

$$
= \frac{E_0 V \sin \alpha}{X_{2q}} + \frac{V^2 \sin \alpha \cos \alpha}{X_{2d}X_{2q}} (X_{2d} - X_{2q}) - \frac{E_0 V \sin \alpha}{X_{2d}X_{2q}} (X_{2d} - X_{2q})
$$

$$
= \frac{E_0 V \sin \alpha}{X_{2d}} + \frac{V^2 \sin \alpha \cos \alpha}{X_{2d}} \left(\frac{X_{2d}}{X_{2q}} - 1 \right)
$$

$$
= \frac{V^2}{X_{2d}} \left[\frac{E_0}{V} \sin \alpha + \tfrac{1}{2} \left(\frac{X_{2d}}{X_{2q}} - 1 \right) \sin 2\alpha \right] \tag{22.8}
$$

Equation (22.8) and Eq. (9.40) of *Electrical Machines* are the same.

22.2 The motional impedance matrix of a machine

The mechanical variables of a machine are only two, speed and torque. There is value in combining the equations relating to these quantities, together with the equations relating to the electrical variables.

The electrical equations are

$$
[v] = [Z] \cdot [i] \tag{16.22}
$$

from which the torque developed by motoring currents is derived

$$
f_m = [i_t] \cdot [G] \cdot [i] \tag{16.38}
$$

If in addition a torque f is applied from an external source, and J is the moment of inertia of the rotor, the equation for the motion of the rotor will be:

$$f + [i_t][G][i] = Jp^2\theta$$

or
$$f = -[i_t][G][i] + Jp^2\theta \tag{22.9}$$

Thus
$$\left[\begin{array}{c} [v] \\ f \end{array} \right] = \left[\begin{array}{c|c} [Z] & 0 \\ \hline -[i_t][G] & Jp^2 \end{array} \right] \cdot \left[\begin{array}{c} [i] \\ \theta \end{array} \right] \tag{22.10}$$

These equations are not linear since $[Z]$ includes terms which are functions of θ and $[i_t]$ appears in the bottom row, but nevertheless Eq. (22.10) summarises the total performance of a primitive machine.

The effect of small disturbances in the applied torque is of importance and can be investigated in the following way.*

Suppose a small disturbance in applied torque is δf so that

$$f \text{ becomes } f + \delta f \tag{22.11}$$
$$[i] \text{ becomes } [i] + \delta[i] \tag{22.12}$$
$$p\theta \text{ becomes } p\theta + \delta p\theta \tag{22.13}$$
$$[v] \text{ becomes } [v] + \delta[v] \tag{22.14}$$

Substituting in Eq. (16.22) and ignoring the terms including $\delta\theta\delta i$

$$[v] = [R + pL + pM + G(p\theta)] \cdot [i] \tag{22.15}$$
$$[v + \delta v] = [R + pL + pM][i + \delta i] + G(p\theta)[i]$$
$$+ G(p\delta\theta)[i]$$
$$+ G(p\theta)[\delta i] \tag{22.16}$$

hence
$$[\delta v] = [R + pL + pM + Gp\theta][\delta i]$$
$$+ [G][i]p(\delta\theta)$$
$$= [Z][\delta i] + [G][i]p(\delta\theta) \tag{22.17}$$

Similarly substituting in Eq. (22.9)

$$f + \delta f = -[(i + \delta i)_t][G][i + \delta i] + Jp^2(\theta + \delta\theta) \tag{22.18}$$
$$\delta f = -[\delta i_t][G][i]$$
$$-[i_t][G][\delta i] + Jp^2\delta\theta$$
$$= -[i_t][G + G_t][\delta i] + Jp^2\delta\theta \tag{22.19}$$

since
$$[\delta i_t][G][i] = [i_t][G_t][\delta i] \tag{22.20}$$

Combining (22.17) and (22.19)

$$\left[\begin{array}{c} [\delta v] \\ \delta f \end{array} \right] = \left[\begin{array}{cc} [Z] & [G][i]p \\ -[i_t][G + G_t] & Jp^2 \end{array} \right] \cdot \left[\begin{array}{c} [\delta i] \\ \delta\theta \end{array} \right] \tag{22.21}$$

* A particular example of this which has practical importance is a synchronous generator driven by a diesel engine, or a motor driving a reciprocating compressor.

These equations are linear. For given conditions, assuming $[i]$ is known, the effect of small changes of $[\delta v]$ or δf can be determined.

Applying Eq. (22.21) to the primitive machine

$$[G][i] \doteq \left[\begin{array}{c|c|c|c|c} & | & | & | +G_{qr} & | +G_q \\ -G_f & | -G_d & | -G_{dr} & | & | \end{array} \right] \cdot \begin{bmatrix} I_f \\ I_{ds} \\ I_{dr} \\ I_{qr} \\ I_{qs} \end{bmatrix} \qquad (22.22)$$

$$\qquad \qquad \qquad \qquad \qquad \qquad \text{from } (16.39)$$

$$= \begin{bmatrix} I_{qr}G_{qr} + I_{qs}G_{qs} \\ -I_fG_f \ -I_{ds}G_d \ -I_{dr}G_{dr} \end{bmatrix} \qquad (22.23)$$

$$[G+G_t] = \left[\begin{array}{c|c|c|c|c} & | & | & | & -G_f & | \\ & | & | & | & -G_d & | \\ & | & | & | & G_{qr}-G_{dr} & | & G_q \\ -G_f & | -G_d & | G_{qr}-G_{dr} & | & | \\ & | & | & G_q & | & | \end{array} \right] \qquad (22.24)$$

$[i_t] \cdot [G+G_t]$

$\quad = [I_f \mid I_{ds} \mid I_{dr} \mid I_{qr} \mid I_{qs}] \cdot [G+G_t] \ =$

$[-I_{qr}G_f \mid -I_{qr}G_d \mid I_{qr}(G_{qr}-G_{dr})+I_{qs}G_q \mid -I_fG_f-I_{ds}G_d+I_{dr}(G_{qr}-G_{dr}) \mid I_{dr}G_q]$
$\qquad \qquad \qquad \qquad \qquad \qquad \qquad \qquad \qquad \qquad \qquad \qquad (22.25)$

and Eq. (22.21) for the primitive machine becomes (see 22.26 on facing page).

The matrix Eq. (22.26) is now linear and the solution of these simultaneous equations proceeds according to the conditions imposed by a particular problem. If for example we are concerned with small amplitude oscillations in applied torque of frequency ω', δf can be represented by a phasor, p replaced by $j\omega'$ and the matrix inverted by a matrix inversion programme including a complex-number sub-routine.

22.3 Faults in power networks

Instead of developing equivalent circuits showing the connections of the sequence networks to represent various types of fault, as was done in Sec. 13.10, the equations for the currents and voltages produced by the fault can be derived by purely matrix methods in the following manner.

22.3.1 Single line-to-earth fault

Commencing with Eq. (13.48) and premultiplying by $[h]$

$$[E_s] = [V_s] + [Z_s][I_s] \qquad (13.48)$$

$$[V_s] = [E_s] - [Z_s][I_s] \qquad (22.27)$$

$$[V] = [h][V_s] = [h][[E_s]-[Z_s][h]^{-1}[I]] \qquad (22.28)$$

$$
\begin{bmatrix}
\delta v_{\mathrm{f}} \\
\delta v_{\mathrm{ds}} \\
\delta v_{\mathrm{dr}} \\
\delta v_{\mathrm{qr}} \\
dv_{\mathrm{qs}} \\
\delta f
\end{bmatrix}
=
\begin{bmatrix}
R_{\mathrm{f}}+L_{\mathrm{f}}\mathrm{p} & M_{\mathrm{fd}}\mathrm{p} & M_{\mathrm{fd}}\mathrm{p} & 0 & 0 & 0 \\
M_{\mathrm{fd}}\mathrm{p} & R_{\mathrm{ds}}+L_{\mathrm{ds}}\mathrm{p} & M_{\mathrm{d}}\mathrm{p} & 0 & 0 & 0 \\
M_{\mathrm{fd}}\mathrm{p} & M_{\mathrm{d}}\mathrm{p} & R_{\mathrm{dr}}+L_{\mathrm{dr}}\mathrm{p} & G_{\mathrm{qr}}(\mathrm{p}\theta) & G_{\mathrm{q}}(\mathrm{p}\theta) & (I_{\mathrm{qr}}G_{\mathrm{qr}}+I_{\mathrm{qs}}G_{\mathrm{qs}})\mathrm{p} \\
-G_{\mathrm{f}}(\mathrm{p}\theta) & -G_{\mathrm{d}}(\mathrm{p}\theta) & -G_{\mathrm{dr}}(\mathrm{p}\theta) & R_{\mathrm{qr}}+L_{\mathrm{qr}}\mathrm{p} & M_{\mathrm{q}}\mathrm{p} & (-I_{\mathrm{f}}G_{\mathrm{f}}-I_{\mathrm{ds}}G_{\mathrm{d}}-I_{\mathrm{dr}}G_{\mathrm{dr}})\mathrm{p} \\
0 & 0 & 0 & M_{\mathrm{q}}\mathrm{p} & R_{\mathrm{qs}}+L_{\mathrm{qs}}\mathrm{p} & 0 \\
-I_{\mathrm{qr}}G_{\mathrm{f}} & -I_{\mathrm{qr}}G_{\mathrm{d}} & \begin{array}{l}I_{\mathrm{qr}}(G_{\mathrm{qr}}-G_{\mathrm{dr}}) \\ +I_{\mathrm{qs}}G_{\mathrm{q}}\end{array} & \begin{array}{l}I_{\mathrm{dr}}(G_{\mathrm{qr}}-G_{\mathrm{dr}}) \\ -I_{\mathrm{f}}G_{\mathrm{f}}-I_{\mathrm{ds}}G_{\mathrm{d}}\end{array} & I_{\mathrm{dr}}G_{\mathrm{q}} & J\mathrm{p}^2
\end{bmatrix}
\cdot
\begin{bmatrix}
\delta i_{\mathrm{f}} \\
\delta i_{\mathrm{ds}} \\
\delta i_{\mathrm{dr}} \\
\delta i_{\mathrm{qr}} \\
\delta i_{\mathrm{qs}} \\
\delta\theta
\end{bmatrix}
$$

(22.26)

The constraints imposed by a single line-to-earth fault as shown in Fig. 13.14 are expressed by

$$[V] = \begin{bmatrix} 0 \\ V_b \\ V_c \end{bmatrix}$$

(22.29)

and

$$[I] = \begin{bmatrix} I_a \\ 0 \\ 0 \end{bmatrix}$$

(22.30)

consequently

$$\begin{bmatrix} 0 \\ V_b \\ V_c \end{bmatrix} = [h]\left[\begin{bmatrix} E_1 \\ 0 \\ 0 \end{bmatrix} - \begin{bmatrix} Z_1 & 0 & 0 \\ 0 & Z_2 & 0 \\ 0 & 0 & Z_0 \end{bmatrix} [h]^{-1} \begin{bmatrix} I_a \\ 0 \\ 0 \end{bmatrix}\right]$$

(22.31)

$$= [h]\left[\begin{bmatrix} E_1 \\ 0 \\ 0 \end{bmatrix} - \begin{bmatrix} Z_1 & 0 & 0 \\ 0 & Z_2 & 0 \\ 0 & 0 & Z_0 \end{bmatrix} \tfrac{1}{3} \begin{bmatrix} I_a \\ I_a \\ I_a \end{bmatrix}\right]$$

$$= \begin{bmatrix} 1 & 1 & 1 \\ h^2 & h & 1 \\ h & h^2 & 1 \end{bmatrix} \begin{bmatrix} E_1 - \tfrac{1}{3} I_a Z_1 \\ - \tfrac{1}{3} I_a Z_2 \\ - \tfrac{1}{3} I_a Z_0 \end{bmatrix}$$

$$= \begin{bmatrix} E_1 & - \tfrac{1}{3} I_a & (Z_1 + Z_2 + Z_0) \\ h^2 E_1 & - \tfrac{1}{3} I_a & (h^2 Z_1 + h Z_2 + Z_0) \\ h E_1 & - \tfrac{1}{3} I_a & (h Z_1 + h^2 Z_2 + Z_0) \end{bmatrix}$$

(22.32)

I_a can be obtained from the first line of Eq. (22.32), V_b and V_c come from the second and third lines.

22.3.2 Double line-to-earth fault (Fig. 13.16)
From Eqs. (13.45) we obtain

$$[E_s] = [V_s] + [Z_s][I_s]$$

(22.33)

$$[Z_s][I_s] = [E_s] - [V_s]$$

(22.34)

$$[I_s] = [Z_s]^{-1}[[E_s] - [V_s]]$$

(22.35)

$$[I] = [h][Z_s]^{-1}[[E_s] - [h]^{-1}[V]]$$

(22.36)

The constraints imposed by the double line-to-earth fault are

$$[I] = \begin{bmatrix} 0 \\ I_b \\ I_c \end{bmatrix}$$

(22.37)

and $\quad [V] = \begin{bmatrix} V_a \\ 0 \\ 0 \end{bmatrix}$

$$(22.38)$$

Hence $\begin{bmatrix} 0 \\ I_b \\ I_c \end{bmatrix} = [h] \begin{bmatrix} 1/Z_1 & 0 & 0 \\ 0 & 1/Z_2 & 0 \\ 0 & 0 & 1/Z_0 \end{bmatrix} \left(\begin{bmatrix} E_1 \\ 0 \\ 0 \end{bmatrix} - [h]^{-1} \begin{bmatrix} V_a \\ 0 \\ 0 \end{bmatrix} \right)$

$$= [h] \begin{bmatrix} 1/Z_1 & . & 0 \\ 0 & 1/Z_2 & 0 \\ 0 & 0 & 1/Z_0 \end{bmatrix} \begin{bmatrix} E - V_a/3 \\ - V_a/3 \\ - V_a/3 \end{bmatrix}$$

$$= \begin{bmatrix} 1 & 1 & 1 \\ h^2 & h & 1 \\ h & h^2 & 1 \end{bmatrix} \begin{bmatrix} E_1/Z_1 - (V_a/3)(1/Z_1) \\ - (V_a/3)(1/Z_2) \\ - (V_a/3)(1/Z_0) \end{bmatrix}$$

$$\begin{bmatrix} 0 \\ I_b \\ I_c \end{bmatrix} = \begin{bmatrix} E_1/Z_1 - (V_a/3)(1/Z_1 + 1/Z_2 + 1/Z_0) \\ h^2 E_1/Z_1 - (V_a/3)(h^2/Z_1 \quad h/Z_2 + 1/Z_0) \\ h E_1/Z_1 - (V_a/3)(h/Z_1 + h^2/Z_2 + 1/Z_0) \end{bmatrix}$$

$$(22.39)$$

V_a can be obtained from the first line of Eq. (22.39) and I_b and I_c from the second and third lines.

22.3.3 Line-to-line fault (unearthed) (Fig. 13.18)

The effects of a fault of this type can be deduced from those of the previous Section by putting $1/Z_0 = 0$ in Eq. (22.39) since the zero-sequence impedance is now infinite.

22.3.4 Symmetrical three-phase fault (Fig. 13.20)

With this type of fault V_a is also zero, but I_a exists and Eq. (22.39) becomes

$$\begin{bmatrix} I_a \\ I_b \\ I_c \end{bmatrix} = \begin{bmatrix} E_1/Z_1 \\ h^2 E_1/Z_1 \\ h E_1/Z_1 \end{bmatrix}$$

$$(22.40)$$

22.4 Complex power

Ambiguity exists in the literature with reference to the precise definition of Complex power. The adoption of SI units and the latest I.E.C. definitions clear up the position. The relevant I.E.C. definitions are as follows:

22.4.1 Active current

The component of the alternating current which is in phase with the voltage.

22.4.2 Reactive current
The component of a current in quadrature with the voltage.

22.4.3 Active power
The mean power in an alternating current circuit. With sinusoidal currents it is equal to the product of the voltage by the active current.

22.4.4 Reactive power
The product of the voltage by the reactive current. The reactive power of an inductive load is positive. Definitions (22.4.1), (22.4.2), (22.4.4) can be applied only to sinusoidal current.

The reason for the choice of sign is because inductive loads are more prevalent than capacitive loads on public supply systems and consequently it is reasonable to speak of generating stations exporting both active and reactive power (kilo watts and kVAr respectively) both values being positive for inductive loads.

The corresponding I.E.C. definition of the term *Complex power* leads to interesting implications.

22.4.5 Complex power
An expression in which the active power is the real part and the reactive power *with sign reversed* is the imaginary part.

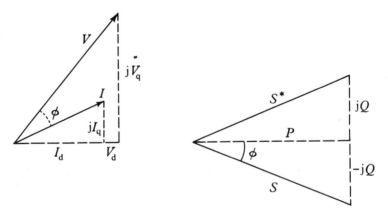

FIG. 22.4 Components of phasors V and I illustrating complex power

Thus if V and I are the voltage and current phasors corresponding to an inductive load having a power factor $\cos \phi$ (Fig. 21.3) and if

$$V = V_d + jV_q \tag{22.41}$$

$$I = I_d + jI_q \tag{22.42}$$

it will be seen that

$$S = V^*I \tag{22.43}$$
$$= (V_d - jV_q)(I_d + jI_q)$$
$$= (V_dI_d + V_qI_q) - j(V_qI_d - V_dI_q)$$
$$= P - jQ \tag{22.44}$$

where
$$P = (V_dI_d + V_qI_q) \tag{22.45}$$

and
$$Q = (V_qI_d - V_dI_q) \tag{22.46}$$

Equations (22.44), (22.45) and (22.46) all correspond to the I.E.C. definition of these quantities and consequently so does Eq. (22.43).

The phasors drawn in Fig. 21.3 also include S^* the *conjugate complex power*

where
$$S^* = P + jQ \tag{22.47}$$

The occasional use of S^* arises

since
$$S^* = I^*V$$
$$= I^*ZI \tag{22.48}$$

just as
$$S = V^*I$$
$$= V^*YV \tag{22.49}$$

Equations (22.48) and (22.49) are seen to be duals. Both arise in circuit analysis and are important since both complex power and its conjugate are invariant in a.c. circuits.*

The I.E.C. definition of complex power does not conflict with the phasor diagrams in Figs. 15.1 and 15.2 of the Appendix Chap. 15 of *Electrical Machines*.

The same phasor diagram (with different scaling) can be still used for Y, I and S as shown in Fig. 15.2 but in interpreting the components of S it must be remembered that the value of the reactive power lagging is that of *minus* the imaginary component of S in accordance with the I.E.C. definition.

22.5 Kron's method of tearing

If an attempt is to be made to analyse the load flow in a large power system and ultimately to determine the currents flowing in all parts of the network, together with the voltages at all busbars, the procedure to be used is based on the nodal method described in Chap. 6. In the first place the admittance matrix of the system has to be established. This procedure is straightforward but the next step is to determine its inverse. The admittance matrix will be a large one with very many terms and consequently a computer having a very large store will be needed and the process of inversion will take a comparatively long time. Errors

* Sec. 4.3.3.

arising from mistakes in presenting input data and transcribing the output data are difficult to trace and add to the cost of the process.

A simplification is introduced by Kron's method of tearing whereby a complicated system is viewed as a number of constituent parts or zones connected together by a superimposed system of interconnectors. This is a realistic attitude to take since actual systems are frequently geographically arranged in precisely this way. By considering the constituent parts separately before interconnection, and then taking into account the effect of the interconnections, the whole connected system is built up.

Kron's method is developed in the following way and illustrated by the system shown in Fig. 22.5. The power network shown there has seven nodes,

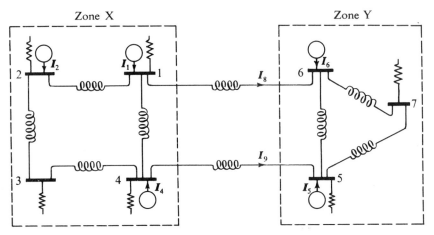

FIG. 22.5 A power system divided into two zones X and Y

five being generator nodes. The system obviously divides into two separate zones X and Y with interconnectors P and Q joining them together. If the method of tearing is to be used there must be no mutual induction between the zones, and the only interconnection being the elements specified. The original system can be 'torn' into the constituent parts as shown in Fig. 22.6 with an additional network showing the voltages and currents in the 'torn' conductors. It is unlikely in practice there will be any voltage source in a conductor which has been cut but if there is one it must be included.

For identification purposes the busbars **1, 2, 3, 4, 5, 6** and **7** in Fig. 22.5 become nodes **a, b, c, d, e, f** and **g** respectively in Fig. 22.6.

The nodal equations for the original system is given by

$$[I_A] = [Y_A] \cdot [V_A] \tag{22.50}$$

where $[Y]$ is a 7×7 matrix in this case corresponding to seven nodes. We wish to solve these equations, that is to determine $[Z_A] = [Y_A]^{-1}$ without actually inverting the matrix $[Y_A]$ by conventional methods. To these equations (22.50) which represent the system before tearing, we add additional equations representing the branches to be cut, the currents in them and the series voltages if any exist. This gives

$$\begin{bmatrix} [I_A] \\ [I_B] \end{bmatrix} = \begin{bmatrix} [Y_A] & 0 \\ 0 & [Y_B] \end{bmatrix} \begin{bmatrix} [V_A] \\ [V_B] \end{bmatrix} \qquad (22.51)$$

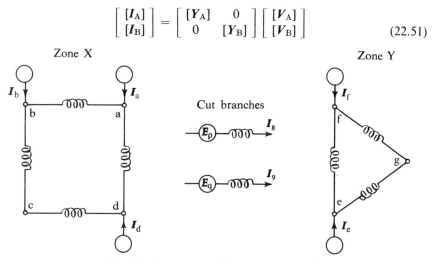

Zone X   Zone Y

Cut branches

FIG. 22.6 Separation of the zones X **and** Y

In the example we are discussing

$[I_A]$ includes all the nodal source currents $(I_1 \ I_2 \ I_4 \ I_5 \ I_6 \ I_7)$

$[I_B]$ includes the cut branch currents $(I_p \ I_q)$

$[V_A]$ includes the nodal voltages $(V_1 \ V_2 \ V_3 \ V_4 \ V_5 \ V_6 \ V_7)$

$[V_B]$ includes the cut branch series voltages $(E_p \text{ and } E_q)$ but as these are usually zero $[V_B]$ is usually ze:o

and we now have an enlarged admittance matrix 9×9.

The next step is to construct a second related system representing the original one after tearing and after adjustments have been made to ensure that the currents of the new system (the *primitive* or *separated* system) correspond to those of the original (the *connected* system).

The new system will have adjusted values of the nodal source currents to compensate for the loss of the currents in the cut branches. A connection matrix is then established relating the new (primitive) source currents to those of the connected system

$$[I'] = [c'] . [I] \qquad (22.52)$$

These changes ensure that the system power is unchanged hence to comply with the principle of power invariance in the connected and separated systems we note that

$$[V_t][I] = [V'_t][I'] = [V'_t][c'][I]$$

hence

$$[V_t] = [V'_t][c'] \qquad (22.53)$$

and
$$[V] = [c't][V']$$
(22.54)

In the example

$$[I_A'] = \begin{bmatrix} I_a \\ I_b \\ I_c \\ I_d \\ I_e \\ I_f \\ I_g \end{bmatrix} = \begin{bmatrix} I_1 & -I_8 \\ I_2 & \\ I_3 & \\ I_4 & -I_9 \\ I_5 & +I_9 \\ I_6 & +I_8 \\ I_7 & \end{bmatrix} \qquad \begin{matrix} \\ \\ I_c = 0 \\ \\ \\ \\ I_g = 0 \end{matrix}$$

$$[I_B'] = \begin{bmatrix} I_p \\ I_q \end{bmatrix} = \begin{bmatrix} I_8 \\ I_9 \end{bmatrix} \qquad \begin{matrix} E_p = 0 \\ E_q = 0 \end{matrix}$$
(22.55)

The form of the connection matrix is seen to be

$$\begin{bmatrix} [I'_A] \\ [I'_B] \end{bmatrix} = \begin{bmatrix} [1] & [c] \\ [0] & [1] \end{bmatrix} \cdot \begin{bmatrix} [I_A] \\ [I_B] \end{bmatrix}$$
(22.56)

where $[c]$ is a simpler connection matrix consisting of elements which are $+1$, -1 or 0 and shows the nodes to which the cut branches are connected.

In the example

$$[c] = \begin{bmatrix} -1 & 0 \\ 0 & 0 \\ 0 & 0 \\ 0 & -1 \\ 0 & 1 \\ 1 & 0 \\ 0 & 0 \end{bmatrix}$$

The admittance matrix of the separated network will be

$$[Y'] = \begin{bmatrix} [Y'_A] & 0 \\ 0 & [Y'_B] \end{bmatrix}$$
(22.57)

where $[Y'_A]$ refers to the sections and $[Y'_B]$ to the cut conductors. $[Y'_B]$ will be a diagonal matrix and $[Y'_A]$ will be a diagonal grouping of matrices each relating to a separate section.

Thus

$$\begin{bmatrix} [Y'_A] & 0 \\ 0 & [Y'_B] \end{bmatrix} = \begin{bmatrix} [Y'_{A1}] & \cdot & \cdot & \cdot \\ \cdot & [Y'_{A2}] & \cdot & \cdot \\ \cdot & \cdot & [Y'_{An}] & \cdot \\ \cdot & \cdot & \cdot & [Y'_B] \end{bmatrix}$$
(22.58)

The currents in the primitive system are related to the nodal voltages by

$$[I'] = [Y'] \cdot [V']$$
(22.59)

hence $[V'] = [Z'] . [I']$ (22.60)

where $[Z'] = [Y']^{-1} = \begin{bmatrix} [Z'_{A1}] & . & . & . \\ . & [Z'_{A2}] & . & . \\ . & . & [Z'_{An}] & . \\ . & . & . & [Z'_B] \end{bmatrix}$ (22.61)

where $[Z'_{A1}] = [Y'_{A1}]^{-1}$ (22.62)

$[Z'_{A2}] = [Y'_{A2}]^{-1}$ (22.63)

$[Z'_B] = [Y'_B]^{-1} = [Y_B]^{-1}$ (22.64)

The inversion of $[Y']$ is thus obtained by inverting the admittance matrices of the separated sections and then assembling these $[Z]$ matrices about the diagonal, together with $[Z'_B]$ thus forming $[Z']$.

Since each section matrix is much smaller than the whole connected matrix, the time taken by the inversion process is considerably reduced.

It should be noted that $[Y_B] = [Y'_B]$ and so $[Z'_B] = [Z_B]$ (22.65)

The matrix $[Z_B]$ can probably be written down by inspection of the network more readily than the corresponding admittance matrix $[Y_B]$.

We can now proceed from the study of the unconnected network to that of the connected system. From the principle of power invariance the impedance matrix of the connected matrix

$$[Z] = [c'_t] . [Z'] . [c']$$ (22.66)

$$= \begin{bmatrix} 1 & 0 \\ [c_t] & 1 \end{bmatrix} . \begin{bmatrix} [Z'_A] & 0 \\ 0 & [Z'_B] \end{bmatrix} . \begin{bmatrix} 1 & [c] \\ 0 & 1 \end{bmatrix}$$ (22.67)

$$\begin{bmatrix} [Z'_A] & [Z'_A][c] \\ [c_t][Z'_A] & [c_t][Z'_A][c] + [Z'_B] \end{bmatrix}$$ (22.68)

The enlarged performance equations of the connected network are therefore given by

$$[V] = [Z] . [I]$$ (22.69)

or

$$\begin{bmatrix} [V_A] \\ [V_B] \end{bmatrix} = \begin{bmatrix} [Z_1] & [Z_2] \\ [Z_3] & [Z_4] \end{bmatrix} . \begin{bmatrix} [I_A] \\ [I_B] \end{bmatrix}$$ (22.70)

where, from Eq. (21.68)

$$[Z_1] = [Z'_A]$$
$$[Z_2] = [Z'_A][c]$$
$$[Z_3] = [c_t][Z'_A]$$ (22.71)
$$[Z_4] = [c_t][Z'_A][c] + [Z'_B]$$

Finally these equations can be reduced to determine the nodal voltages $[V_A]$ in terms of the original source currents $[I_A]$ of the original connected network.

Since $[V_B] = 0$, the matrix in Eq. (21.68) can be partitioned giving

$$[V_A] = [Z_A][I_A] \tag{22.72}$$

where $$[Z_A] = [Z_1] - [Z_2][Z_4]^{-1}[Z_3] \tag{22.73}$$

Thus $$[Z_A] = [Z'_A] - [Z'_A][c] [[c_t][Z'_A][c] + [Z'_B]]^{-1} [c_t][Z'_A] \tag{22.74}$$

or $$[Z_A] = [Z'_A] [1 - [c][Y'_A][c_t][Z'_A]] \tag{22.75}$$

where $$[Y'_A] = [[c_t][Z'_A][c] + [Z'_B]]^{-1} \tag{22.76}$$

22.5.1 Program to determine the nodal voltages of a large system

The method of tearing can be summarised in the following steps.

1. Tear the network into N convenient sub-networks. There must be no mutual inductance between these sub-networks.
2. Establish the admittance matrices $[Y'_{A1}][Y'_{A2}] - - - - [Y'_{An}]$ of these sub-networks.
3. Establish the impedance matrix $[Z'_B]$ of the cut branches.
4. Establish the connection matrix $[c]$ showing the relationship between the nodal currents in the separated networks and the corresponding values in the connected networks.
5. Obtain the transpose $[c_t]$.
6. Invert the admittance matrices and obtain $[Z'_{A1}][Z'_{A2}] - - - - [Z'_{An}]$
7. Obtain $[Z'_A]$ by assembling $[Z'_{A1}] [Z'_{A2}] - - - - [Z'_{An}]$ about the diagonal.

$$[Z'_A] = [Y'_A]^{-1} \tag{22.77}$$

8. Program the following equations in a matrix interpretive scheme

$$[B] = [c_t][Z'_A] \tag{22.78}$$

$$[Y'_A] = [[B][c] + [Z'_B]]^{-1} \tag{22.79}$$

$$[Z_A] = [Z'_A][1 - [c][Y'_A][B]] \tag{22.80}$$

$$[V_A] = [Z_A][I_A] \tag{22.81}$$

Equation (21.81) thus yields the column matrix of nodal voltages.

22.5.2 Advantages of the method of tearing

The advantages of the method of tearing can be summarised as follows:

1. The admittance matrix $[Y'_A]$ has all its elements grouped along the diagonal and consequently can be inverted in sections.
2. If any changes are made to any of these sections it is not necessary to re-invert the whole matrix, only the sections concerned.
3. Frequently the sub-divisions are identical sub-networks and so their matrices can be repeated.
4. Many of the matrix manipulations are simple additions. Multiplication by a connection matrix only involves addition or subtraction of elements and the computation time is short.

5. If additions or subtractions are made to the network, only those matrices referring to the section concerned need changing.
6. The final $[\mathbf{Z_A}]$ matrix for the system can be re-used if the whole system becomes part of a still larger system. A new connection matrix will be needed to account for the new interconnectors and the total system matrix will be enlarged in much the same way as the system itself develops.

22.6 Tables of functions

Many of the calculations described in earlier chapters are simplified by the use of the following tables of functions (pages 348–376).

They are particularly important in problems involving complex numbers, enabling results to be obtained by simple slide-rule manipulation. Their use is illustrated in the following examples:

22.6.1 Calculation of the inductance coefficient c due to short circuit
Since
$$c = 1 - k^2 \qquad (22.82)$$
$$\text{also (20.36)}$$

the inductance coefficient corresponding to any coefficient of coupling can be obtained from Table 10.

Example:
   If
$$k = 0.412$$
from Table 10
$$c = 0.8303$$

22.6.2 Calculation of the magnitude of Z given R and X
$$Z = \sqrt{(R^2 + X^2)} \qquad (22.83)$$
hence
$$Z = R\sqrt{(1 + (X/R)^2)} \qquad (22.84)$$
or
$$= X\sqrt{(1 + (R/X)^2)} \qquad (22.85)$$

The ratio R/X or X/R (whichever is less than one) is first obtained by slide rule and the corresponding value of $\sqrt{(1 + (R/X)^2)}$ or $\sqrt{(1 + X/R)^2)}$ is obtained from Table 11. The value of Z is determined from Eq. (22.84) or (22.85) by slide rule.

Example:
   If
$$R = 54.3$$

and

$$X = 46\cdot9$$
$$X/R = 0\cdot864$$

from Table 11
$$Z = 54\cdot3 \times 1\cdot3216$$
$$= 71\cdot8.$$

22.6.3 Conversion from cartesian to polar coordinates
If

$$V \exp(j\phi) = a+jb \tag{22.86}$$

then

$$V = \sqrt{(a^2+b^2)} \tag{22.87}$$

and

$$\phi = \arctan b/a \tag{22.88}$$

V is found from a and b in the same way that Z was found from R and X

$$V = a\sqrt{(1+(b/a)^2)} \tag{22.89}$$
$$= b\sqrt{(1+(a/b)^2)} \tag{22.90}$$

The ratio a/b or b/a (whichever is less than one) is determined by slide rule and V obtained from Eq. (22.89) or (22.90), together with Table 11.

Arctan a/b or arctan b/a is next obtained from Table 5. The angle ϕ depends on the signs of a and b and on which ratio has been used.

The relationship between ϕ and arctan b/a or arctan a/b is given below:

$a+$	$b+$	$b < a$	$\phi = \arctan b/a$
$a+$	$b+$	$a < b$	$\phi = 90 - \arctan a/b$
$a-$	$b+$	$a < b$	$\phi = 90 + \arctan a/b$
$a-$	$b+$	$b < a$	$\phi = 180 - \arctan b/a$
$a-$	$b-$	$b < a$	$\phi = 180 + \arctan b/a$
$a-$	$b-$	$a < b$	$\phi = 270 - \arctan a/b$
$a+$	$b-$	$a < b$	$\phi = 270 + \arctan a/b$
$a+$	$b-$	$b < a$	$\phi = 360 - \arctan b/a$

The tables enable the conversion to be made to slide-rule accuracy and this transform is particularly useful when complex numbers have to be multiplied or divided.

Example:
If

$$V \exp(j\phi) = -10+j2\cdot34$$
$$b/a = 0\cdot234$$

from Table 11
$$V = 10 \times 1 \cdot 0270$$
$$= 10 \cdot 27$$
$$\phi = 180 - \arctan 0 \cdot 234$$
from Table 5
$$= 180 - 13 \cdot 17$$
$$= 166 \cdot 83 \text{ degrees.}$$

22.6.4 Conversion from polar to cartesian coordinates
Since
$$V \exp (j\phi) = V (\cos \phi + j \sin \phi)$$
$$= a + jb \tag{22.86}$$
$$a = V \cos \phi \tag{22.91}$$
$$b = V \sin \phi \tag{22.92}$$

a and b can thus be found with the aid of Tables 6 and 7.

Example:
  If
$$a + jb = 100 \exp (j\ 79 \cdot 6 \text{ deg})$$
from Table 7
$$a = 100 \times 0 \cdot 1805$$
$$= 18 \cdot 05$$
from Table 6
$$b = 100 \times 0 \cdot 9836$$
$$= 98 \cdot 36$$

22.6.5 Conversion from impedance to admittance
Since
$$Y = 1/Z \tag{22.93}$$
if
$$Z = R + jX \tag{22.94}$$
then
$$Y = 1/(R + jX)$$
$$= \frac{R - jX}{(R + jX)(R - jX)}$$
$$= \frac{R}{R^2 + X^2} - j \frac{X}{R^2 + X^2}$$
$$= g + jb \tag{22.95}$$
where
$$g = R/(R^2 + X^2)$$
$$= \frac{1}{R} \frac{1}{1 + (X/R)^2} \tag{22.96}$$
$$= \frac{1}{X} \frac{R/X}{1 + (R/X)^2} \tag{22.97}$$

and

$$b = -X/(R^2 + X^2)$$

$$= -\frac{1}{R}\frac{X/R}{1+(X/R)^2} \tag{22.98}$$

$$= -\frac{1}{X}\frac{1}{1+(R/X)^2} \tag{22.99}$$

The ratio R/X or X/R (whichever is less than one) is obtained by slide rule and the corresponding values of

$$1/(1+(R/X)^2) \quad \text{and} \quad (R/X)/(1+(R/X)^2)$$

or

$$1/(1+(X/R)^2) \quad \text{and} \quad (X/R)/(1+(X/R)^2)$$

are found from Tables 12 and 13.

Substitution in Eqs. (22.96), (22.98) or (22.97), (22.99), are simple slide rule multiplication exercises.

Example:

If

$$Z = 0{\cdot}418 + j0{\cdot}320$$

$$X/R = 0{\cdot}766$$

from Table 12 $$g = \frac{1}{0{\cdot}418}\,0{\cdot}6302$$

$$= 1{\cdot}51$$

from Table 13 $$b = \frac{1}{0{\cdot}418}\,0{\cdot}4827$$

$$= 1{\cdot}15$$

22.6.6 Complex power and power factor

If the 'Complex power' S is known, the power factor can be determined from the components of S^*.

$$S^* = P + jQ$$

$$\cos\phi = P/S$$

$$= P/\sqrt{(P^2 + Q^2)}$$

$$= \frac{1}{\sqrt{(1+(Q/P)^2)}} \tag{22.100}$$

$$= \frac{P/Q}{\sqrt{(1+(P/Q)^2)}} \tag{22.101}$$

Cos ϕ is thus obtained from the ratio Q/P or P/Q, whichever is less than one, in conjunction with Tables 14 and 15.

Similarly (if required)

$$\sin \phi = Q/S$$
$$= Q/\sqrt{(P^2+Q^2)}$$
$$= \frac{1}{\sqrt{(1+(P/Q)^2)}} \tag{22.102}$$
$$= \frac{Q/P}{\sqrt{(1+(Q/P)^2)}} \tag{22.103}$$

and thus $\sin \phi$ can also be found from the ratio of the components of Complex power, using the Tables 14 and 15.

Example:

If

$$S^* = 100+j56\cdot2$$
$$Q/P = 0\cdot562$$

from Table 14 $\qquad \cos \phi = 0\cdot8718$

from Table 15 $\qquad \sin \phi = 0\cdot4899.$

22.6.7 Transmission line calculations

To obtain the complex constants M and N (Sec. 9.8.2) given the value of the complex product YZ.

The values of M and N are given in the tables forming a matrix, each element being a value M and N, the one being immediately above the other. The rows correspond to the value of a and the columns to the value of b in steps of $0\cdot01$ where

$$YZ = a+jb \tag{22.104}$$

Thus if $\qquad YZ = -0\cdot28+j0\cdot03$

the corresponding values M and N obtained from Table 16 are

$$M = 0\cdot8632+j0\cdot0143$$
$$N = 0\cdot954+j0\cdot0049$$

In order to interpolate between points given in the tables it is necessary to remember that a given point W will lie intermediate between four adjacent points. In other words W will be within a square $P\,Q\,R\,S$ as shown in Fig. 22.7.

If $\qquad\qquad\qquad P = a_1+jb_1 \tag{22.105}$

we have $\qquad\qquad Q = (a_1+0\cdot01)+jb_1 \tag{22.106}$

$\qquad\qquad\qquad R = a_1+j(b_1+0\cdot01) \tag{22.107}$

$\qquad\qquad\qquad S = (a_1+0\cdot01)+j(b_1+0\cdot01) \tag{22.108}$

$0\cdot01$ being the standard increment used in the table. The coordinates of W can be expressed as

$$W = a_3 + jb_3 \qquad (22.109)$$

or

$$W = (a_1 + 0 \cdot 01x) + j(b_1 + 0 \cdot 01y) \qquad (22.110)$$

where

$$x = 100(a_3 - a_1) \qquad (22.111)$$

and

$$y = 100(b_3 - b_1) \qquad (22.112)$$

The transform to the M and N plane, transforms the square $P\,Q\,R\,S$ to the quadrilateral $P'\,Q'\,R'\,S'$ and the point W to W' which has the same relative position with respect to the angular points $P'\,Q'\,R'$ and S' that W has to $P\,Q\,R$ and S.

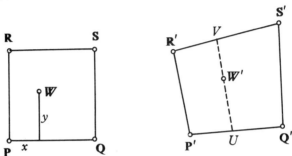

FIG. 22.7 Interpolation between points $PQRS$

The co-ordinates of $P'\,Q'\,R'$ and S' are

$$P' = p_1 + jq_1 \qquad (22.113)$$
$$Q' = p_2 + jq_2 \qquad (22.114)$$
$$R' = p_3 + jq_3 \qquad (22.115)$$
$$S' = p_4 + jq_4 \qquad (22.116)$$

A method of determining the approximate position of W' is to divide the lines $P'\,Q'$ and $R'\,S'$ in the ratio x at U and V and then to divide the line UV in the ratio y.

From this we are able to deduce that the co-ordinates of W' will be given by

$$W' = p_W + jq_W$$

where

$$p_W = p_1 + x(p_2 - p_1)$$
$$+ y(p_3 - p_1)$$
$$+ xy[(p_4 - p_3) - (p_2 - p_1)] \qquad (22.117)$$
$$q_W = q_1 + x(q_2 - q_1)$$
$$+ y(q_3 - q_1)$$
$$+ xy[(q_4 - q_3) - (q_2 - q_1)] \qquad (22.118)$$

Equations (22.117) and (22.118) apply in general but for these tables the xy terms are very small and may be neglected.

Example:

If

$$Z = -0.327 + j0.143$$
$$a_1 = 0.32 \quad b_1 = 0.14$$
$$a_2 = 0.33 \quad b_2 = 0.15$$
$$a_3 = 0.327 \quad b_3 = 0.143$$
$$x = 0.007/0.01 = 0.7 \quad y = 0.003/0.01 = 0.3$$

From Table 16
$$P' = 0.8434 + j0.0663$$
$$Q' = 0.8387 + j0.0662$$
$$R' = 0.8433 + j0.0711$$
$$S' = 0.8386 + j0.0709$$
$$p_\text{w} = 0.8434 + 0.7(0.8387 - 0.8434)$$
$$+ 0.3(0.8433 - 0.8434)$$
$$= 0.8434 + 0.7(0.0053) = 0.8471$$
$$q_\text{w} = 0.0663 + 0.7(0.0662 - 0.0663)$$
$$+ 0.3(0.0711 - 0.0663)$$
$$= 0.0663 + 0.3(0.0048) = 0.0677$$
$$M = 0.8471 + j0.0677$$

22.7 Phasor diagrams

22.7.1 Conventions

In this book and in *Electrical Machines*, consistent and unambiguous conventions have been attached to all phasor diagrams used to represent steady-state alternating quantities. Attention has been drawn to the fact that this does not obtain in all text books,* and a student must tend to be confused when he goes from one book to another and finds different diagrams purporting to mean the same thing. Everyone is bound to deplore such a state of affairs and the present author can do no more than reiterate the reasons for the adoption of the conventions he has used in these text books and to hope that eventually this system will be universally recommended.

22.7.2 Voltage and current phasor diagrams

The quantities normally displayed in phasor diagrams are voltage and current. Other alternating quantities (e.g. flux) are occasionally introduced but the main purpose of the diagram is to show:

(a) the way in which the various currents in a circuit are related to each other,

* M. G. Scroggie, *Phasor Diagrams*, (Iliffe), 1967.

(*b*) the way in which the voltage drops in the circuit are related to each other, and

(*c*) the phase relationship between currents and voltages.

In general, phasor diagrams consist of the superposition of two separate diagrams, one for voltage and the other for current.

22.7.3 What is a phasor?

A phasor is defined in Sec. 7.1.2 as a complex number used to represent an alternating quantity and in Fig. 7.3 we see a phasor $a+jb$ drawn on a diagram (which mathematicians speak of as an Argand diagram) or as electrical engineers sometimes say 'drawn in the complex plane'. This phasor is used to represent a voltage V and it should be noted that the line used in the diagram has its ends labelled **O** and **V** or alternatively an arrow head is placed on the non-zero end of the line. In Fig. 7.3 both conventions have been used to make certain there is no ambiguity.

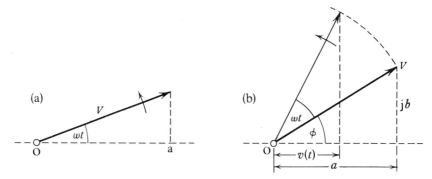

**FIG. 22.7 (a) The geometrical interpretation of $v(t) = \mathrm{Re}\, V \exp j\omega t$.
From 7.3 (b) Geometrical interpretation of $v(t) = \mathrm{Re}\, V \exp j\phi \exp j\omega t$**

However, it will be appreciated that we are assuming that we know precisely what is meant by 'the voltage V' in the circuit from which it has been taken. This is quite another matter—nothing to do with phasor diagrams—but entirely a matter of *circuit conventions*.

The circuit convention for voltage used in this book is defined in Sec. 1.2.2 introducing the double-subscript notation for potential difference.

22.7.4 Voltage phasor diagrams

We have used a single-subscript notation for nodal voltage in Chap. 6. This is quite sufficient and self evident. V_A means the potential at node **A** with respect to some common datum which has to be specified. In order to refer to the potential difference between nodes we must use double subscripts and for this we need to adopt a convention. The one used in this book is that

$$V_{AB} \equiv V_A - V_B \qquad (22.119)$$

It implies that if V_A is greater than V_B, V_{AB} will be positive or in other words when V_{AB} is positive, the node referred to by the first subscript is positive with respect to the second. This convention is the one adopted by the I.E.C. and Eq. (22.119) is quite logical. But it is a matter of convention and unfortunately some authorities reverse the subscript order.

The convention expressed in Eq. (22.119) is immediately applicable to voltage phasor diagrams. The phasor represented by the line **OV** in Fig. 22.7 can represent the nodal voltage at a node **A** (with reference to datum **O**) by using label **A** instead of **V** at the opposite end of the phasor to zero.

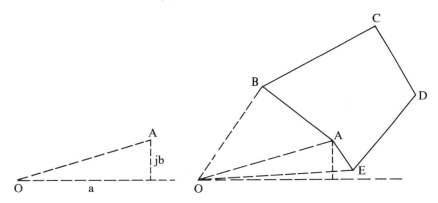

FIG. 22.8 A phasor polygon of voltages relating to nodes $A\ B\ C\ D$ and E

Since the nodal voltage of other nodes **B C D E** can be similarly expressed, a voltage phasor diagram such as Fig. 22.8 can be drawn. On this diagram the nodal-voltage phasors are shown by dotted lines and the nodal-pair voltages or the phasors representing voltage between nodes $V_{AB}\ V_{BC}\ V_{CD}\ V_{DE}\ V_{EA}$ are shown by full lines.

These phasors form a polygon and a phasor diagram for a particular network may consist of this polygon only without drawing in the dotted lines to the datum. In fact the complete information relating to the nodal voltage is there if we simply plot the nodal points on the complex plane without connecting them. The lines joining pairs of points merely add to the visualisation of $V_{AB}\ V_{AC}\ V_{BC}$ as required.

It will be seen that one single line represents two phasors, for example V_{AB} and V_{BA}, and from Eq. (22.119)

$$V_{AB} \equiv -V_{BA} \qquad (22.120)$$

If the convention described in this section is strictly enforced there is no need to use arrowheads on a voltage phasor diagram. However, a reader may feel happier to use an occasional arrowhead to pick out a particular nodal-pair voltage in which case the convention used in *Electrical Machines*, Sec. 1.2.4 can also be applied.

22.7.5 Current phasor diagrams

In exactly the same manner, phasor diagrams for currents in networks can be constructed and properly labelled points can be plotted in the complex plane for this purpose eliminating the use of arrowheads. The logic of this will be understood if we have correctly interpreted the principle of duality discussed in Chap. 8.

The dual of *nodal voltage* is *loop current*, consequently loop currents can be represented by phasors in exactly the same way as nodal-voltage phasors are drawn.

For example Fig. 22.9 shows a circuit which has two loops **1** and **2**, and three nodes **A B** and **C**.

The reference directions of the two loop currents must be specified and these shown by arrows on the network tracing out the loops. This decision is quite arbitrary and is taken according to convenience as discussed in Chap. 5.

These currents are designated I_1 and I_2 and each current can be represented by a phasor

$$I_1 = a_1 + jb_1 \qquad (22.121)$$
$$I_2 = a_2 + jb_2$$

In the complex plane, points **1** and **2** are plotted with respect to the origin **O**. Note that we are using exactly the same procedure as we did for plotting nodal-voltage phasors. (Fig. 22.9(*b*)).

In a similar way the difference between two loop currents can be found if we adopt the convention.

$$I_{12} \equiv I_1 - I_2 \qquad (22.122)$$

or

$$I_{21} \equiv I_2 - I_1$$

So we see that the points **1**, **2** and **0** form a triangle in the complex plane. The third side of the triangle represents the difference between the two loop currents, that is the current in the central branch of the diagram. The current in this branch is either I_{12} flowing from **B** to **C** in the branch or I_{21} flowing from **C** to **B** and obviously

$$I_{21} \equiv -I_{12} \qquad (22.123)$$

The same line on the diagram represents both these difference currents and which we wish to specify is entirely a matter of convenience but one of the equations (22.122) must be used.

The corresponding voltage phasor diagram showing the nodal voltages **A B** and **C** is drawn in Fig. 22.9(*c*).

To clarify the current diagram the phasors representing the branch currents I_a I_b and I_c can be drawn in. This can be done by adding correctly pointing arrowheads to the diagram and labelling each arrowhead by the symbol for a branch current.

Since
$$I_a = I_1$$
$$I_b = I_{12} = I_1 - I_2$$
$$I_c = I_2$$

Fig. 22.9(d) shows these currents. Checking the phasor diagrams of Fig. 22.9 against the circuit elements used it will be noted that

(a) V_{ab} and I_a are in phase
(b) I_c leads V_{BC} by a right angle
(c) I_b lags V_{BC} by approximately 60°

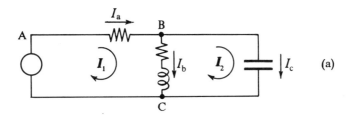

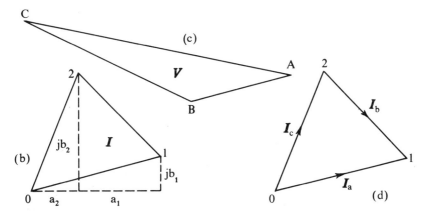

FIG. 22.9 (a) A simple a.c. circuit. (b) Loop current phasors I_1 and I_2. (c) Nodal voltage phasor polygon. (d) Branch current phasor polygon

22.7.6 Transformer phasor diagram

The phasor diagram corresponding to a transformer connected to constant voltage a.c. mains and feeding a lagging power-factor load is a particular case where diagrams are drawn differently in different text books.

Figure 22.10(a) shows a transformer of turns ratio 2:1 with terminals marked in accordance with Sec. 4.4 and Fig. 22.10(b) is its equivalent circuit when connected to supply and load.

Here we see there are three loops slightly complicated by the ratio change of the idealized transformer. Terminals **B** and **D** are taken as the reference nodes for the primary and secondary circuits respectively.

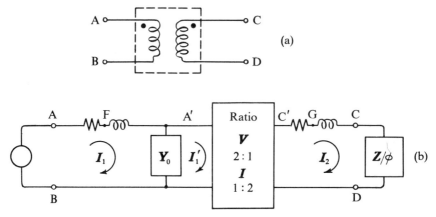

(a)

(b)

FIG. 22.10 (a) Conventional representation of a transformer. (b) Equivalent circuit of a transformer

The phasor diagram of the transformer (Fig. 22.11(a)) consists of four super-imposed phasor polygons for primary voltages and currents and secondary voltages and currents respectively.

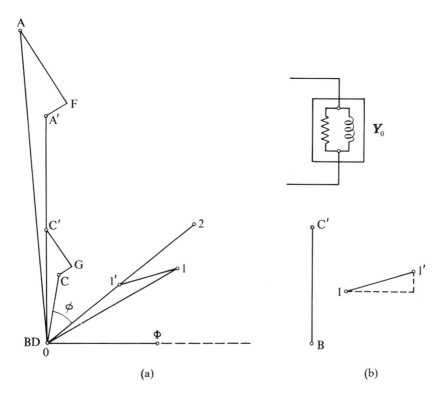

(a)   (b)

FIG. 22.11 (a) Phasor diagram of a transformer. (b) Components of magnetising current

The phasor diagram shows the following relationships

(a) $V_{\text{A'B}}$ and $V_{\text{C'B}}$ are in quadrature with flux Φ

(b) I_2 and V_{CB} are related by the load impedance Z and I_2 lags V_{CB} by angle ϕ.

(c) $V_{\text{AB}} = V_{\text{A'B}} + V_{\text{AA'}}$.

(d) $V_{\text{C'D}} = V_{\text{CD}} + V_{\text{C'C}}$.

(e) $I_1 = I'_1 + I_{11'}$.

(f) $I_{11'}$ is the current in Y the magnetising admittance and leads the flux phasor by a small angle. It is sometimes divided into components in phase and in quadrature with the voltage across it (Fig. 22.11(b)).

Figure 22.11(a) is identical with Fig. 2.23 in *Electrical Machines* but the latter uses arrowheads to help to indicate individual currents and voltages.

Table 1 Exp X

	0·000	0·002	0·004	0·006	0·008
0·00	1·0000	1·0020	1·0040	1·0060	1·0080
0·01	1·0101	1·0121	1·0141	1·0161	1·0182
0·02	1·0202	1·0222	1·0243	1·0263	1·0284
0·03	1·0305	1·0325	1·0346	1·0367	1·0387
0·04	1·0408	1·0429	1·0450	1·0471	1·0492
0·05	1·0513	1·0534	1·0555	1·0576	1·0597
0·06	1·0618	1·0640	1·0661	1·0682	1·0704
0·07	1·0725	1·0747	1·0768	1.0790	1·0811
0·08	1·0833	1·0855	1·0876	1·0898	1·0920
0·09	1·0942	1·0964	1·0986	1·1008	1·1030
0·10	1·1052	1·1074	1·1096	1·1118	1·1140
0·11	1·1163	1·1185	1·1208	1·1230	1·1252
0·12	1·1275	1·1298	1·1320	1·1343	1·1366
0·13	1·1388	1·1411	1·1434	1·1457	1·1480
0·14	1·1503	1·1526	1·1549	1·1572	1·1595
0·15	1·1618	1·1642	1·1665	1·1688	1·1712
0·16	1·1735	1·1759	1·1782	1·1806	1·1829
0·17	1·1853	1·1877	1·1901	1·1924	1·1948
0·18	1·1972	1·1996	1·2020	1·2044	1·2068
0·19	1·2092	1·2117	1·2141	1·2165	1·2190
0·20	1·2214	1·2238	1·2263	1·2288	1·2312
0·21	1·2337	1·2361	1·2386	1·2411	1·2436
0·22	1·2461	1·2486	1·2511	1·2536	1·2561
0·23	1·2586	1·2611	1·2636	1·2662	1·2687
0·24	1·2712	1·2738	1·2763	1·2789	1·2815
0·25	1·2840	1·2866	1·2892	1·2918	1·2943
0·26	1·2969	1·2995	1·3021	1·3047	1·3073
0·27	1·3100	1·3126	1·3152	1·3178	1·3205
0·28	1·3231	1·3258	1·3284	1·3311	1·3338
0·29	1·3364	1·3391	1·3418	1·3445	1·3472
0·30	1·3499	1·3526	1·3553	1·3580	1·3607
0·31	1·3634	1·3662	1·3689	1·3716	1·3744
0·32	1·3771	1·3799	1·3826	1·3854	1·3882
0·33	1·3910	1·3938	1·3965	1·3993	1·4021
0·34	1·4049	1·4078	1·4106	1·4134	1·4162
0·35	1·4191	1·4219	1·4248	1·4276	1·4305
0·36	1·4333	1·4362	1·4391	1·4420	1·4448
0·37	1·4477	1·4506	1·4535	1·4564	1·4594
0·38	1·4623	1·4652	1·4681	1·4711	1·4740
0·39	1·4770	1·4799	1·4829	1·4859	1·4888
0·40	1·4918	1·4948	1·4978	1·5008	1·5038
0·41	1·5068	1·5098	1·5129	1·5159	1·5189
0·42	1·5220	1·5250	1·5281	1·5311	1·5342
0·43	1·5373	1·5403	1·5434	1·5465	1·5496
0·44	1·5527	1·5558	1·5589	1·5621	1·5652
0·45	1·5683	1·5715	1·5746	1·5778	1·5809
0·46	1·5841	1·5872	1·5904	1·5936	1·5968
0·47	1·6000	1·6032	1·6064	1·6096	1·6128
0·48	1·6161	1·6193	1·6226	1·6258	1·6291
0·49	1·6323	1·6356	1·6389	1·6421	1·6454
0·50	1·6487	1·6520	1·6553	1·6586	1·6620

Table 1—continued   349

	0·000	0·002	0·004	0·006	0·008
0·51	1·6653	1·6686	1·6720	1·6753	1·6787
0·52	1·6820	1·6854	1·6888	1·6922	1·6955
0·53	1·6989	1·7023	1·7057	1·7092	1·7126
0·54	1·7160	1·7194	1·7229	1·7263	1·7298
0·55	1·7333	1·7367	1·7402	1·7437	1·7472
0·56	1·7507	1·7542	1·7577	1·7612	1·7647
0·57	1·7683	1·7718	1·7754	1·7789	1·7825
0·58	1·7860	1·7896	1·7932	1·7968	1·8004
0·59	1·8040	1·8076	1·8112	1·8148	1·8185
0·60	1·8221	1·8258	1·8294	1·8331	1·8368
0·61	1·8404	1·8441	1·8478	1·8515	1·8552
0·62	1·8589	1·8626	1·8664	1·8701	1·8739
0·63	1·8776	1·8814	1·8851	1·8889	1·8927
0·64	1·8965	1·9003	1·9041	1·9079	1·9117
0·65	1·9155	1·9194	1·9232	1·9271	1·9309
0·66	1·9348	1·9387	1·9425	1·9464	1·9503
0·67	1·9542	1·9581	1·9621	1·9660	1·9699
0·68	1·9739	1·9778	1·9818	1·9858	1·9897
0·69	1·9937	1·9977	2·0017	2·0057	2·0097
0·70	2·0138	2·0178	2·0218	2·0259	2·0299
0·71	2·0340	2·0381	2·0421	2·0462	2·0503
0·72	2·0544	2·0585	2·0627	2·0668	2·0709
0·73	2·0751	2·0792	2·0834	2·0876	2·0917
0·74	2·0959	2·1001	2·1043	2·1085	2·1128
0·75	2·1170	2·1212	2·1255	2·1297	2·1340
0·76	2·1383	2·1426	2·1468	2·1511	2·1555
0·77	2·1598	2·1641	2·1684	2·1728	2·1771
0·78	2·1815	2·1858	2·1902	2·1946	2·1990
0·79	2·2034	2·2078	2·2122	2·2167	2·2211
0·80	2·2255	2·2300	2·2345	2·2389	2·2434
0·81	2·2479	2·2524	2·2569	2·2614	2·2660
0·82	2·2705	2·2750	2·2796	2·2842	2·2887
0·83	2·2933	2·2979	2·3025	2·3071	2·3117
0·84	2·3164	2·3210	2·3257	2·3303	2·3350
0·85	2·3396	2·3443	2·3490	2·3537	2·3584
0·86	2·3632	2·3679	2·3726	2·3774	2·3821
0·87	2·3869	2·3917	2·3965	2·4013	2·4061
0·88	2·4109	2·4157	2·4206	2·4254	2·4303
0·89	2·4351	2·4400	2·4449	2·4498	2·4547
0·90	2·4596	2·4645	2·4695	2·4744	2·4794
0·91	2·4843	2·4893	2·4943	2·4993	2·5043
0·92	2·5093	2·5143	2·5193	2·5244	2·5294
0·93	2·5345	2·5396	2·5447	2·5498	2·5549
0·94	2·5600	2·5651	2·5702	2·5754	2·5805
0·95	2·5857	2·5909	2·5961	2·6013	2·6065
0·96	2·6117	2·6169	2·6222	2·6274	2·6327
0·97	2·6379	2·6432	2·6485	2·6538	2·6591
0·98	2·6645	2·6698	2·6751	2·6805	2·6859
0·99	2·6912	2·6966	2·7020	2·7074	2·7129
1·00	2·7183	2·7237	2·7292	2·7346	2·7401

Table 2 Exp $(-X)$

	0·00	0·02	0·04	0·06	0·08
0·00	1·0000	0·9802	0·9608	0·9418	0·9231
0·10	0·9048	0·8869	0·8694	0·8521	0·8353
0·20	0·8187	0·8025	0·7866	0·7711	0·7558
0·30	0·7408	0·7261	0·7118	0·6977	0·6839
0·40	0·6703	0·6570	0·6440	0·6313	0·6188
0·50	0·6065	0·5945	0·5827	0·5712	0·5599
0·60	0·5488	0·5379	0·5273	0·5169	0·5066
0·70	0·4966	0·4868	0·4771	0·4677	0·4584
0·80	0·4493	0·4404	0·4317	0·4232	0·4148
0·90	0·4066	0·3985	0·3906	0·3829	0·3753
1·00	0·3679	0·3606	0·3535	0·3465	0·3396
1·10	0·3329	0·3263	0·3198	0·3135	0·3073
1·20	0·3012	0·2952	0·2894	0·2837	0·2780
1·30	0·2725	0·2671	0·2618	0·2567	0·2516
1·40	0·2466	0·2417	0·2369	0·2322	0·2276
1·50	0·2231	0·2187	0·2144	0·2101	0·2060
1·60	0·2019	0·1979	0·1940	0·1901	0·1864
1·70	0·1827	0·1791	0·1755	0·1720	0·1686
1·80	0·1653	0·1620	0·1588	0·1557	0·1526
1·90	0·1496	0·1466	0·1437	0·1409	0·1381
2·00	0·1353	0·1327	0·1300	0·1275	0·1249
2·10	0·1225	0·1200	0·1177	0·1153	0·1130
2·20	0·1108	0·1086	0·1065	0·1044	0·1023
2·30	0·1003	0·0983	0·0963	0·0944	0·0926
2·40	0·0907	0·0889	0·0872	0·0854	0·0837
2·50	0·0821	0·0805	0·0789	0·0773	0·0758
2·60	0·0743	0·0728	0·0714	0·0699	0·0686
2·70	0·0672	0·0659	0·0646	0·0633	0·0620
2·80	0·0608	0·0596	0·0584	0·0573	0·0561
2·90	0·0550	0·0539	0·0529	0·0518	0·0508
3·00	0·0498	0·0488	0·0478	0·0469	0·0460
3·10	0·0450	0·0442	0·0433	0·0424	0·0416
3·20	0·0408	0·0400	0·0392	0·0384	0·0376
3·30	0·0369	0·0362	0·0354	0·0347	0·0340
3·40	0·0334	0·0327	0·0321	0·0314	0·0308
3·50	0·0302	0·0296	0·0290	0·0284	0·0279
3·60	0·0273	0·0268	0·0263	0·0257	0·0252
3·70	0·0247	0·0242	0·0238	0·0233	0·0228
3·80	0·0224	0·0219	0·0215	0·0211	0·0207
3·90	0·0202	0·0198	0·0194	0·0191	0·0187
4·00	0·0183	0·0180	0·0176	0·0172	0·0169
4·10	0·0166	0·0162	0·0159	0·0156	0·0153
4·20	0·0150	0·0147	0·0144	0·0141	0·0138
4·30	0·0136	0·0133	0·0130	0·0128	0·0125
4·40	0·0123	0·0120	0·0118	0·0116	0·0113
4·50	0·0111	0·0109	0·0107	0·0105	0·0103
4·60	0·0101	0·0099	0·0097	0·0095	0·0093
4·70	0·0091	0·0089	0·0087	0·0086	0·0084
4·80	0·0082	0·0081	0·0079	0·0078	0·0076
4·90	0·0074	0·0073	0·0072	0·0070	0·0069
5·00	0·0067	0·0066	0·0065	0·0063	0·0062

Table 3   Exp $(-X)$ 351

	0·000	0·002	0·004	0·006	0·008
0·00	1·0000	0·9980	0·9960	0·9940	0·9920
0·01	0·9900	0·9881	0·9861	0·9841	0·9822
0·02	0·9802	0·9782	0·9763	0·9743	0·9724
0·03	0·9704	0·9685	0·9666	0·9646	0·9627
0·04	0·9608	0·9589	0·9570	0·9550	0·9531
0·05	0·9512	0·9493	0·9474	0·9455	0·9436
0·06	0·9418	0·9399	0·9380	0·9361	0·9343
0·07	0·9324	0·9305	0·9287	0·9268	0·9250
0·08	0·9231	0·9213	0·9194	0·9176	0·9158
0·09	0·9139	0·9121	0·9103	0·9085	0·9066
0·10	0·9048	0·9030	0·9012	0·8994	0·8976
0·11	0·8958	0·8940	0·8923	0·8905	0·8887
0·12	0·8869	0·8851	0·8834	0·8816	0·8799
0·13	0·8781	0·8763	0·8746	0·8728	0·8711
0·14	0·8694	0·8676	0·8659	0·8642	0·8624
0·15	0·8607	0·8590	0·8573	0·8556	0·8538
0·16	0·8521	0·8504	0·8487	0·8470	0·8454
0·17	0·8437	0·8420	0·8403	0·8386	0·8369
0·18	0·8353	0·8336	0·8319	0·8303	0·8286
0·19	0·8270	0·8253	0·8237	0·8220	0·8204
0·20	0·8187	0·8171	0·8155	0·8138	0·8122
0·21	0·8106	0·8090	0·8073	0·8057	0·8041
0·22	0·8025	0·8009	0·7993	0·7977	0·7961
0·23	0·7945	0·7929	0·7914	0·7898	0·7882
0·24	0·7866	0·7851	0·7835	0·7819	0·7804
0·25	0·7788	0·7772	0·7757	0·7741	0·7726
0·26	0·7711	0·7695	0·7680	0·7664	0·7649
0·27	0·7634	0·7619	0·7603	0·7588	0·7573
0·28	0·7558	0·7543	0·7528	0·7513	0·7498
0·29	0·7483	0·7468	0·7453	0·7438	0·7423
0·30	0·7408	0·7393	0·7379	0·7364	0·7349
0·31	0·7334	0·7320	0·7305	0·7291	0·7276
0·32	0·7261	0·7247	0·7233	0·7218	0·7204
0·33	0·7189	0·7175	0·7161	0·7146	0·7132
0·34	0·7118	0·7103	0·7089	0·7075	0·7061
0·35	0·7047	0·7033	0·7019	0·7005	0·6991
0·36	0·6977	0·6963	0·6949	0·6935	0·6921
0·37	0·6907	0·6894	0·6880	0·6866	0·6852
0·38	0·6839	0·6825	0·6811	0·6798	0·6784
0·39	0·6771	0·6757	0·6744	0·6730	0.6717
0·40	0·6703	0·6690	0·6676	0·6663	0·6650
0·41	0·6637	0·6623	0·6610	0·6597	0·6584
0·42	0·6570	0·6557	0·6544	0·6531	0·6518
0·43	0·6505	0·6492	0·6479	0·6466	0·6453
0·44	0·6440	0·6427	0·6415	0·6402	0·6389
0·45	0·6376	0·6364	0·6351	0·6338	0·6325
0·46	0·6313	0·6300	0·6288	0·6275	0·6263
0·47	0·6250	0·6238	0·6225	0·6213	0·6200
0·48	0·6188	0·6175	0·6163	0·6151	0·6139
0·49	0·6126	0·6114	0·6102	0·6090	0·6077
0·50	0·6065	0·6053	0·6041	0·6029	0·6017

	0·000	0·002	0·004	0·006	0·008
0·51	0·6005	0·5993	0·5981	0·5969	0·5957
0·52	0·5945	0·5933	0·5921	0·5910	0·5898
0·53	0·5886	0·5874	0·5863	0·5851	0·5839
0·54	0·5827	0·5816	0·5804	0·5793	0·5781
0·55	0·5769	0·5758	0·5746	0·5735	0·5724
0·56	0·5712	0·5701	0·5689	0·5678	0·5667
0·57	0·5655	0·5644	0·5633	0·5621	0·5610
0·58	0·5599	0·5588	0·5577	0·5565	0·5554
0·59	0·5543	0·5532	0·5521	0·5510	0·5499
0·60	0·5488	0·5477	0·5466	0·5455	0·5444
0·61	0·5434	0·5423	0·5412	0·5401	0·5390
0·62	0·5379	0·5369	0·5358	0·5347	0·5337
0·63	0·5326	0·5315	0·5305	0·5294	0·5283
0·64	0·5273	0·5262	0·5252	0·5241	0·5231
0·65	0·5220	0·5210	0·5200	0·5189	0·5179
0·66	0·5169	0·5158	0.5148	0.5138	0.5127
0·67	0·5117	0·5107	0·5097	0·5086	0·5076
0·68	0·5066	0·5056	0·5046	0·5036	0·5026
0·69	0·5016	0·5006	0·4996	0·4986	0·4976
0·70	0·4966	0·4956	0·4946	0·4936	0·4926
0·71	0·4916	0·4907	0·4897	0·4887	0·4877
0·72	0·4868	0·4858	0·4848	0·4838	0·4829
0·73	0·4819	0·4809	0·4800	0·4790	0·4781
0·74	0·4771	0·4762	0·4752	0·4743	0·4733
0·75	0.4724	0.4714	0.4705	0.4695	0.4686
0·76	0·4677	0·4667	0·4658	0·4649	0·4639
0·77	0·4630	0·4621	0·4612	0·4602	0·4593
0·78	0·4584	0·4575	0·4566	0·4557	0·4548
0·79	0·4538	0·4529	0·4520	0·4511	0·4502
0·80	0·4493	0·4484	0·4475	0·4466	0·4457
0·81	0·4449	0·4440	0·4431	0·4422	0·4413
0·82	0·4404	0·4396	0·4387	0·4378	0·4369
0·83	0·4360	0·4352	0·4343	0·4334	0·4326
0·84	0·4317	0·4308	0·4300	0·4291	0·4283
0·85	0·4274	0·4266	0·4257	0·4249	0·4240
0·86	0·4232	0·4223	0·4215	0·4206	0·4198
0·87	0·4190	0·4181	0·4173	0·4164	0·4156
0·88	0·4148	0·4140	0·4131	0·4123	0·4115
0·89	0·4107	0·4098	0·4090	0·4082	0·4074
0·90	0·4066	0·4058	0·4049	0·4041	0·4033
0·91	0·4025	0·4017	0·4009	0·4001	0·3993
0·92	0·3985	0·3977	0·3969	0·3961	0·3953
0·93	0·3946	0·3938	0·3930	0·3922	0·3914
0·94	0·3906	0·3898	0·3891	0·3883	0·3875
0·95	0·3867	0·3860	0·3852	0·3844	0·3837
0·96	0·3829	0·3821	0·3814	0·3806	0·3798
0·97	0·3791	0·3783	0·3776	0·3768	0·3761
0·98	0·3753	0·3746	0·3738	0·3731	0·3723
0·99	0·3716	0·3708	0·3701	0·3694	0·3686
1·00	0·3679	0·3671	0·3664	0·3657	0·3649

Table 4   Log X 353

	0·00	0·02	0·04	0·06	0·08
1·00	0·0000	0·0198	0·0392	0·0583	0·0770
1·10	0·0953	0·1133	0·1310	0·1484	0·1655
1·20	0·1823	0·1989	0·2151	0·2311	0·2469
1·30	0·2624	0·2776	0·2927	0·3075	0·3221
1·40	0·3365	0·3507	0·3646	0·3784	0·3920
1·50	0·4055	0·4187	0·4318	0·4447	0·4574
1·60	0·4700	0·4824	0·4947	0·5068	0·5188
1·70	0·5306	0·5423	0·5539	0·5653	0·5766
1·80	0·5878	0·5988	0·6098	0·6206	0·6313
1·90	0·6419	0·6523	0·6627	0·6729	0·6831
2·00	0·6931	0·7031	0·7129	0·7227	0·7324
2·10	0·7419	0·7514	0·7608	0·7701	0·7793
2·20	0·7885	0·7975	0·8065	0·8154	0·8242
2·30	0·8329	0·8416	0·8502	0·8587	0·8671
2·40	0·8755	0·8838	0·8920	0·9002	0.9083
2·50	0·9163	0·9243	0·9322	0·9400	0·9478
2·60	0·9555	0·9632	0·9708	0·9783	0·9858
2·70	0·9933	1·0006	1·0080	1·0152	1·0225
2·80	1·0296	1·0367	1·0438	1·0508	1·0578
2·90	1·0647	1·0716	1·0784	1·0852	1·0919
3·00	1·0986	1·1053	1·1119	1·1184	1·1249
3·10	1·1314	1·1378	1·1442	1·1506	1·1569
3·20	1·1632	1·1694	1·1756	1·1817	1·1878
3·30	1·1939	1·2000	1·2060	1·2119	1·2179
3·40	1·2238	1·2296	1·2355	1·2413	1·2470
3·50	1·2528	1·2585	1·2641	1·2698	1·2754
3·60	1·2809	1·2865	1·2920	1·2975	1·3029
3·70	1·3083	1·3137	1·3191	1·3244	1·3297
3·80	1·3350	1·3403	1·3455	1·3507	1·3558
3·90	1·3610	1·3661	1·3712	1·3762	1·3813
4·00	1·3863	1·3913	1·3962	1·4012	1·4061
4·10	1·4110	1·4159	1·4207	1·4255	1·4303
4·20	1·4351	1·4398	1·4446	1·4493	1·4540
4·30	1·4586	1·4633	1·4679	1·4725	1·4770
4·40	1·4816	1·4861	1·4907	1·4951	1·4996
4·50	1·5041	1·5085	1·5129	1·5173	1·5217
4·60	1·5261	1·5304	1·5347	1·5390	1·5433
4·70	1·5476	1·5518	1·5560	1·5602	1·5644
4·80	1·5686	1·5728	1·5769	1·5810	1·5851
4·90	1·5892	1·5933	1·5974	1·6014	1·6054
5·00	1·6094	1·6134	1·6174	1·6214	1·6253
5·10	1·6292	1·6332	1·6371	1·6409	1·6448
5·20	1·6487	1·6525	1·6563	1·6601	1·6639
5·30	1·6677	1·6715	1·6752	1·6790	1·6827
5·40	1·6864	1·6901	1·6938	1·6974	1·7011
5·50	1·7047	1·7084	1·7120	1·7156	1·7192
5·60	1·7228	1·7263	1·7299	1.7334	1·7370
5·70	1·7405	1·7440	1·7475	1·7509	1·7544
5·80	1·7579	1·7613	1·7647	1·7681	1·7716
5·90	1·7750	1·7783	1·7817	1·7851	1·7884
6·00	1·7918	1·7951	1·7984	1·8017	1·8050

Table 4—continued

	0·00	0·02	0·04	0·06	0·08
6·10	1·8083	1·8116	1·8148	1·8181	1·8213
6·20	1·8245	1·8278	1·8310	1·8342	1·8374
6·30	1·8405	1·8437	1·8469	1·8500	1·8532
6·40	1·8563	1·8594	1·8625	1·8656	1·8687
6·50	1·8718	1·8749	1·8779	1·8810	1·8840
6·60	1·8871	1·8901	1·8931	1·8961	1·8991
6·70	1·9021	1·9051	1·9081	1·9110	1·9140
6·80	1·9169	1·9199	1·9228	1·9257	1·9286
6·90	1·9315	1·9344	1·9373	1·9402	1·9430
7·00	1·9459	1·9488	1·9516	1·9544	1·9573
7·10	1·9601	1·9629	1·9657	1·9685	1·9713
7·20	1·9741	1·9769	1·9796	1·9824	1·9851
7·30	1·9879	1·9906	1·9933	1·9961	1·9988
7·40	2·0015	2·0042	2·0069	2·0096	2·0122
7·50	2·0149	2·0176	2·0202	2·0229	2·0255
7·60	2·0281	2·0308	2·0334	2·0360	2·0386
7·70	2·0412	2·0438	2·0464	2·0490	2·0516
7·80	2·0541	2·0567	2·0592	2·0618	2·0643
7·90	2·0669	2·0694	2·0719	2·0744	2·0769
8·00	2·0794	2·0819	2·0844	2·0869	2·0894
8·10	2·0919	2·0943	2·0968	2·0992	2·1017
8·20	2·1041	2·1066	2·1090	2·1114	2·1138
8·30	2·1163	2·1187	2·1211	2·1235	2·1258
8·40	2·1282	2·1306	2·1330	2·1353	2·1377
8·50	2·1401	2·1424	2·1448	2·1471	2·1494
8·60	2·1518	2·1541	2·1564	2·1587	2·1610
8·70	2·1633	2·1656	2·1679	2·1702	2·1725
8·80	2·1748	2·1770	2·1793	2·1815	2·1838
8·90	2·1861	2·1883	2·1905	2·1928	2·1950
9·00	2·1972	2·1994	2·2017	2·2039	2·2061
9·10	2·2083	2·2105	2·2127	2·2148	2·2170
9·20	2·2192	2·2214	2·2235	2·2257	2·2279
9·30	2·2300	2·2322	2·2343	2·2364	2·2386
9·40	2·2407	2·2428	2·2450	2·2471	2·2492
9·50	2·2513	2·2534	2·2555	2·2576	2·2597
9·60	2·2618	2·2638	2·2659	2·2680	2·2701
9·70	2·2721	2·2742	2·2762	2·2783	2·2803
9·80	2·2824	2·2844	2·2865	2·2885	2·2905
9·90	2·2925	2·2946	2·2966	2·2986	2·3006
10·00	2·3026	2·3046	2·3066	2·3086	2·3106

Table 5   Arctan X (deg.)   355

	0·000	0·002	0·004	0·006	0·008
0·00	0·000	0·115	0·229	0·344	0·458
0·01	0·573	0·688	0·802	0·917	1·031
0·02	1·146	1·260	1·375	1·489	1·604
0·03	1·718	1·833	1·947	2·062	2·176
0·04	2·291	2·405	2·519	2·634	2·748
0·05	2·862	2·977	3·091	3·205	3·319
0·06	3·434	3·548	3·662	3·776	3·890
0·07	4·004	4·118	4·232	4·346	4·460
0·08	4·574	4·688	4·802	4·915	5·029
0·09	5·143	5·256	5·370	5·484	5·597
0·10	5·711	5·824	5·937	6·051	6·164
0·11	6·277	6·390	6·504	6·617	6·730
0·12	6·843	6·956	7·069	7·181	7·294
0·13	7·407	7·520	7·632	7·745	7·857
0·14	7·970	8·082	8·194	8·306	8·419
0·15	8·531	8·643	8·755	8·867	8·979
0·16	9·090	9·202	9·314	9·425	9·537
0·17	9·648	9·759	9·871	9·982	10·093
0·18	10·204	10·315	10·426	10·537	10·647
0·19	10·758	10·869	10·979	11·089	11·200
0·20	11·310	11·420	11·530	11·640	11·750
0·21	11·860	11·969	12·079	12·189	12·298
0·22	12·407	12·517	12·626	12·735	12·844
0·23	12·953	13·062	13·170	13·279	13·387
0·24	13·496	13·604	13·712	13·820	13·928
0·25	14·036	14·144	14·252	14·359	14·467
0·26	14·574	14·681	14·789	14·896	15·003
0·27	15·110	15·216	15·323	15·430	15·536
0·28	15·642	15·748	15·855	15·961	16·066
0·29	16·172	16·278	16·383	16·489	16·594
0·30	16·699	16·804	16·909	17·014	17·119
0·31	17·223	17·328	17·432	17·537	17·641
0·32	17·745	17·849	17·952	18·056	18·159
0·33	18·263	18·366	18·469	18·572	18·675
0·34	18·778	18·881	18·983	19·086	19·188
0·35	19·290	19·392	19·494	19·596	19·697
0·36	19·799	19·900	20·002	20·103	20·204
0·37	20·304	20·405	20·506	20·606	20·707
0·38	20·807	20·907	21·007	21·107	21·206
0·39	21·306	21·405	21·504	21·604	21·703
0·40	21·801	21·900	21·999	22·097	22·195
0·41	22·294	22·392	22·490	22·587	22·685
0·42	22·782	22·880	22·977	23·074	23·171
0·43	23·268	23·364	23·461	23·557	23·653
0·44	23·749	23·845	23·941	24·037	24·132
0·45	24·228	24·323	24·418	24·513	24·608
0·46	24·702	24·797	24·891	24·986	25·080
0·47	25·174	25·267	25·361	25·454	25·548
0·48	25·641	25·734	25·827	25·920	26·012
0·49	26·105	26·197	26·289	26·381	26·473
0·50	26·565	26·657	26·748	26·839	26·931

Table 5—continued

	0·000	0·002	0·004	0·006	0·008
0·51	27·022	27·112	27·203	27·294	27·384
0·52	27·474	27·565	27·655	27·744	27·834
0·53	27·924	28·013	28·102	28·191	28·280
0·54	28·369	28·458	28·546	28·635	28·723
0·55	28·811	28·899	28·986	29·074	29·162
0·56	29·249	29·336	29·423	29·510	29·597
0·57	29·683	29·770	29·856	29·942	30·028
0·58	30·114	30·199	30·285	30·370	30·456
0·59	30·541	30·626	30·710	30·795	30·879
0·60	30·964	31·048	31·132	31·216	31·300
0·61	31·383	31·467	31·550	31·633	31·716
0·62	31·799	31·882	31·964	32·047	32·129
0·63	32·211	32·293	32·375	32·456	32·538
0·64	32·619	32·700	32·782	32·862	32·943
0·65	33·024	33·104	33·185	33·265	33·345
0·66	33·425	33·505	33·584	33·664	33·743
0·67	33·822	33·901	33·980	34·059	34·137
0·68	34·216	34·294	34·372	34·450	34·528
0·69	34·606	34·683	34·761	34·838	34·915
0·70	34·992	35·069	35·146	35·222	35·298
0·71	35·375	35·451	35·527	35·603	35·678
0·72	35·754	35·829	35·905	35·980	36·055
0·73	36·129	36·204	36·279	36·353	36·427
0·74	36·501	36·575	36·649	36·723	36·796
0·75	36·870	36·943	37·016	37·089	37·162
0·76	37·235	37·307	37·380	37·452	37·524
0·77	37·596	37·668	37·740	37·811	37·883
0·78	37·954	38·025	38·096	38·167	38·238
0·79	38·309	38·379	38·450	38·520	38·590
0·80	38·660	38·730	38·799	38·869	38·938
0·81	39·007	39·077	39·146	39·214	39·283
0·82	39·352	39·420	39·489	39·557	39·625
0·83	39·693	39·760	39·828	39·896	39·963
0·84	40·030	40·097	40·164	40·231	40·298
0·85	40·365	40·431	40·497	40·564	40·630
0·86	40·696	40·761	40·827	40·893	40·958
0·87	41·023	41·088	41·153	41·218	41·283
0·88	41·348	41·412	41·477	41·541	41·605
0·89	41·669	41·733	41·797	41·860	41·924
0·90	41·987	42·050	42·114	42·177	42·239
0·91	42·302	42·365	42·427	42·490	42·552
0·92	42·614	42·676	42·738	42·800	42·861
0·93	42·923	42·984	43·045	43·107	43·168
0·94	43·229	43·289	43·350	43·410	43·471
0·95	43·531	43·591	43·651	43·711	43·771
0·96	43·831	43·890	43·950	44·009	44·068
0·97	44·128	44·187	44·245	44·304	44·363
0·98	44·421	44·480	44·538	44·596	44·654
0·99	44·712	44·770	44·828	44·885	44·943
1·00	45·000	45·057	45·114	45·171	45·228

Table 6   Sin X 357

	0·0	0·2	0·4	0·6	0·8
0·00	0·0000	0·0035	0·0070	0·0105	0·0140
1·00	0·0175	0·0209	0·0244	0·0279	0·0314
2·00	0·0349	0·0384	0·0419	0·0454	0·0488
3·00	0·0523	0·0558	0·0593	0·0628	0·0663
4·00	0·0698	0·0732	0·0767	0·0802	0·0837
5·00	0·0872	0·0906	0·0941	0·0976	0·1011
6·00	0·1045	0·1080	0·1115	0·1149	0·1184
7·00	0·1219	0·1253	0·1288	0·1323	0·1357
8·00	0·1392	0·1426	0·1461	0·1495	0·1530
9·00	0·1564	0·1599	0·1633	0·1668	0·1702
10·00	0·1736	0·1771	0·1805	0·1840	0·1874
11·00	0·1908	0·1942	0·1977	0·2011	0·2045
12·00	0·2079	0·2133	0·2147	0·2181	0·2215
13·00	0·2250	0·2284	0·2317	0·2351	0·2385
14·00	0·2419	0·2453	0·2487	0·2521	0·2554
15·00	0·2588	0·2622	0·2656	0·2689	0·2723
16·00	0·2756	0·2790	0·2823	0·2857	0·2890
17·00	0·2924	0·2957	0·2990	0·3024	0·3057
18·00	0·3090	0·3123	0·3156	0·3190	0·3223
19·00	0·3256	0·3289	0·3322	0·3355	0·3387
20·00	0·3420	0·3453	0·3486	0·3518	0·3551
21·00	0·3584	0·3616	0·3649	0·3681	0·3714
22·00	0·3746	0·3778	0·3811	0·3843	0·3875
23·00	0·3907	0·3939	0·3971	0·4003	0·4035
24·00	0·4067	0·4099	0·4131	0·4163	0·4195
25·00	0·4226	0·4258	0·4289	0·4321	0·4352
26·00	0·4384	0·4415	0·4446	0·4478	0·4509
27·00	0·4540	0·4571	0·4602	0·4633	0·4664
28·00	0·4695	0·4726	0·4756	0·4787	0·4818
29·00	0·4848	0·4879	0·4909	0·4939	0·4970
30·00	0·5000	0·5030	0·5060	0·5090	0·5120
31·00	0·5150	0·5180	0·5210	0·5240	0·5270
32·00	0·5299	0·5329	0·5358	0·5388	0·5417
33·00	0·5446	0·5476	0·5505	0·5534	0·5563
34·00	0·5592	0·5621	0·5650	0·5678	0·5707
35·00	0·5736	0·5764	0·5793	0·5821	0·5850
36·00	0·5878	0·5906	0·5934	0·5962	0·5990
37·00	0·6018	0·6046	0·6074	0·6101	0·6129
38·00	0·6157	0·6184	0·6211	0·6239	0·6266
39·00	0·6293	0·6320	0·6347	0·6374	0·6401
40·00	0·6428	0·6455	0·6481	0·6508	0·6534
41·00	0·6561	0·6587	0·6613	0·6639	0·6665
42·00	0·6691	0·6717	0·6743	0·6769	0·6794
43·00	0·6820	0·6845	0·6871	0·6896	0·6921
44·00	0·6947	0·6972	0·6997	0·7022	0·7046
45·00	0·7071	0·7096	0·7120	0·7145	0·7169
46·00	0·7193	0·7218	0·7242	0·7266	0·7290
47·00	0·7314	0·7337	0·7361	0·7385	0·7408
48·00	0·7431	0·7455	0·7478	0·7501	0·7524
49·00	0·7547	0·7570	0·7593	0·7615	0·7638
50·00	0·7660	0·7683	0·7705	0·7727	0·7749

Table 6—continued

	0·0	0·2	0·4	0·6	0·8
51·00	0·7771	0·7793	0·7815	0·7837	0·7859
52·00	0·7880	0·7902	0·7923	0·7944	0·7965
53·00	0·7986	0·8007	0·8028	0·8049	0·8070
54·00	0·8090	0·8111	0·8131	0·8151	0·8171
55·00	0·8192	0·8211	0·8231	0·8251	0·8271
56·00	0·8290	0·8310	0·8329	0·8348	0·8368
57·00	0·8387	0·8406	0·8425	0·8443	0·8462
58·00	0·8480	0·8499	0·8517	0·8536	0·8554
59·00	0·8572	0·8590	0·8607	0·8625	0·8643
60·00	0·8660	0·8678	0·8695	0·8712	0·8729
61·00	0·8746	0·8763	0·8780	0·8796	0·8813
62·00	0·8829	0·8846	0·8862	0·8878	0·8894
63·00	0·8910	0·8926	0·8942	0·8957	0·8973
64·00	0·8988	0·9003	0·9018	0·9033	0·9048
65·00	0·9063	0·9078	0·9092	0·9107	0·9121
66·00	0·9135	0·9150	0·9164	0·9178	0·9191
67·00	0·9205	0·9219	0·9232	0·9245	0·9259
68·00	0·9272	0·9285	0·9298	0·9311	0·9323
69·00	0·9336	0·9348	0·9361	0·9373	0·9385
70·00	0·9397	0·9409	0·9421	0·9432	0·9444
71·00	0·9455	0·9466	0·9478	0·9489	0·9500
72·00	0·9511	0·9521	0·9532	0·9542	0·9553
73·00	0·9563	0·9573	0·9583	0·9593	0·9603
74·00	0·9613	0·9622	0·9632	0·9641	0·9650
75·00	0·9659	0·9668	0·9677	0·9686	0·9694
76·00	0·9703	0·9711	0·9720	0·9728	0·9736
77·00	0·9744	0·9751	0·9759	0·9767	0·9774
78·00	0·9781	0·9789	0·9796	0·9803	0·9810
79·00	0·9816	0·9823	0·9829	0·9836	0·9842
80·00	0·9848	0·9854	0·9860	0·9866	0·9871
81·00	0·9877	0·9882	0·9888	0·9893	0·9898
82·00	0·9903	0·9907	0·9912	0·9917	0·9921
83·00	0·9925	0·9930	0·9934	0·9938	0·9942
84·00	0·9945	0·9949	0·9952	0·9956	0·9959
85·00	0·9962	0·9965	0·9968	0·9971	0·9973
86·00	0·9976	0·9978	0·9980	0·9982	0·9984
87·00	0·9986	0·9988	0·9990	0·9991	0·9993
88·00	0·9994	0·9995	0·9996	0·9997	0·9998
89·00	0·9998	0·9999	0·9999	1·0000	1·0000
90·00	1·0000	1·0000	1·0000	0·9999	0·9999

Table 7   Cos X 359

	0·0	0·2	0·4	0·6	0·8
0·00	1·0000	1·0000	1·0000	0·9999	0·9999
1·00	0·9998	0·9998	0·9997	0·9996	0·9995
2·00	0·9994	0·9993	0·9991	0·9990	0·9988
3·00	0·9986	0·9984	0·9982	0·9980	0·9978
4·00	0·9976	0·9973	0·9971	0·9968	0·9965
5·00	0·9962	0·9959	0·9956	0·9952	0·9949
6·00	0·9945	0·9942	0·9938	0·9934	0·9930
7·00	0·9925	0·9921	0·9917	0·9912	0·9907
8·00	0·9903	0·9898	0·9893	0·9888	0·9882
9·00	0·9877	0·9871	0·9866	0·9860	0·9854
10·00	0·9848	0·9842	0·9836	0·9829	0·9823
11·00	0·9816	0·9810	0·9803	0·9796	0·9789
12·00	0·9781	0·9774	0·9767	0·9759	0·9751
13·00	0·9744	0·9736	0·9728	0·9720	0·9711
14·00	0·9703	0·9694	0·9686	0·9677	0·9668
15·00	0·9659	0·9650	0·9641	0·9632	0·9622
16·00	0·9613	0·9603	0·9593	0·9583	0·9573
17·00	0·9563	0·9553	0·9542	0·9532	0·9521
18·00	0·9511	0·9500	0·9489	0·9478	0·9466
19·00	0·9455	0·9444	0·9432	0·9421	0·9409
20·00	0·9397	0·9385	0·9373	0·9361	0·9348
21·00	0·9336	0·9323	0·9311	0·9298	0·9285
22·00	0·9272	0·9259	0·9245	0·9232	0·9219
23·00	0·9205	0·9191	0·9178	0·9164	0·9150
24·00	0·9135	0·9121	0·9107	0·9092	0·9078
25·00	0·9063	0·9048	0·9033	0·9018	0·9003
26·00	0·8988	0·8973	0·8957	0·8942	0·8926
27·00	0·8910	0·8894	0·8878	0·8862	0·8846
28·00	0·8829	0·8813	0·8796	0·8780	0·8763
29·00	0·8746	0·8729	0·8712	0·8695	0·8678
30·00	0·8660	0·8643	0·8625	0·8607	0·8590
31·00	0·8572	0·8554	0·8536	0·8517	0·8499
32·00	0·8480	0·8462	0·8443	0·8425	0·8406
33·00	0·8387	0·8368	0·8348	0·8329	0·8310
34·00	0·8290	0·8271	0·8251	0·8231	0·8211
35·00	0·8192	0·8171	0·8151	0·8131	0·8111
36·00	0·8090	0·8070	0·8049	0·8028	0·8007
37·00	0·7986	0·7965	0·7944	0·7923	0·7902
38·00	0·7880	0·7859	0·7837	0·7815	0·7793
39·00	0·7771	0·7749	0·7727	0·7705	0·7683
40·00	0·7660	0·7638	0·7615	0·7593	0·7570
41·00	0·7547	0·7524	0·7501	0·7478	0·7455
42·00	0·7431	0·7408	0·7385	0·7361	0·7337
43·00	0·7314	0·7290	0·7266	0·7242	0·7218
44·00	0·7193	0·7169	0·7145	0·7120	0·7096
45·00	0·7071	0·7046	0·7022	0·6997	0·6972
46·00	0·6947	0·6921	0·6896	0·6871	0·6845
47·00	0·6820	0·6794	0·6769	0·6743	0·6717
48·00	0·6691	0·6665	0·6639	0·6613	0·6587
49·00	0·6561	0·6534	0·6508	0·6481	0·6455
50·00	0·6428	0·6401	0·6374	0·7347	0·6320

Table 7—continued

	0·0	0·2	0·4	0·6	0·8
51·00	0·6239	0·6266	0·6239	0·6211	0·6184
52·00	0·6157	0·6129	0·6101	0·6074	0·6046
53·00	0·6018	0·5990	0·5962	0·5934	0·5096
54·00	0·5878	0·5850	0·5821	0·5793	0·5764
55·00	0·5736	0·5707	0·5678	0·5650	0·5621
56·00	0·5592	0·5563	0·5534	0·5505	0·5476
57·00	0·5446	0·5417	0·5388	0·5358	0·5329
58·00	0·5299	0·5270	0·5240	0·5210	0·5180
59·00	0·5150	0·5120	0·5090	0·5060	0·5030
60·00	0·5000	0·4970	0·4939	0·4909	0·4879
61·00	0·4848	0·4818	0·4787	0·4756	0·4726
62·00	0·4695	0·4664	0·4633	0·4602	0·4571
63·00	0·4540	0·4509	0·4478	0·4446	0·4415
64·00	0·4384	0·4352	0·4321	0·4289	0·4258
65·00	0·4226	0·4195	0·4163	0·4131	0·4099
66·00	0·4067	0·4035	0·4003	0·3971	0·3939
67·00	0·3907	0·3875	0·3843	0·3811	0·3778
68·00	0·3746	0·3714	0·3681	0·3649	0·3616
69·00	0·3584	0·3551	0·3518	0·3486	0·3453
70·00	0·3420	0·3387	0·3355	0·3322	0·3289
71·00	0·3256	0·3223	0·3190	0·3156	0·3123
72·00	0·3090	0·3057	0·3024	0·2990	0·2957
73·00	0·2924	0·2890	0·2857	0·2823	0·2790
74·00	0·2756	0·2723	0·2689	0·2656	0·2622
75·00	0·2588	0·2554	0·2521	0·2487	0·2453
76·00	0·2419	0·2385	0·2351	0·2317	0·2284
77·00	0·2250	0·2215	0·2181	0·2147	0·2113
78·00	0·2079	0·2045	0·2011	0·1977	0·1942
79·00	0·1908	0·1874	0·1840	0·1805	0·1771
80·00	0·1736	0·1702	0·1668	0·1633	0·1599
81·00	0·1564	0·1530	0·1495	0·1461	0·1426
82·00	0·1392	0·1357	0·1323	0·1288	0·1253
83·00	0·1219	0·1184	0·1149	0·1115	0·1080
84·00	0·1045	0·1011	0·0976	0·0941	0·0906
85·00	0·0872	0·0837	0·0802	0·0767	0·0732
86·00	0·0698	0·0663	0·0628	0·0593	0·0558
87·00	0·0523	0·0488	0·0454	0·0419	0·0384
88·00	0·0349	0·0314	0·0279	0·0244	0·0209
89·00	0·0175	0·0140	0·0105	0·0070	0·0035
90·00	0·0000	−0·0035	−0·0070	−0·0105	−0·0140

Table 8   1/X   361

	0·00	0·02	0·04	0·06	0·08
1·00	1·0000	0·9804	0·9615	0·9434	0·9259
1·10	0·9091	0·8929	0·8772	0·8621	0·8475
1·20	0·8333	0·8197	0·8065	0·7937	0·7812
1·30	0·7692	0·7576	0·7463	0·7353	0·7246
1·40	0·7143	0·7042	0·6944	0·6849	0·6757
1·50	0·6667	0·6579	0·6494	0·6410	0·6329
1·60	0·6250	0·6173	0·6098	0·6024	0·5952
1·70	0·5882	0·5814	0·5747	0·5682	0·5618
1·80	0·5556	0·5495	0·5435	0·5376	0·5319
1·90	0·5263	0·5208	0·5155	0·5102	0·5051
2·00	0·5000	0·4950	0·4902	0·4854	0·4808
2·10	0·4762	0·4717	0·4673	0·4630	0·4587
2·20	0·4545	0·4505	0·4464	0·4425	0·4386
2·30	0·4348	0·4310	0·4274	0·4237	0·4202
2·40	0·4167	0·4132	0·4098	0·4065	0·4032
2·50	0·4000	0·3968	0·3937	0·3906	0·3876
2·60	0·3846	0·3817	0·3788	0·3759	0·3731
2·70	0·3704	0·3676	0·3650	0·3623	0·3597
2·80	0·3571	0·3546	0·3521	0·3497	0·3472
2·90	0·3448	0·3425	0·3401	0·3378	0·3356
3·00	0·3333	0·3311	0·3289	0·3268	0·3247
3·10	0·3226	0·3205	0·3185	0·3165	0·3145
3·20	0·3125	0·3106	0·3086	0·3067	0·3049
3·30	0·3030	0·3012	0·2994	0·2976	0·2959
3·40	0·2941	0·2924	0·2907	0·2890	0·2874
3·50	0·2857	0·2841	0·2825	0·2809	0·2793
3·60	0·2778	0·2762	0·2747	0·2732	0·2717
3·70	0·2703	0·2688	0·2674	0·2660	0·2646
3·80	0·2632	0·2618	0·2604	0·2591	0·2577
3·90	0·2564	0·2551	0·2538	0·2525	0·2513
4·00	0·2500	0·2488	0·2475	0·2463	0·2451
4·10	0·2439	0·2427	0·2415	0·2404	0·2392
4·20	0·2381	0·2370	0·2358	0·2347	0·2336
4·30	0·2326	0·2315	0·2304	0·2294	0·2283
4·40	0·2273	0·2262	0·2252	0·2242	0·2232
4·50	0·2222	0·2212	0·2203	0·2193	0·2183
4·60	0·2174	0·2165	0·2155	0·2146	0·2137
4·70	0·2128	0·2119	0·2110	0·2101	0·2092
4·80	0·2083	0·2075	0·2066	0·2058	0·2049
4·90	0·2041	0·2033	0·2024	0·2016	0·2008
5·00	0·2000	0·1992	0·1984	0·1976	0·1969
5·10	0·1961	0·1953	0·1946	0·1938	0·1931
5·20	0·1923	0·1916	0·1908	0·1901	0·1894
5·30	0·1887	0·1880	0·1873	0·1866	0·1859
5·40	0·1852	0·1845	0·1838	0·1832	0·1825
5·50	0·1818	0·1812	0·1805	0·1799	0·1792
5·60	0·1786	0·1779	0·1773	0·1767	0·1761
5·70	0·1754	0·1748	0·1742	0·1736	0·1730
5·80	0·1724	0·1718	0·1712	0·1706	0·1701
5·90	0·1695	0·1689	0·1684	0·1678	0·1672
6·00	0·1667	0·1661	0·1656	0·1650	0·1645

Table 8—continued

	0·00	0·02	0·04	0·06	0·08
6·10	0·1639	0·1634	0·1629	0·1623	0·1618
6·20	0·1613	0·1608	0·1603	0·1597	0·1592
6·30	0·1587	0·1582	0·1577	0·1572	0·1567
6·40	0·1563	0·1558	0·1553	0·1548	0·1543
6·50	0·1538	0·1534	0·1529	0·1524	0·1520
6·60	0·1515	0·1511	0·1506	0·1502	0·1497
6·70	0·1493	0·1488	0·1484	0·1479	0·1475
6·80	0·1471	0·1466	0·1462	0·1458	0·1453
6·90	0·1449	0·1445	0·1441	0·1437	0·1433
7·00	0·1429	0·1425	0·1420	0·1416	0·1412
7·10	0·1408	0·1404	0·1401	0·1397	0·1393
7·20	0·1389	0·1385	0·1381	0·1377	0·1374
7·30	0·1370	0·1366	0·1362	0·1359	0·1355
7·40	0·1351	0·1348	0·1344	0·1340	0·1337
7·50	0·1333	0·1330	0·1326	0·1323	0·1319
7·60	0·1316	0·1312	0·1309	0·1305	0·1302
7·70	0·1299	0·1295	0·1292	0·1289	0·1285
7·80	0·1282	0·1279	0·1276	0·1272	0·1269
7·90	0·1266	0·1263	0·1259	0·1256	0·1253
8·00	0·1250	0·1247	0·1244	0·1241	0·1238
8·10	0·1235	0·1232	0·1229	0·1225	0·1222
8·20	0·1220	0·1217	0·1214	0·1211	0·1208
8·30	0·1205	0·1202	0·1199	0·1196	0·1193
8·40	0·1190	0·1188	0·1185	0·1182	0·1179
8·50	0·1176	0·1174	0·1171	0·1168	0·1166
8·60	0·1163	0·1160	0·1157	0·1155	0·1152
8·70	0·1149	0·1147	0·1144	0·1142	0·1139
8·80	0·1136	0·1134	0·1131	0·1129	0·1126
8·90	0·1124	0·1121	0·1119	0·1116	0·1114
9·00	0·1111	0·1109	0·1106	0·1104	0·1101
9·10	0·1099	0·1096	0·1094	0·1092	0·1089
9·20	0·1087	0·1085	0·1082	0·1080	0·1078
9·30	0·1075	0·1073	0·1071	0·1068	0·1066
9·40	0·1064	0·1062	0·1059	0·1057	0·1055
9·50	0·1053	0·1050	0·1048	0·1046	0·1044
9·60	0·1042	0·1040	0·1037	0·1035	0·1033
9·70	0·1031	0·1029	0·1027	0·1025	0·1022
9·80	0·1020	0·1018	0·1016	0·1014	0·1012
9·90	0·1010	0·1008	0·1006	0·1004	0·1002
10·00	0·1000	0·0998	0·0996	0·0994	0·0992

Table 9 X^2 363

	0·00	0·02	0·04	0·06	0·08
1·00	1·0000	1·0404	1·0816	1·1236	1·1664
1·10	1·2100	1·2544	1·2996	1·3456	1·3924
1·20	1·4400	1·4884	1·5376	1·5876	1·6384
1·30	1·6900	1·7424	1·7956	1·8496	1·9044
1·40	1·9600	2·0164	2·0736	2·1316	2·1904
1·50	2·2500	2·3104	2·3716	2·4336	2·4964
1·60	2·5600	2·6244	2·6896	2·7556	2·8224
1·70	2·8900	2·9584	3·0276	3·0976	3·1684
1·80	3·2400	3·3124	3·3856	3·4596	3·5344
1·90	3·6100	3·6864	3·7636	3·8416	3·9204
2·00	4·0000	4·0804	4·1616	4·2436	4·3264
2·10	4·4100	4·4944	4·5796	4·6656	4·7524
2·20	4·8400	4·9284	5·0176	5·1076	5·1984
2·30	5·2900	5·3824	5·4756	5·5696	5·6644
2·40	5·7600	5·8564	5·9536	6·0516	6·1504
2·50	6·2500	6·3504	6·4516	6·5536	6·6564
2·60	6·7600	6·8644	6·9696	7·0756	7·1824
2·70	7·2900	7·3984	7·5076	7·6176	7·7284
2·80	7·8400	7·9524	8·0656	8·1796	8·2944
2·90	8·4100	8·5264	8·6436	8·7616	8·8804
3·00	9·0000	9·1204	9·2416	9·3636	9·4864
3·10	9·6100	9·7344	9·8596	9·9856	10·112
3·20	10·240	10·368	10·498	10·628	10·758
3·30	10·890	11·022	11·156	11·290	11·424
3·40	11·560	11·696	11·834	11·972	12·110
3·50	12·250	12·390	12·532	12·674	12·816
3·60	12·960	13·104	13·250	13·396	13·542
3·70	13·690	13·838	13·988	14·138	14·288
3·80	14·440	14·592	14·746	14·900	15·054
3·90	15·210	15·366	15·524	15·682	15·840
4·00	16·000	16·160	16·322	16·484	16·646
4·10	16·810	16·974	17·140	17·306	17·472
4·20	17·640	17·808	17·978	18·148	18·318
4·30	18·490	18·662	18·836	19·010	19·184
4·40	19·360	19·536	19·714	19·892	20·070
4·50	20·250	20·430	20·612	20·794	20·976
4·60	21·160	21·344	21·530	21·716	21·902
4·70	22·090	22·278	22·468	22·658	22·848
4·80	23·040	23·232	23·426	23·620	23·814
4·90	24·010	24·206	24·404	24·602	24·800
5·00	25·000	25·200	25·402	25·604	25·806
5·10	26·010	26·214	26·420	26·626	26·832
5·20	27·040	27·248	27·458	27·668	27·878
5·30	28·090	28·302	28·516	28·730	28·944
5·40	29·160	29·376	29·594	29·812	30·030
5·50	30·250	30·470	30·692	30·914	31·136
5·60	31·360	31·584	31·810	32·036	32·262
5·70	32·490	32·718	32·948	33·178	33·408
5·80	33·640	33·872	34·106	34·340	34·574
5·90	34·810	35·046	35·284	35·522	35·760
6·00	36·000	36·240	36·482	36·724	36·966

Table 9—continued

	0·00	0·02	0·04	0·06	0·08
6·10	37·210	37·454	37·700	37·946	38·192
6·20	38·440	38·688	38·938	39·188	39·438
6·30	39·690	39·942	40·196	40·450	40·704
6·40	40·960	41·216	41·474	41·732	41·990
6·50	42·250	42·510	42·772	43·034	43·296
6·60	43·560	43·824	44·090	44·356	44·622
6·70	44·890	45·158	45·428	45·698	45·968
6·80	46·240	46·512	46·786	47·060	47·334
6·90	47·610	47·886	48·164	48·442	48·720
7·00	49·000	49·280	49·562	49·844	50·126
7·10	50·410	50·694	50·980	51·266	51·552
7·20	51·840	52·128	52·418	52·708	52·998
7·30	53·290	53·582	53·876	54·170	54·464
7·40	54·760	55·056	55·354	55·652	55·950
7·50	56·250	56·550	56·852	57·154	57·456
7·60	57·760	58·064	58·370	58·676	58·982
7·70	59·290	59·598	59·908	60·218	60·528
7·80	60·840	61·152	61·466	61·780	62·094
7·90	62·410	62·726	63·044	63·362	63·680
8·00	64·000	64·320	64·642	64·964	65·286
8·10	65·610	65·934	66·260	66·586	66·912
8·20	67·240	67·568	67·898	68·228	68·558
8·30	68·890	69·222	69·556	69·890	70·224
8·40	70·560	70·896	71·234	71·572	71·910
8·50	72·250	72·590	72·932	73·274	73·616
8·60	73·960	74·304	74·650	74·996	75·342
8·70	75·690	76·038	76·388	76·738	77·088
8·80	77·440	77·792	78·146	78·500	78·854
8·90	79·210	79·566	79·924	80·282	80·640
9·00	81·000	81·360	81·722	82·084	82·446
9·10	82·810	83·174	83·540	83·906	84·272
9·20	84·640	85·008	85·378	85·748	86·118
9·30	86·490	86·862	87·236	87·610	87·984
9·40	88·360	88·736	89·114	89·492	89·870
9·50	90·250	90·630	91·012	91·394	91·776
9·60	92·160	92·544	92·930	93·316	93·702
9·70	94·090	94·478	94·868	95·258	95·648
9·80	96·040	96·432	96·826	97·220	97·614
9·90	98·010	98·406	98·804	99·202	99·600
10·00	100·00	100·40	100·80	101·20	101·610

Table 10 $1 - X^2$ 365

	0·000	0·002	0·004	0·006	0·008
0·00	1·0000	1·0000	1·0000	1·0000	0·9999
0·01	0·9999	0·9999	0·9998	0·9997	0·9997
0·02	0·9996	0·9995	0·9994	0·9993	0·9992
0·03	0·9991	0·9990	0·9988	0·9987	0·9986
0·04	0·9984	0·9982	0·9981	0·9979	0·9977
0·05	0·9975	0·9973	0·9971	0·9969	0·9966
0·06	0·9964	0·9962	0·9959	0·9956	0·9954
0·07	0·9951	0·9948	0·9945	0·9942	0·9939
0·08	0·9936	0·9933	0·9929	0·9926	0·9923
0·09	0·9919	0·9915	0·9912	0·9908	0·9904
0·10	0·9900	0·9896	0·9892	0·9888	0·9883
0·11	0·9879	0·9875	0·9870	0·9865	0·9861
0·12	0·9856	0·9851	0·9846	0·9841	0·9836
0·13	0·9831	0·9826	0·9820	0·9815	0·9810
0·14	0·9804	0·9798	0·9793	0·9787	0·9781
0·15	0·9775	0·9769	0·9763	0·9757	0·9750
0·16	0·9744	0·9738	0·9731	0·9724	0·9718
0·17	0·9711	0·9704	0·9697	0·9690	0·9683
0·18	0·9676	0·9669	0·9661	0·9654	0·9647
0·19	0·9639	0·9631	0·9624	0·9616	0·9608
0·20	0·9600	0·9592	0·9484	0·9576	0·9567
0·21	0·9559	0·9551	0·9542	0·9533	0·9525
0·22	0·9516	0·9507	0·9498	0·9489	0·9480
0·23	0·9471	0·9462	0·9452	0·9443	0·9434
0·24	0·9424	0·9414	0·9405	0·9395	0·9385
0·25	0·9375	0·9365	0·9355	0·9345	0·9334
0·26	0·9324	0·9314	0·9303	0·9292	0·9282
0·27	0·9271	0·9260	0·9249	0·9238	0·9227
0·28	0·9216	0·9205	0·9193	0·9182	0·9171
0·29	0·9159	0·9147	0·9136	0·9124	0·9112
0·30	0·9100	0·9088	0·9076	0·9064	0·9051
0·31	0·9039	0·9027	0·9014	0·9001	0·8989
0·32	0·8976	0·8963	0·8950	0·8937	0·8924
0·33	0·8911	0·8898	0·8884	0·8871	0·8858
0·34	0·8844	0·8830	0·8817	0·8803	0·8789
0·35	0·8775	0·8761	0·8747	0·8733	0·8718
0·36	0·8704	0·8690	0·8675	0·8660	0·8646
0·37	0·8631	0·8616	0·8601	0·8586	0·8571
0·38	0·8556	0·8541	0·8525	0·8510	0·8495
0·39	0·8479	0·8463	0·8448	0·8432	0·8416
0·40	0·8400	0·8384	0·8368	0·8352	0·8335
0·41	0·8319	0·8303	0·8286	0·8269	0·8253
0·42	0·8236	0·8219	0·8202	0·8185	0·8168
0·43	0·8151	0·8134	0·8116	0·8099	0·8082
0·44	0·8064	0·8046	0·8029	0·8011	0·7993
0·45	0·7975	0·7957	0·7939	0·7921	0·7902
0·46	0·7884	0·7866	0·7847	0·7828	0·7810
0·47	0·7791	0·7772	0·7753	0·7734	0·7715
0·48	0·7696	0·7677	0·7657	0·7638	0·7619
0·49	0·7599	0·7579	0·7560	0·7540	0·7520
0·50	0·7500	0·7480	0·7460	0·7440	0·7419

Table 10—continued

	0·000	0·002	0·004	0·006	0·008
0·51	0·7399	0·7379	0·7358	0·7337	0·7317
0·52	0·7296	0·7275	0·7254	0·7233	0·7212
0·53	0·7191	0·7170	0·7148	0·7127	0·7106
0·54	0·7084	0·7062	0·7041	0·7019	0·6997
0·55	0·6975	0·6953	0·6931	0·6909	0·6886
0·56	0·6864	0·6842	0·6819	0·6796	0·6774
0·57	0·6751	0·6728	0·6705	0·6682	0·6659
0·58	0·6636	0·6613	0·6589	0·6566	0·6543
0·59	0·6519	0·6495	0·6472	0·6448	0·6424
0·60	0·6400	0·6376	0·6352	0·6328	0·6303
0·61	0·6279	0·6255	0·6230	0·6205	0·6181
0·62	0·6156	0·6131	0·6106	0·6081	0·6056
0·63	0·6031	0·6006	0·5980	0·5955	0·5930
0·64	0·5904	0·5878	0·5853	0·5827	0·5801
0·65	0·5775	0·5749	0·5723	0·5697	0·5670
0·66	0·5644	0·5618	0·5591	0·5564	0·5538
0·67	0·5511	0·5484	0·5457	0·5430	0·5403
0·68	0·5376	0·5349	0·5321	0·5294	0·5267
0·69	0·5239	0·5211	0·5184	0·5156	0·5128
0·70	0·5100	0·5072	0·5044	0·5016	0·4987
0·71	0·4959	0·4931	0·4902	0·4873	0·4845
0·72	0·4816	0·4787	0·4758	0·4729	0·4700
0·73	0·4671	0·4642	0·4612	0·4583	0·4554
0·74	0·4524	0·4494	0·4465	0·4435	0·4405
0·75	0·4375	0·4345	0·4315	0·4285	0·4254
0·76	0·4224	0·4194	0·4163	0·4132	0·4102
0·77	0·4071	0·4040	0·4009	0·3978	0·3947
0·78	0·3916	0·3885	0·3853	0·3822	0·3791
0·79	0·3759	0·3727	0·3696	0·3664	0·3692
0·80	0·3600	0·3568	0·3536	0·3504	0·3471
0·81	0·3439	0·3407	0·3374	0·3341	0·3309
0·82	0·3276	0·3243	0·3210	0·3177	0·3144
0·83	0·3111	0·3078	0·3044	0·3011	0·2978
0·84	0·2944	0·2910	0·2877	0·2843	0·2809
0·85	0·2775	0·2741	0·2707	0·2673	0·2638
0·86	0·2604	0·2570	0·2535	0·2500	0·2466
0·87	0·2431	0·2396	0·2361	0·2326	0·2291
0·88	0·2256	0·2221	0·2185	0·2150	0·2115
0·89	0·2079	0·2043	0·2008	0·1972	0·1936
0·90	0·1900	0·1864	0·1828	0·1792	0·1755
0·91	0·1719	0·1683	0·1646	0·1609	0·1573
0·92	0·1536	0·1499	0·1462	0·1425	0·1388
0·93	0·1351	0·1314	0·1276	0·1239	0·1202
0·94	0·1164	0·1126	0·1089	0·1051	0·1013
0·95	0·0975	0·0937	0·0899	0·0861	0·0822
0·96	0·0784	0·0746	0·0707	0·0668	0·0630
0·97	0·0591	0·0552	0·0513	0·0474	0·0435
0·98	0·0396	0·0357	0·0317	0·0278	0·0239
0·99	0·0199	0·0159	0·0120	0·0080	0·0040
1·00	0·0000	−0·0040	−0·0080	−0·0120	−0·0161

Table 11 $\sqrt{(1+X^2)}$ 367

	0·000	0·002	0·004	0·006	0·008
0·00	1·0000	1·0000	1·0000	1·0000	1·0000
0·01	1·0000	1·0001	1·0001	1·0001	1·0002
0·02	1·0002	1·0002	1·0003	1·0003	1·0004
0·03	1·0004	1·0005	1·0006	1·0006	1·0007
0·04	1·0008	1·0009	1·0010	1·0011	1·0012
0·05	1·0012	1·0014	1·0015	1·0016	1·0017
0·06	1·0018	1·0019	1·0020	1·0022	1·0023
0·07	1·0024	1·0026	1·0027	1·0029	1·0030
0·08	1·0032	1·0034	1·0035	1·0037	1·0039
0·09	1·0040	1·0042	1·0044	1·0046	1·0048
0·10	1·0050	1·0052	1·0054	1·0056	1·0058
0·11	1·0060	1·0063	1·0065	1·0067	1·0069
0·12	1·0072	1·0074	1·0077	1·0079	1·0082
0·13	1·0084	1·0087	1·0089	1·0092	1·0095
0·14	1·0098	1·0100	1·0103	1·0106	1·0109
0·15	1·0112	1·0115	1·0118	1·0121	1·0124
0·16	1·0127	1·0130	1·0134	1·0137	1·0140
0·17	1·0143	1·0147	1·0150	1·0154	1·0157
0·18	1·0161	1·0164	1·0168	1·0172	1·0175
0·19	1·0179	1·0183	1·0186	1·0190	1·0194
0·20	1·0198	1·0202	1·0206	1·0210	1·0214
0·21	1·0218	1·0222	1·0226	1·0231	1·0235
0·22	1·0239	1·0243	1·0248	1·0252	1·0257
0·23	1·0261	1·0266	1·0270	1·0275	1·0279
0·24	1·0284	1·0289	1·0293	1·0298	1·0303
0·25	1·0308	1·0313	1·0318	1·0322	1·0327
0·26	1·0332	1·0338	1·0343	1·0348	1·0353
0·27	1·0358	1·0363	1·0369	1·0374	1·0379
0·28	1·0385	1·0390	1·0395	1·0401	1·0406
0·29	1·0412	1·0418	1·0423	1·0429	1·0435
0·30	1·0440	1·0446	1·0452	1·0458	1·0464
0·31	1·0469	1·0475	1·0481	1·0487	1·0493
0·32	1·0500	1·0506	1·0512	1·0518	1·0524
0·33	1·0530	1·0537	1·0543	1·0549	1·0556
0·34	1·0562	1·0569	1·0575	1·0582	1·0588
0·35	1·0595	1·0601	1·0608	1·0615	1·0622
0·36	1·0628	1·0635	1·0642	1·0649	1·0656
0·37	1·0663	1·0670	1·0676	1·0684	1·0691
0·38	1·0698	1·0705	1·0712	1·0719	1·0726
0·39	1·0734	1·0741	1·0748	1·0756	1·0763
0·40	1·0770	1·0778	1·0785	1·0793	1·0800
0·41	1·0808	1·0815	1·0823	1·0831	1·0838
0·42	1·0846	1·0854	1·0862	1·0870	1·0877
0·43	1·0885	1·0893	1·0901	1·0909	1·0917
0·44	1·0925	1·0933	1·0941	1·0950	1·0958
0·45	1·0966	1·0974	1·0982	1·0991	1·0999
0·46	1·1007	1·1016	1·1024	1·1032	1·1041
0·47	1·1049	1·1058	1·1067	1·1075	1·1084
0·48	1·1092	1·1101	1·1110	1·1118	1·1127
0·49	1·1136	1·1145	1·1154	1·1163	1·1171
0·50	1·1180	1·1189	1·1198	1·1207	1·1216

Table 11—continued

	0·000	0·002	0·004	0·006	0·008
0·51	1·1225	1·1235	1·1244	1·1253	1·1262
0·52	1·1271	1·1280	1·1290	1·1299	1·1308
0·53	1·1318	1·1327	1·1336	1·1346	1·1355
0·54	1·1365	1·1374	1·1384	1·1393	1·1403
0·55	1·1413	1·1422	1·1432	1·1442	1·1451
0·56	1·1461	1·1471	1·1481	1·1491	1·1501
0·57	1·1510	1·1520	1·1530	1·1540	1·1550
0·58	1·1560	1·1570	1·1580	1·1590	1·1601
0·59	1·1611	1·1621	1·1631	1·1641	1·1652
0·60	1·1662	1·1672	1·1683	1·1693	1·1703
0·61	1·1714	1·1724	1·1735	1·1745	1·1756
0·62	1·1766	1·1777	1·1787	1·1798	1·1808
0·63	1·1819	1·1830	1·1840	1·1851	1·1862
0·64	1·1873	1·1883	1·1894	1·1905	1·1916
0·65	1·1927	1·1938	1·1949	1·1960	1·1971
0·66	1·1982	1·1993	1·2004	1·2015	1·2026
0·67	1·2037	1·2048	1·2059	1·2071	1·2082
0·68	1·2093	1·2104	1·2116	1·2127	1·2138
0·69	1·2149	1·2161	1·2172	1·2184	1·2195
0·70	1·2207	1·2218	1·2230	1·2241	1·2253
0·71	1·2264	1·2276	1·2287	1·2299	1·2311
0·72	1·2322	1·2334	1·2346	1·2357	1·2369
0·73	1·2381	1·2393	1·2405	1·2417	1·2428
0·74	1·2440	1·2452	1·2464	1·2476	1·2488
0·75	1·2500	1·2512	1·2524	1·2536	1·2548
0·76	1·2560	1·2572	1·2584	1·2597	1·2609
0·77	1·2621	1·2633	1·2645	1·2658	1·2670
0·78	1·2682	1·2695	1·2707	1·2719	1·2732
0·79	1·2744	1·2756	1·2769	1·2781	1·2794
0·80	1·2806	1·2819	1·2831	1·2844	1·2856
0·81	1·2869	1·2882	1·2894	1·2907	1·2919
0·82	1·2932	1·2945	1·2958	1·2970	1·2983
0·83	1·2996	1·3009	1·3021	1·3034	1·3047
0·84	1·3060	1·3073	1·3086	1·3099	1·3111
0·85	1·3124	1·3137	1·3150	1·3163	1·3176
0·86	1·3189	1·3202	1·3216	1·3229	1·3242
0·87	1·3255	1·3268	1·3281	1·3294	1·3307
0·88	1·3321	1·3334	1·3347	1·3360	1·3374
0·89	1·3387	1·3400	1·3414	1·3427	1·3440
0·90	1·3454	1·3467	1·3480	1·3494	1·3507
0·91	1·3521	1·3534	1·3548	1·3561	1·3575
0·92	1·3588	1·3602	1·3615	1·3629	1·3643
0·93	1·3656	1·3670	1·3683	1·3697	1·3711
0·94	1·3724	1·3738	1·3752	1·3766	1·3779
0·95	1·3793	1·3807	1·3821	1·3835	1·3848
0·96	1·3862	1·3876	1·3890	1·3904	1·3918
0·97	1·3932	1·3946	1·3959	1·3973	1·3987
0·98	1·4001	1·4015	1·4029	1·4043	1·4058
0·99	1·4072	1·4086	1·4100	1·4114	1·4128
1·00	1·4142	1·4156	1·4170	1·4185	1·4199

Table 12 $1/(1 + X^2)$

	0·000	0·002	0·004	0·006	0·008
0·00	1·0000	1·0000	1·0000	1·0000	0·9999
0·01	0·9999	0·9999	0·9998	0·9997	0·9997
0·02	0·9996	0·9995	0·9994	0·9993	0·9992
0·03	0·9991	0·9990	0·9988	0·9987	0·9986
0·04	0·9984	0·9982	0·9981	0·9979	0·9977
0·05	0·9975	0·9973	0·9971	0·9969	0·9966
0·06	0·9964	0·9962	0·9959	0·9957	0·9954
0·07	0·9951	0·9948	0·9946	0·9943	0·9940
0·08	0·9936	0·9933	0·9930	0·9927	0·9923
0·09	0·9920	0·9916	0·9912	0·9909	0·9905
0·10	0·9901	0·9897	0·9893	0·9889	0·9885
0·11	0·9880	0·9876	0·9872	0·9867	0·9863
0·12	0·9858	0·9853	0·9849	0·9844	0·9839
0·13	0·9834	0·9829	0·9824	0·9818	0·9813
0·14	0·9808	0·9802	0·9797	0·9791	0·9786
0·15	0·9780	0·9774	0·9768	0·9762	0·9756
0·16	0·9750	0·9744	0·9738	0·9732	0·9726
0·17	0·9719	0·9713	0·9706	0·9700	0·9693
0·18	0·9686	0·9679	0·9673	0·9666	0·9659
0·19	0·9652	0·9644	0·9637	0·9630	0·9623
0·20	0·9615	0·9608	0·9600	0·9593	0·9585
0·21	0·9578	0·9570	0·9562	0·9554	0·9546
0·22	0·9538	0·9530	0·9522	0·9514	0·9506
0·23	0·9498	0·9489	0·9481	0·9472	0·9464
0·24	0·9455	0·9447	0·9438	0·9429	0·9421
0·25	0·9412	0·9403	0·9394	0·9385	0·9376
0·26	0·9367	0·9358	0·9348	0·9339	0·9330
0·27	0·9321	0·9311	0·9302	0·9292	0·9283
0·28	0·9273	0·9263	0·9254	0·9244	0·9234
0·29	0·9224	0·9214	0·9204	0·9194	0·9184
0·30	0·9174	0·9164	0·9154	0·9144	0·9134
0·31	0·9123	0·9113	0·9103	0·9092	0·9082
0·32	0·9071	0·9061	0·9050	0·9039	0·9029
0·33	0·9018	0·9007	0·8996	0·8986	0·8975
0·34	0·8964	0·8953	0·8942	0·8931	0·8920
0·35	0·8909	0·8898	0·8886	0·8875	0·8864
0·36	0·8853	0·8841	0·8830	0·8819	0·8807
0·37	0·8796	0·8784	0·8773	0·8761	0·8750
0·38	0·8738	0·8727	0·8715	0·8703	0·8692
0·39	0·8680	0·8668	0·8656	0·8644	0·8633
0·40	0·8621	0·8609	0·8597	0·8585	0·8573
0·41	0·8561	0·8549	0·8537	0·8525	0·8513
0·42	0·8501	0·8488	0·8476	0·8464	0·8452
0·43	0·8440	0·8427	0·8415	0·8403	0·8390
0·44	0·8378	0·8366	0·8353	0·8341	0·8328
0·45	0·8316	0·8304	0·8291	0·8279	0·8266
0·46	0·8254	0·8241	0·8228	0·8216	0·8203
0·47	0·8191	0·8178	0·8165	0·8153	0·8140
0·48	0·8127	0·8115	0·8102	0·8089	0·8077
0·49	0·8064	0·8051	0·8038	0·8026	0·8013
0·50	0·8000	0·7987	0·7974	0·7962	0·7949

Table 12—continued

	0·000	0·002	0·004	0·006	0·008
0·51	0·7936	0·7923	0·7910	0·7897	0·7884
0·52	0·7872	0·7859	0·7846	0·7833	0·7820
0·53	0·7807	0·7794	0·7781	0·7768	0·7755
0·54	0·7742	0·7729	0·7716	0·7703	0·7691
0·55	0·7678	0·7665	0·7652	0·7639	0·7626
0·56	0·7613	0·7600	0·7587	0·7574	0·7561
0·57	0·7548	0·7535	0·7522	0·7509	0·7496
0·58	0·7483	0·7470	0·7457	0·7444	0·7431
0·59	0·7418	0·7405	0·7392	0·7379	0·7366
0·60	0·7353	0·7340	0·7327	0·7314	0·7301
0·61	0·7288	0·7275	0·7262	0·7249	0·7236
0·62	0·7223	0·7210	0·7197	0·7185	0·7172
0·63	0·7159	0·7146	0·7133	0·7120	0·7107
0·64	0·7094	0·7081	0·7068	0·7056	0·7043
0·65	0·7030	0·7017	0·7004	0·6991	0·6979
0·66	0·6966	0·6953	0·6940	0·6927	0·6915
0·67	0·6902	0·6889	0·6876	0·6864	0·6851
0·68	0·6838	0·6825	0·6813	0·6800	0·6787
0·69	0·6775	0·6762	0·6749	0·6737	0·6724
0·70	0·6711	0·6699	0·6686	0·6674	0·6661
0·71	0·6648	0·6636	0·6623	0·6611	0·6598
0·72	0·6586	0·6573	0·6561	0·6548	0·6536
0·73	0·6524	0·6511	0·6499	0·6486	0·6474
0·74	0·6462	0·6449	0·6437	0·6425	0·6412
0·75	0·6400	0·6388	0·6375	0·6363	0·6351
0·76	0·6339	0·6327	0·6314	0·6302	0·6290
0·77	0·6278	0·6266	0·6254	0·6242	0·6229
0·78	0·6217	0·6205	0·6193	0·6181	0·6169
0·79	0·6157	0·6145	0·6133	0·6121	0·6109
0·80	0·6098	0·6086	0·6074	0·6062	0·6050
0·81	0·6038	0·6026	0·6015	0·6003	0·5991
0·82	0·5979	0·5968	0·5956	0·5944	0·5933
0·83	0·5921	0·5909	0·5898	0·5886	0·5875
0·84	0·5863	0·5851	0·5840	0·5828	0·5817
0·85	0·5806	0·5794	0·5783	0·5771	0·5760
0·86	0·5748	0·5737	0·5726	0·5714	0·5703
0·87	0·5692	0·5681	0·5669	0·5658	0·5647
0·88	0·5636	0·5625	0·5613	0·5602	0·5591
0·89	0·5580	0·5569	0·5558	0·5547	0·5536
0·90	0·5525	0·5514	0·5503	0·5492	0·5481
0·91	0·5470	0·5459	0·5448	0·5438	0·5427
0·92	0·5416	0·5405	0·5394	0·5384	0·5373
0·93	0·5362	0·5352	0·5341	0·5330	0·5320
0·94	0·5309	0·5298	0·5288	0·5277	0·5267
0·95	0·5256	0·5246	0·5235	0·5225	0·5214
0·96	0·5204	0·5194	0·5183	0·5173	0·5163
0·97	0·5152	0·5142	0·5132	0·5121	0·5111
0·98	0·5101	0·5091	0·5081	0·5070	0·5060
0·99	0·5050	0·5040	0·5030	0·5020	0·5010
1·00	0·5000	0·4990	0·4980	0·4970	0·4960

Table 13 $X/(1+X^2)$ 371

	0·000	0·002	0·004	0·006	0·008
0·00	0·0000	0·0020	0·0040	0·0060	0·0080
0·01	0·0100	0·0120	0·0140	0·0160	0·0180
0·02	0·0200	0·0220	0·0240	0·0260	0·0280
0·03	0·0030	0·0320	0·0340	0·0360	0·0379
0·04	0·0399	0·0419	0·0439	0·0459	0·0479
0·05	0·0499	0·0519	0·0538	0·0558	0·0578
0·06	0·0598	0·0618	0·0637	0·0657	0·0677
0·07	0·0697	0·0716	0·0736	0·0756	0·0775
0·08	0·0795	0·0815	0·0834	0·0854	0·0873
0·09	0·0893	0·0912	0·0932	0·0951	0·0971
0·10	0·0990	0·1009	0·1029	0·1048	0·1068
0·11	0·1087	0·1106	0·1125	0·1145	0·1164
0·12	0·1183	0·1202	0·1221	0·1240	0·1259
0·13	0·1278	0·1297	0·1316	0·1335	0·1354
0·14	0·1373	0·1392	0·1411	0·1430	0·1448
0·15	0·1467	0·1486	0·1504	0·1523	0·1542
0·16	0·1560	0·1579	0·1597	0·1615	0·1634
0·17	0·1652	0·1671	0·1689	0·1707	0·1725
0·18	0·1744	0·1762	0·1780	0·1798	0·1816
0·19	0·1834	0·1852	0·1870	0·1887	0·1905
0·20	0·1923	0·1941	0·1958	0·1976	0·1994
0·21	0·2011	0·2029	0·2046	0·2064	0·2081
0·22	0·2098	0·2116	0·2133	0·2150	0·2167
0·23	0·2184	0·2202	0·2219	0·2235	0·2252
0·24	0·2269	0·2286	0·2303	0·2320	0·2336
0·25	0·2353	0·2370	0·2386	0·2403	0·2419
0·26	0·2435	0·2452	0·2468	0·2484	0·2500
0·27	0·2517	0·2533	0·2549	0·2565	0·2581
0·28	0·2596	0·2612	0·2628	0·2644	0·2659
0·29	0·2675	0·2691	0·2706	0·2722	0·2737
0·30	0·2752	0·2768	0·2783	0·2798	0·2813
0·31	0·2828	0·2843	0·2858	0·2873	0·2888
0·32	0·2903	0·2918	0·2932	0·2947	0·2961
0·33	0·2976	0·2990	0·3005	0·3019	0·3033
0·34	0·3048	0·3062	0·3076	0·3090	0·3104
0·35	0·3118	0·3132	0·3146	0·3160	0·3173
0·36	0·3187	0·3201	0·3214	0·3228	0·3241
0·37	0·3254	0·3268	0·3281	0·3294	0·3307
0·38	0·3321	0·3334	0·3347	0·3359	0·3372
0·39	0·3385	0·3398	0·3411	0·3423	0·3436
0·40	0·3448	0·3461	0·3473	0·3485	0·3498
0·41	0·3510	0·3522	0·3534	0·3546	0·3558
0·42	0·3570	0·3582	0·3594	0·3606	0·3617
0·43	0·3629	0·3641	0·3652	0·3664	0·3675
0·44	0·3686	0·3698	0·3709	0·3720	0·3731
0·45	0·3742	0·3753	0·3764	0·3775	0·3786
0·46	0·3797	0·3807	0·3818	0·3829	0·3839
0·47	0·3850	0·3860	0·3870	0·3881	0·3891
0·48	0·3901	0·3911	0·3921	0·3931	0·3941
0·49	0·3951	0·3961	0·3971	0·3981	0·3990
0·50	0·4000	0·4010	0·4019	0·4029	0·4038

Table 13—continued

	0·000	0·002	0·004	0·006	0·008
0·51	0·4047	0·4057	0·4066	0·4075	0·4084
0·52	0·4093	0·4102	0·4111	0·4120	0·4129
0·53	0·4138	0·4146	0·4155	0·4164	0·4172
0·54	0·4181	0·4189	0·4198	0·4206	0·4214
0·55	0·4223	0·4231	0·4239	0·4247	0·4255
0·56	0·4263	0·4271	0·4279	0·4287	0·4294
0·57	0·4302	0·4310	0·4317	0·4325	0·4333
0·58	0·4340	0·4347	0·4355	0·4362	0·4369
0·59	0·4377	0·4384	0·4391	0·4398	0·4405
0·60	0·4412	0·4419	0·4426	0·4432	0·4439
0·61	0·4446	0·4452	0·4459	0·4466	0·4472
0·62	0·4478	0·4485	0·4491	0·4498	0·4504
0·63	0·4510	0·4516	0·4522	0·4528	0·4534
0·64	0·4540	0·4546	0·4552	0·4558	0·4564
0·65	0·4569	0·4575	0·4581	0·4586	0·4592
0·66	0·4597	0·4603	0·4608	0·4614	0·4619
0·67	0·4624	0·4629	0·4635	0·4640	0·4645
0·68	0·4650	0·4655	0·4660	0·4665	0·4670
0·69	0·4674	0·4679	0·4684	0·4689	0·4693
0·70	0·4698	0·4703	0·4707	0·4712	0·4716
0·71	0·4720	0·4725	0·4729	0·4733	0·4738
0·72	0·4742	0·4746	0·4750	0·4754	0·4758
0·73	0·4762	0·4766	0·4770	0·4774	0·4778
0·74	0·4782	0·4785	0·4789	0·4793	0·4796
0·75	0·4800	0·4804	0·4807	0·4811	0·4814
0·76	0·4817	0·4821	0·4824	0·4827	0·4831
0·77	0·4834	0·4837	0·4840	0·4843	0·4846
0·78	0·4850	0·4853	0·4856	0·4858	0·4861
0·79	0·4864	0·4867	0·4870	0·4873	0·4875
0·80	0·4878	0·4881	0·4883	0·4886	0·4888
0·81	0·4891	0·4894	0·4896	0·4898	0·4901
0·82	0·4903	0·4905	0·4908	0·4910	0·4912
0·83	0·4914	0·4917	0·4919	0·4921	0·4923
0·84	0·4925	0·4927	0·4929	0·4931	0·4933
0·85	0·4935	0·4937	0·4938	0·4940	0·4942
0·86	0·4944	0·4945	0·4947	0·4949	0·4950
0·87	0·4952	0·4953	0·4955	0·4957	0·4958
0·88	0·4959	0·4961	0·4962	0·4964	0·4965
0·89	0·4966	0·4968	0·4969	0·4970	0·4971
0·90	0·4972	0·4974	0·4975	0·4976	0·4977
0·91	0·4978	0·4979	0·4980	0·4981	0·4982
0·92	0·4983	0·4984	0·4984	0·4985	0·4986
0·93	0·4987	0·4988	0·4988	0·4989	0·4990
0·94	0·4990	0·4991	0·4992	0·4992	0·4993
0·95	0·4993	0·4994	0·4994	0·4995	0·4995
0·96	0·4996	0·4996	0·4997	0·4997	0·4997
0·97	0·4998	0·4998	0·4998	0·4999	0·4999
0·98	0·4999	0·4999	0·4999	0·5000	0·5000
0·99	0·5000	0·5000	0·5000	0·5000	0·5000
1·00	0·5000	0·5000	0·5000	0·5000	0·5000

Table 14 $1/\sqrt{(1+X^2)}$ 373

	0·000	0·002	0·004	0·006	0·008
0·00	1·0000	1·0000	1·0000	1·0000	1·0000
0·01	1·0000	0·9999	0·9999	0·9999	0·9998
0·02	0·9998	0·9998	0·9997	0·9997	0·9996
0·03	0·9996	0·9995	0·9994	0·9994	0·9993
0·04	0·9992	0·9991	0·9990	0·9989	0·9988
0·05	0·9988	0·9987	0·9985	0·9984	0·9983
0·06	0·9982	0·9981	0·9980	0·9978	0·9977
0·07	0·9976	0·9974	0·9973	0·9971	0·9970
0·08	0·9968	0·9967	0·9965	0·9963	0·9962
0·09	0·9960	0·9958	0·9956	0·9954	0·9952
0·10	0·9950	0·9948	0·9946	0·9944	0·9942
0·11	0·9940	0·9938	0·9936	0·9933	0·9931
0·12	0·9929	0·9926	0·9924	0·9922	0·9919
0·13	0·9917	0·9914	0·9911	0·9909	0·9906
0·14	0·9903	0·9901	0·9898	0·9895	0·9892
0·15	0·9889	0·9886	0·9883	0·9880	0·9877
0·16	0·9874	0·9871	0·9868	0·9865	0·9862
0·17	0·9859	0·9855	0·9852	0·9849	0·9845
0·18	0·9842	0·9838	0·9835	0·9831	0·9828
0·19	0·9824	0·9821	0·9817	0·9813	0·9810
0·20	0·9806	0·9802	0·9798	0·9794	0·9790
0·21	0·9787	0·9783	0·9779	0·9775	0·9771
0·22	0·9766	0·9762	0·9758	0·9754	0·9750
0·23	0·9746	0·9741	0·9737	0·9733	0·9728
0·24	0·9724	0·9719	0·9715	0·9710	0·9706
0·25	0·9701	0·9697	0·9692	0·9688	0·9683
0·26	0·9678	0·9673	0·9669	0·9664	0·9659
0·27	0·9654	0·9649	0·9645	0·9640	0·9635
0·28	0·9630	0·9625	0·9620	0·9615	0·9609
0·29	0·9604	0·9599	0·9594	0·9589	0·9584
0·30	0·9578	0·9573	0·9568	0·9562	0·9557
0·31	0·9552	0·9546	0·9541	0·9535	0·9530
0·32	0·9524	0·9519	0·9513	0·9508	0·9502
0·33	0·9496	0·9491	0·9485	0·9479	0·9473
0·34	0·9468	0·9462	0·9456	0·9450	0·9444
0·35	0·9439	0·9433	0·9427	0·9421	0·9415
0·36	0·9409	0·9403	0·9397	0·9391	0·9385
0·37	0·9379	0·9373	0·9366	0·9360	0·9354
0·38	0·9348	0·9342	0·9335	0·9329	0·9323
0·39	0·9317	0·9310	0·9304	0·9298	0·9291
0·40	0·9285	0·9278	0·9272	0·9265	0·9259
0·41	0·9253	0·9246	0·9239	0·9233	0·9226
0·42	0·9220	0·9213	0·9207	0·9200	0·9193
0·43	0·9187	0·9180	0·9173	0·9167	0·9160
0·44	0·9153	0·9146	0·9140	0·9133	0·9126
0·45	0·9119	0·9112	0·9106	0·9099	0·9092
0·46	0·9085	0·9078	0·9071	0·9064	0·9057
0·47	0·9050	0·9043	0·9036	0·9029	0·9022
0·48	0·9015	0·9008	0·9001	0·8994	0·8987
0·49	0·8980	0·8973	0·8966	0·8959	0·8951
0·50	0·8944	0·8937	0·8930	0·8923	0·8916

Table 14—continued

	0·000	0·002	0·004	0·006	0·008
0·51	0·8908	0·8901	0·8894	0·8887	0·8879
0·52	0·8872	0·8865	0·8858	0·8850	0·8843
0·53	0·8836	0·8828	0·8821	0·8814	0·8806
0·54	0·8799	0·8792	0·8784	0·8777	0·8770
0·55	0·8762	0·8755	0·8747	0·8740	0·8732
0·56	0·8725	0·8718	0·8710	0·8703	0·8695
0·57	0·8688	0·8680	0·8673	0·8665	0·8658
0·58	0·8650	0·8643	0·8635	0·8628	0·8620
0·59	0·8613	0·8605	0·8598	0·8590	0·8582
0·60	0·8575	0·8567	0·8560	0·8552	0·8545
0·61	0·8537	0·8529	0·8522	0·8514	0·8507
0·62	0·8499	0·8491	0·8484	0·8476	0·8469
0·63	0·8461	0·8453	0·8446	0·8438	0·8430
0·64	0·8423	0·8415	0·8407	0·8400	0·8392
0·65	0·8384	0·8377	0·8369	0·8361	0·8354
0·66	0·8346	0·8338	0·8331	0·8323	0·8315
0·67	0·8308	0·8300	0·8292	0·8285	0·8277
0·68	0·8269	0·8262	0·8254	0·8246	0·8238
0·69	0·8231	0·8223	0·8215	0·8208	0·8200
0·70	0·8192	0·8185	0·8177	0·8169	0·8162
0·71	0·8154	0·8146	0·8138	0·8131	0·8123
0·72	0·8115	0·8108	0·8100	0·8092	0·8085
0·73	0·8077	0·8069	0·8061	0·8054	0·8046
0·74	0·8038	0·8031	0·8023	0·8015	0·8008
0·75	0·8000	0·7992	0·7985	0·7977	0·7969
0·76	0·7962	0·7954	0·7946	0·7939	0·7931
0·77	0·7923	0·7916	0·7908	0·7900	0·7893
0·78	0·7885	0·7877	0·7870	0·7862	0·7854
0·79	0·7847	0·7839	0·7832	0·7824	0·7816
0·80	0·7809	0·7801	0·7793	0·7786	0·7778
0·81	0·7771	0·7763	0·7755	0·7748	0·7740
0·82	0·7733	0·7725	0·7718	0·7710	0·7702
0·83	0·7695	0·7687	0·7680	0·7672	0·7665
0·84	0·7657	0·7650	0·7642	0·7634	0·7627
0·85	0·7619	0·7612	0·7604	0·7597	0·7589
0·86	0·7582	0·7574	0·7567	0·7559	0·7552
0·87	0·7544	0·7537	0·7529	0·7522	0·7515
0·88	0·7507	0·7500	0·7492	0·7485	0·7477
0·89	0·7470	0·7463	0·7455	0·7448	0·7440
0·90	0·7433	0·7426	0·7418	0·7411	0·7403
0·91	0·7396	0·7389	0·7381	0·7374	0·7367
0·92	0·7359	0·7352	0·7345	0·7337	0·7330
0·93	0·7323	0·7315	0·7308	0·7301	0·7294
0·94	0·7286	0·7279	0·7272	0·7264	0·7257
0·95	0·7250	0·7243	0·7236	0·7228	0·7221
0·96	0·7214	0·7207	0·7199	0·7192	0·7185
0·97	0·7178	0·7171	0·7164	0·7156	0·7149
0·98	0·7142	0·7135	0·7128	0·7121	0·7114
0·99	0·7107	0·7099	0·7092	0·7085	0·7078
1·00	0·7071	0·7064	0·7057	0·7050	0·7043

Table 15 $X/\sqrt{(1+X^2)}$ 375

	0·000	0·002	0·004	0·006	0·008
0·00	0·0000	0·0020	0·0040	0·0060	0·0080
0·01	0·0100	0·0120	0·0140	0·0160	0·0180
0·02	0·0200	0·0220	0·0240	0·0260	0·0280
0·03	0·0300	0·0320	0·0340	0·0360	0·0380
0·04	0·0400	0·0420	0·0440	0·0460	0·0479
0·05	0·0499	0·0519	0·0539	0·0559	0·0579
0·06	0·0599	0·0619	0·0639	0·0659	0·0678
0·07	0·0698	0·0718	0·0738	0·0758	0·0778
0·08	0·0797	0·0817	0·0837	0·0857	0·0877
0·09	0·0896	0·0916	0·0936	0·0956	0·0975
0·10	0·0995	0·1015	0·1034	0·1054	0·1074
0·11	0·1093	0·1113	0·1133	0·1152	0·1172
0·12	0·1191	0·1211	0·1231	0·1250	0·1270
0·13	0·1289	0·1309	0·1328	0·1348	0·1367
0·14	0·1386	0·1406	0·1425	0·1445	0·1464
0·15	0·1483	0·1503	0·1522	0·1541	0·1561
0·16	0·1580	0·1599	0·1618	0·1638	0·1657
0·17	0·1676	0·1695	0·1714	0·1733	0·1752
0·18	0·1772	0·1791	0·1810	0·1829	0·1848
0·19	0·1867	0·1886	0·1904	0·1923	0·1942
0·20	0·1961	0·1980	0·1999	0·2018	0·2036
0·21	0·2055	0·2074	0·2093	0·2111	0·2130
0·22	0·2149	0·2167	0·2186	0·2204	0·2223
0·23	0·2241	0·2260	0·2278	0·2297	0·2315
0·24	0·2334	0·2352	0·2370	0·2389	0·2407
0·25	0·2425	0·2444	0·2462	0·2480	0·2498
0·26	0·2516	0·2534	0·2553	0·2571	0·2589
0·27	0·2607	0·2625	0·2643	0·2661	0·2678
0·28	0·2696	0·2714	0·2732	0·2750	0·2768
0·29	0·2785	0·2803	0·2821	0·2838	0·2856
0·30	0·2873	0·2891	0·2909	0·2926	0·2944
0·31	0·2961	0·2978	0·2996	0·3013	0·3030
0·32	0·3048	0·3065	0·3082	0·3099	0·3117
0·33	0·3134	0·3151	0·3168	0·3185	0·3202
0·34	0·3219	0·3236	0·3253	0·3270	0·3287
0·35	0·3304	0·3320	0·3337	0·3354	0·3371
0·36	0·3387	0·3404	0·3420	0·3437	0·3454
0·37	0·3470	0·3487	0·3503	0·3519	0·3536
0·38	0·3552	0·3568	0·3585	0·3601	0·3617
0·39	0·3633	0·3650	0·3666	0·3682	0·3698
0·40	0·3714	0·3730	0·3746	0·3762	0·3778
0·41	0·3794	0·3809	0·3825	0·3841	0·3857
0·42	0·3872	0·3888	0·3904	0·3919	0·3935
0·43	0·3950	0·3966	0·3981	0·3997	0·4012
0·44	0·4027	0·4043	0·4058	0·4073	0·4088
0·45	0·4104	0·4119	0·4134	0·4149	0·4164
0·46	0·4179	0·4194	0·4209	0·4224	0·4239
0·47	0·4254	0·4268	0·4283	0·4298	0·4313
0·48	0·4327	0·4342	0·4357	0·4371	0·4386
0·49	0·4400	0·4415	0·4429	0·4443	0·4458
0·50	0·4472	0·4486	0·4501	0·4515	0·4529

Table 15—continued

	0·000	0·002	0·004	0·006	0·008
0·51	0·4543	0·4557	0·4571	0·4586	0·4600
0·52	0·4614	0·4627	0·4641	0·4655	0·4669
0·53	0·4683	0·4697	0·4710	0·4724	0·4738
0·54	0·4751	0·4765	0·4779	0·4792	0·4806
0·55	0·4819	0·4833	0·4846	0·4859	0·4873
0·56	0·4886	0·4899	0·4913	0·4926	0·4939
0·57	0·4952	0·4965	0·4978	0·4991	0·5004
0·58	0·5017	0·5030	0·5043	0·5056	0·5069
0·59	0·5081	0·5094	0·5107	0·5120	0·5132
0·60	0·5145	0·5158	0·5170	0·5183	0·5195
0·61	0·5208	0·5220	0·5232	0·5245	0·5257
0·62	0·5269	0·5282	0·5294	0·5306	0·5318
0·63	0·5330	0·5342	0·5355	0·5367	0·5379
0·64	0·5391	0·5402	0·5414	0·5426	0·5438
0·65	0·5450	0·5462	0·5473	0·5485	0·5497
0·66	0·5508	0·5520	0·5532	0·5543	0·5555
0·67	0·5566	0·5578	0·5589	0·5600	0·5612
0·68	0·5623	0·5634	0·5646	0·5657	0·5668
0·69	0·5679	0·5690	0·5701	0·5713	0·5724
0·70	0·5735	0·5746	0·5757	0·5767	0·5778
0·71	0·5789	0·5800	0·5811	0·5822	0·5832
0·72	0·5843	0·5854	0·5864	0·5875	0·5886
0·73	0·5896	0·5907	0·5917	0·5928	0·5938
0·74	0·5948	0·5959	0·5969	0·5979	0·5990
0·75	0·6000	0·6010	0·6020	0·6031	0·6041
0·76	0·6051	0·6061	0·6071	0·6081	0·6091
0·77	0·6101	0·6111	0·6121	0·6131	0·6140
0·78	0·6150	0·6160	0·6170	0·6180	0·6189
0·79	0·6199	0·6209	0·6218	0·6228	0·6237
0·80	0·6247	0·6256	0·6266	0·6275	0·6285
0·81	0·6294	0·6304	0·6313	0·6322	0·6332
0·82	0·6341	0·6350	0·6359	0·6368	0·6378
0·83	0·6387	0·6396	0·6405	0·6414	0·6423
0·84	0·6432	0·6441	0·6450	0·6459	0·6468
0·85	0·6476	0·6485	0·6494	0·6503	0·6512
0·86	0·6520	0·6529	0·6538	0·6546	0·6555
0·87	0·6564	0·6572	0·6581	0·6589	0·6598
0·88	0·6606	0·6615	0·6623	0·6632	0·6640
0·89	0·6648	0·6657	0·6665	0·6673	0·6681
0·90	0·6690	0·6698	0·6706	0·6714	0·6722
0·91	0·6730	0·6738	0·6747	0·6755	0·6763
0·92	0·6771	0·6779	0·6786	0·6794	0·6802
0·93	0·6810	0·6818	0·6826	0·6834	0·6841
0·94	0·6849	0·6857	0·6865	0·6872	0·6880
0·95	0·6887	0·6895	0·6903	0·6910	0·6918
0·96	0·6925	0·6933	0·6940	0·6948	0·6955
0·97	0·6963	0·6970	0·6977	0·6985	0·6992
0·98	0·6999	0·7007	0·7014	0.7021	0·7028
0·99	0·7035	0·7043	0·7050	0·7057	0·7064
1·00	0·7071	0·7078	0·7085	0·7092	0·7099

Note

For a transmission line the relationship between the input and output voltages and currents is given by

$$\begin{bmatrix} V_1 \\ I_1 \end{bmatrix} = \begin{bmatrix} M & ZN \\ YN & M \end{bmatrix} \cdot \begin{bmatrix} V_2 \\ I_2 \end{bmatrix} \tag{9.69}$$

where

$$M = 1 + \frac{YZ}{2} + \frac{(YZ)^2}{24} + \frac{(YZ)^3}{720} \tag{9.67}$$

and

$$N = 1 + \frac{YZ}{6} + \frac{(YZ)^2}{120} + \frac{(YZ)^3}{5040} \tag{9.68}$$

In Table 16 values of M and N have been computed for a range of typical values of the complex product YZ

$$YZ = a + jb \tag{22.104}$$

a and b have been given values at intervals of 0·01 from

$$a = \text{zero} \quad \text{to} \quad a = -0·35$$
and
$$b = \text{zero} \quad \text{to} \quad b = 0.27$$

To use the table turn to the page where the column heading is the required value of b and choose the figures in this column on the line corresponding to the value of a
e.g.

$$\text{If } YZ = -0.13 + j0·15$$
from page 384
$$M = 0·9348 + j0·0734$$
$$N = 0.9783 + j0·0247$$

The method of interpolation is described in Sec. 22.6.7.

Table 16—continued

b a	0·00	0·01	0·02	0·03
0·00	1·0000 + j0·0000 1·0000 + j0·0000	1·0000 + j0·0050 1·0000 + j0·0017	1·0000 + j0·0100 1·0000 + j0·0033	1·0000 + j0·0150 1·0000 + j0·0050
−0·01	0·9950 + j0·0000 0·9983 + j0·0000	0·9950 + j0·0050 0·9983 + j0·0017	0·9950 + j0·0100 0·9983 + j0·0033	0·9950 + j0·0150 0·9983 + j0·0050
−0·02	0·9900 + j0·0000 0·9967 + j0·0000	0·9900 + j0·0050 0·9967 + j0·0017	0·9900 + j0·0100 0·9967 + j0·0033	0·9900 + j0·0150 0·9967 + j0·0050
−0·03	0·9850 + j0·0000 0·9950 + j0·0000	0·9850 + j0·0050 0·9950 + j0·0017	0·9850 + j0·0100 0·9950 + j0·0033	0·9850 + j0·0149 0·9950 + j0·0050
−0·04	0·9801 + j0·0000 0·9933 + j0·0000	0·9801 + j0·0050 0·9933 + j0·0017	0·9800 + j0·0099 0·9933 + j0·0033	0·9800 + j0·0149 0·9933 + j0·0050
−0·05	0·9751 + j0·0000 0·9917 + j0·0000	0·9751 + j0·0050 0·9917 + j0·0017	0·9751 + j0·0099 0·9917 + j0·0033	0·9751 + j0·0149 0·9917 + j0·0050
−0·06	0·9701 + j0·0000 0·9900 + j0·0000	0·9701 + j0·0050 0·9900 + j0·0017	0·9701 + j0·0099 0·9900 + j0·0033	0·9701 + j0·0149 0·9900 + j0·0050
−0·07	0·9652 + j0·0000 0·9884 + j0·0000	0·9652 + j0·0049 0·9884 + j0·0017	0·9652 + j0·0099 0·9884 + j0·0033	0·9652 + j0·0148 0·9884 + j0·0050
−0·08	0·9603 + j0·0000 0·9867 + j0·0000	0·9603 + j0·0049 0·9867 + j0·0017	0·9602 + j0·0099 0·9867 + j0·0033	0·9602 + j0·0148 0·9867 + j0·0050
−0·09	0·9553 + j0·0000 0·9851 + j0·0000	0·9553 + j0·0049 0·9851 + j0·0017	0·9553 + j0·0099 0·9851 + j0·0033	0·9553 + j0·0148 0·9851 + j0·0050
−0·10	0·9504 + j0·0000 0·9834 + j0·0000	0·9504 + j0·0049 0·9834 + j0·0017	0·9504 + j0·0098 0·9834 + j0·0033	0·9504 + j0·0148 0·9834 + j0·0050
−0·11	0·9455 + j0·0000 0·9818 + j0·0000	0·9455 + j0·0049 0·9818 + j0·0016	0·9455 + j0·0098 0·9818 + j0·0033	0·9455 + j0·0147 0·9818 + j0·0049
−0·12	0·9406 + j0·0000 0·9801 + j0·0000	0·9406 + j0·0049 0·9801 + j0·0016	0·9406 + j0·0098 0·9801 + j0·0033	0·9406 + j0·0147 0·9801 + j0·0049
−0·13	0·9357 + j0·0000 0·9785 + j0·0000	0·9357 + j0·0049 0·9785 + j0·0016	0·9357 + j0·0098 0·9785 + j0·0033	0·9357 + j0·0147 0·9785 + j0·0049
−0·14	0·9308 + j0·0000 0·9768 + j0·0000	0·9308 + j0·0049 0·9768 + j0·0016	0·9308 + j0·0098 0·9768 + j0·0033	0·9308 + j0·0147 0·9768 + j0·0049
−0·15	0·9259 + j0·0000 0·9752 + j0·0000	0·9259 + j0·0049 0·9752 + j0·0016	0·9259 + j0·0098 0·9752 + j0·0033	0·9259 + j0·0146 0·9752 + j0·0049
−0·16	0·9211 + j0·0000 0·9735 + j0·0000	0·9211 + j0·0049 0·9735 + j0·0016	0·9210 + j0·0097 0·9735 + j0·0033	0·9210 + j0·0146 0·9735 + j0·0049
−0·17	0·9162 + j0·0000 0·9719 + j0·0000	0·9162 + j0·0049 0·9719 + j0·0016	0·9162 + j0·0097 0·9719 + j0·0033	0·9162 + j0·0146 0·9719 + j0·0049
−0·18	0·9113 + j0·0000 0·9703 + j0·0000	0·9113 + j0·0049 0·9703 + j0·0016	0·9113 + j0·0097 0·9703 + j0·0033	0·9113 + j0·0146 0·9703 + j0·0049
−0·19	0·9065 + j0·0000 0·9686 + j0·0000	0·9065 + j0·0048 0·9686 + j0·0016	0·9065 + j0·0097 0·9686 + j0·0033	0·9065 + j0·0145 0·9686 + j0·0049

Table 16—continued

379

b a	0·00	0·01	0·02	0·03
−0·20	0·9017 + j0·0000	0·9017 + j0·0048	0·9016 + j0·0097	0·9016 + j0·0145
	0·9670 + j0·0000	0·9670 + j0·0016	0·9670 + j0·0033	0·9670 + j0·0049
−0·21	0·8968 + j0·0000	0·8968 + j0·0048	0·8968 + j0·0097	0·8968 + j0·0145
	0·9654 + j0·0000	0·9654 + j0·0016	0·9654 + j0·0033	0·9654 + j0·0049
−0·22	0·8920 + j0·0000	0·8920 + j0·0048	0·8920 + j0·0096	0·8920 + j0·0145
	0·9637 + j0·0000	0·9637 + j0·0016	0·9637 + j0·0033	0·9637 + j0·0049
−0·23	0·8872 + j0·0000	0·8872 + j0·0048	0·8872 + j0·0096	0·8872 + j0·0144
	0·9621 + j0·0000	0·9621 + j0·0016	0·9621 + j0·0033	0·9621 + j0·0049
−0·24	0·8824 + j0·0000	0·8824 + j0·0048	0·8824 + j0·0096	0·8823 + j0·0144
	0·9605 + j0·0000	0·9605 + j0·0016	0·9605 + j0·0033	0·9605 + j0·0049
−0·25	0·8776 + j0·0000	0·8776 + j0·0048	0·8876 + j0·0096	0·8775 + j0·0144
	0·9589 + j0·0000	0·9589 + j0·0016	0·9588 + j0·0033	0·9588 + j0·0049
−0·26	0·8728 + j0·0000	0·8728 + j0·0048	0·8728 + j0·0096	0·8728 + j0·0144
	0·9572 + j0·0000	0·9572 + j0·0016	0·9572 + j0·0032	0·9572 + j0·0049
−0·27	0·8680 + j0·0000	0·8680 + j0·0048	0·8680 + j0·0096	0·8680 + j0·0143
	0·9556 + j0·0000	0·9556 + j0·0016	0·9556 + j0·0032	0·9556 + j0·0049
−0·28	0·8632 + j0·0000	0·8632 + j0·0048	0·8632 + j0·0095	0·8632 + j0·0143
	0·9540 + j0·0000	0·9540 + j0·0016	0·8540 + j0·0032	0·9540 + j0·0049
−0·29	0·8585 + j0·0000	0·8585 + j0·0048	0·8585 + j0·0095	0·8584 + j0·0143
	0·9524 + j0·0000	0·9524 + j0·0016	0·9524 + j0·0032	0·9524 + j0·0049
−0·30	0·8537 + j0·0000	0·8537 + j0·0048	0·8537 + j0·0095	0·8537 + j0·0143
	0·9507 + j0·0000	0·9507 + j0·0016	0·9507 + j0·0032	0·9507 + j0·0049
−0·31	0·8490 + j0·0000	0·8490 + j0·0047	0·8489 + j0·0095	0·8489 + j0·0142
	0·9491 + j0·0000	0·9491 + j0·0016	0·9491 + j0·0032	0·9491 + j0·0048
−0·32	0·8442 + j0·0000	0·8442 + j0·0047	0·8442 + j0·0095	0·8442 + j0·0142
	0·9475 + j0·0000	0·9475 + j0·0016	0·9475 + j0·0032	0·9475 + j0·0048
−0·33	0·8395 + j0·0000	0·8395 + j0·0047	0·8395 + j0·0095	0·8395 + j0·0142
	0·9459 + j0·0000	0·9459 + j0·0016	0·9459 + j0·0032	0·9459 + j0·0048
−0·34	0·8348 + j0·0000	0·8348 + j0·0047	0·8347 + j0·0094	0·8347 + j0·0142
	0·9443 + j0·0000	0·9443 + j0·0016	0·9443 + j0·0032	0·9443 + j0·0048
−0·35	0·8300 + j0·0000	0·8300 + j0·0047	0·8300 + j0·0094	0·8300 + j0·0141
	0·9427 + j0·0000	0·9427 + j0·0016	0·9427 + j0·0032	0·9427 + j0·0048

Table 16—continued

b / a	0·04	0·05	0·06	0·07
0·00	0·9999 + j0·0200	0·9999 + j0·0250	0·9999 + j0·0300	0·9998 + j0·0350
	1·0000 + j0·0067	1·0000 + j0·0083	1·0000 + j0·0100	1·0000 + j0·0117
−0·01	0·9949 + j0·0200	0·9949 + j0·0250	0·9949 + j0·0299	0·9948 + j0·0349
	0·9983 + j0·0067	0·9983 + j0·0083	0·9983 + j0·0100	0·9983 + j0·0117
−0·02	0·9900 + j0·0199	0·9899 + j0·0249	0·9899 + j0·0299	0·9898 + j0·0349
	0·9967 + j0·0067	0·9966 + j0·0083	0·9966 + j0·0100	0·9966 + j0·0116
−0·03	0·9850 + j0·0199	0·9849 + j0·0249	0·9849 + j0·0298	0·9848 + j0·0348
	0·9950 + j0·0066	0·9950 + j0·0083	0·9950 + j0·0100	0·9950 + j0·0116
−0·04	0·9800 + j0·0199	0·9800 + j0·0248	0·9799 + j0·0298	0·9799 + j0·0348
	0·9933 + j0·0066	0·9933 + j0·0083	0·9933 + j0·0100	0·9933 + j0·0116
−0·05	0·9750 + j0·0198	0·9750 + j0·0248	0·9750 + j0·0298	0·9749 + j0·0347
	0·9917 + j0·0066	0·9917 + j0·0083	0·9917 + j0·0100	0·9916 + j0·0116
−0·06	0·9701 + j0·0198	0·9700 + j0·0248	0·9700 + j0·0297	0·9699 + j0·0347
	0·9900 + j0·0066	0·9900 + j0·0083	0·9900 + j0·0099	0·9900 + j0·0116
−0·07	0·9651 + j0·0198	0·9651 + j0·0247	0·9651 + j0·0297	0·9650 + j0·0346
	0·9884 + j0·0066	0·9884 + j0·0083	0·9883 + j0·0099	0·9883 + j0·0116
−0·08	0·9602 + j0·0197	0·9602 + j0·0247	0·9601 + j0·0296	0·9601 + j0·0345
	0·9867 + j0·0066	0·9867 + j0·0083	0·9867 + j0·0099	0·9867 + j0·0116
−0·09	0·9553 + j0·0197	0·9552 + j0·0246	0·9552 + j0·0296	0·9551 + j0·0345
	0·9851 + j0·0066	0·9850 + j0·0083	0·9850 + j0·0099	0·9850 + j0·0116
−0·10	0·9503 + j0·0197	0·9503 + j0·0246	0·9503 + j0·0295	0·9502 + j0·0344
	0·9834 + j0·0066	0·9834 + j0·0083	0·9834 + j0·0099	0·9834 + j0·0116
−0·11	0·9454 + j0·0196	0·9454 + j0·0245	0·9454 + j0·0295	0·9453 + j0·0344
	0·9818 + j0·0066	0·9817 + j0·0082	0·9817 + j0·0099	0·9817 + j0·0115
−0·12	0·9405 + j0·0196	0·9405 + j0·0245	0·9404 + j0·0294	0·9404 + j0·0343
	0·9801 + j0·0066	0·9801 + j0·0082	0·9801 + j0·0099	0·9801 + j0·0115
−0·13	0·9356 + j0·0196	0·9356 + j0·0245	0·9356 + j0·0294	0·9355 + j0·0342
	0·9785 + j0·0066	0·9785 + j0·0082	0·9784 + j0·0099	0·9784 + j0·0115
−0·14	0·9307 + j0·0195	0·9307 + j0·0244	0·9307 + j0·0293	0·9306 + j0·0342
	0·9768 + j0·0066	0·9768 + j0·0082	0·9768 + j0·0099	0·9768 + j0·0115
−0·15	0·9259 + j0·0195	0·9258 + j0·0244	0·9258 + j0·0293	0·9257 + j0·0341
	0·9752 + j0·0066	0·9752 + j0·0082	0·9752 + j0·0099	0·9751 + j0·0115
−0·16	0·9210 + j0·0195	0·9210 + j0·0243	0·9209 + j0·0292	0·9209 + j0·0341
	0·9735 + j0·0066	0·9735 + j0·0082	0·9735 + j0·0098	0·9735 + j0·0115
−0·17	0·9161 + j0·0194	0·9161 + j0·0243	0·9160 + j0·0292	0·9160 + j0·0340
	0·9719 + j0·0066	0·9719 + j0·0082	0·9719 + j0·0098	0·9719 + j0·0115
−0·18	0·9113 + j0·0194	0·9112 + j0·0243	0·9112 + j0·0291	0·9111 + j0·0340
	0·9703 + j0·0065	0·9702 + j0·0082	0·9702 + j0·0098	0·9702 + j0·0115
−0·19	0·9064 + j0·0194	0·9064 + j0·0242	0·9063 + j0·0291	0·9063 + j0·0339
	0·9686 + j0·0065	0·9686 + j0·0082	0·9686 + j0·0098	0·9686 + j0·0114

Table 16—continued 381

b a	0·04	0·05	0·06	0·07
−0·20	0·9016 + j0·0193	0·9016 + j0·0242	0·9015 + j0·0290	0·9015 + j0·0338
	0·9670 + j0·0065	0·9670 + j0·0082	0·9670 + j0·0098	0·9670 + j0·0114
−0·21	0·8968 + j0·0193	0·8967 + j0·0241	0·8967 + j0·0290	0·8966 + j0·0338
	0·9654 + j0·0065	0·9653 + j0·0082	0·9653 + j0·0098	0·9653 + j0·0114
−0·22	0·8919 + j0·0193	0·8919 + j0·0241	0·8919 + j0·0289	0·8918 + j0·0337
	0·9637 + j0·0065	0·9637 + j0·0082	0·9637 + j0·0098	0·9637 + j0·0114
−0·23	0·8871 + j0·0192	0·8871 + j0·0241	0·8870 + j0·0289	0·8870 + j0·0337
	0·9621 + j0·0065	0·9621 + j0·0081	0·9621 + j0·0098	0·9621 + j0·0114
−0·24	0·8823 + j0·0192	0·8823 + j0·0240	0·8822 + j0·0288	0·8822 + j0·0336
	0·9605 + j0·0065	0·9605 + j0·0081	0·9604 + j0·0098	0·9604 + j0·0114
−0·25	0·8775 + j0·0192	0·8775 + j0·0240	0·8774 + j0·0288	0·8774 + j0·0336
	0·9588 + j0·0065	0·9588 + j0·0081	0·9588 + j0·0098	0·9588 + j0·0114
−0·26	0·8727 + j0·0191	0·8727 + j0·0239	0·8726 + j0·0287	0·8726 + j0·0335
	0·9572 + j0·0065	0·9572 + j0·0081	0·9572 + j0·0097	0·9572 + j0·0114
−0·27	0·8679 + j0·0191	0·8679 + j0·0239	0·8679 + j0·0287	0·8678 + j0·0334
	0·9556 + j0·0065	0·9556 + j0·0081	0·9556 + j0·0097	0·9556 + j0·0114
−0·28	0·8632 + j0·0191	0·9631 + j0·0238	0·9631 + j0·0286	0·8630 + j0·0334
	0·9540 + j0·0065	0·9540 + j0·0081	0·9540 + j0·0097	0·9539 + j0·0113
−0·29	0·8584 + j0·0190	0·8584 + j0·0238	0·8583 + j0·0286	0·8583 + j0·0333
	0·9523 + j0·0065	0·9523 + j0·0081	0·9523 + j0·0097	0·9523 + j0·0113
−0·30	0·8536 + j0·0190	0·8536 + j0·0238	0·8536 + j0·0285	0·8535 + j0·0333
	0·9507 + j0·0065	0·9507 + j0·0081	0·9507 + j0·0097	0·9507 + j0·0113
−0·31	0·8489 + j0·0190	0·8489 + j0·0237	0·8488 + j0·0285	0·8488 + j0·0332
	0·9491 + j0·0065	0·9491 + j0·0081	0·9491 + j0·0097	0·9491 + j0·0113
−0·32	0·8442 + j0·0190	0·8441 + j0·0237	0·8441 + j0·0284	0·8440 + j0·0332
	0·9475 + j0·0065	0·9475 + j0·0081	0·9475 + j0·0097	0·9475 + j0·0113
−0·33	0·8394 + j0·0189	0·8394 + j0·0236	0·8393 + j0·0284	0·8393 + j0·0331
	0·9459 + j0·0064	0·9459 + j0·0081	0·9459 + j0·0097	0·9459 + j0·0113
−0·34	0·8347 + j0·0189	0·8347 + j0·0236	0·8346 + j0·0283	0·8346 + j0·0330
	0·9443 + j0·0064	0·9443 + j0·0081	0·9443 + j0·0097	0·9442 + j0·0113
−0·35	0·8300 + j0·0189	0·8299 + j0·0236	0·8299 + j0·0283	0·8298 + j0·0330
	0·9427 + j0·0064	0·9427 + j0·0080	0·9426 + j0·0097	0·9426 + j0·0113

Table 16—continued

a \ b	0·08	0·09	0·10	0·11
0·00	0·9997 + j0·0400	0·9997 + j0·0450	0·9996 + j0·0500	0·9995 + j0·0550
	0·9999 + j0·0133	0·9999 + j0·0150	0·9999 + j0·0167	0·9999 + j0·0183
−0·01	0·9947 + j0·0399	0·9947 + j0·0449	0·9946 + j0·0499	0·9945 + j0·0549
	0·9983 + j0·0133	0·9983 + j0·0150	0·9983 + j0·0166	0·9982 + j0·0183
−0·02	0·9898 + j0·0399	0·9897 + j0·0448	0·9896 + j0·0498	0·9895 + j0·0548
	0·9966 + j0·0133	0·9966 + j0·0150	0·9966 + j0·0166	0·9966 + j0·0183
−0·03	0·9848 + j0·0398	0·9847 + j0·0448	0·9846 + j0·0497	0·9845 + j0·0547
	0·9950 + j0·0133	0·9949 + j0·0150	0·9949 + j0·0166	0·9949 + j0·0183
−0·04	0·9798 + j0·0397	0·9797 + j0·0447	0·9797 + j0·0497	0·9796 + j0·0546
	0·9933 + j0·0133	0·9933 + j0·0149	0·9933 + j0·0166	0·9932 + j0·0183
−0·05	0·9748 + j0·0397	0·9748 + j0·0446	0·9747 + j0·0496	0·9746 + j0·0545
	0·9916 + j0·0133	0·9916 + j0·0149	0·9916 + j0·0166	0·9916 + j0·0182
−0·06	0·9699 + j0·0396	0·9698 + j0·0446	0·9697 + j0·0495	0·9696 + j0·0544
	0·9900 + j0·0133	0·9900 + j0·0149	0·9899 + j0·0166	0·9899 + j0·0182
−0·07	0·9649 + j0·0395	0·9649 + j0·0445	0·9648 + j0·0494	0·9647 + j0·0544
	0·9883 + j0·0132	0·9883 + j0·0149	0·9883 + j0·0166	0·9883 + j0·0182
−0·08	0·9600 + j0·0395	0·9599 + j0·0444	0·9599 + j0·0493	0·9598 + j0·0543
	0·9867 + j0·0132	0·9867 + j0·0149	0·9866 + j0·0165	0·9866 + j0·0182
−0·09	0·9551 + j0·0394	0·9550 + j0·0443	0·9549 + j0·0493	0·9548 + j0·0542
	0·9850 + j0·0132	0·9850 + j0·0149	0·9850 + j0·0165	0·9850 + j0·0182
−0·10	0·9502 + j0·0393	0·9501 + j0·0443	0·9500 + j0·0492	0·9499 + j0·0541
	0·9834 + j0·0132	0·9833 + j0·0149	0·9833 + j0·0165	0·9833 + j0·0182
−0·11	0·9452 + j0·0393	0·9452 + j0·0442	0·9451 + j0·0491	0·9450 + j0·0540
	0·9817 + j0·0132	0·9817 + j0·0148	0·9817 + j0·0165	0·9817 + j0·0181
−0·12	0·9403 + j0·0392	0·9403 + j0·0441	0·9402 + j0·0490	0·9401 + j0·0539
	0·9801 + j0·0132	0·9801 + j0·0148	0·9800 + j0·0165	0·9800 + j0·0181
−0·13	0·9354 + j0·0391	0·9354 + j0·0440	0·9353 + j0·0489	0·9352 + j0·0538
	0·9784 + j0·0132	0·9784 + j0·0148	0·9784 + j0·0165	0·9784 + j0·0181
−0·14	0·9305 + j0·0391	0·9305 + j0·0440	0·9304 + j0·0488	0·9303 + j0·0537
	0·9768 + j0·0131	0·9768 + j0·0148	0·9767 + j0·0164	0·9767 + j0·0181
−0·15	0·9257 + j0·0390	0·9256 + j0·0439	0·9255 + j0·0488	0·9254 + j0·0536
	0·9751 + j0·0131	0·9751 + j0·0148	0·9751 + j0·0164	0·9751 + j0·0181
−0·16	0·9208 + j0·0389	0·9207 + j0·0438	0·9207 + j0·0487	0·9206 + j0·0535
	0·9735 + j0·0131	0·9735 + j0·0148	0·9735 + j0·0164	0·9734 + j0·0180
−0·17	0·9159 + j0·0389	0·9159 + j0·0437	0·9158 + j0·0486	0·9157 + j0·0535
	0·9719 + j0·0131	0·9718 + j0·0147	0·9718 + j0·0164	0·9718 + j0·0180
−0·18	0·9111 + j0·0388	0·9110 + j0·0437	0·9109 + j0·0485	0·9108 + j0·0534
	0·9702 + j0·0131	0·9702 + j0·0147	0·9702 + j0·0164	0·9702 + j0·0180
−0·19	0·9062 + j0·0387	0·9062 + j0·0436	0·9061 + j0·0484	0·9060 + j0·0533
	0·9686 + j0·0131	0·9686 + j0·0147	0·9686 + j0·0164	0·9685 + j0·0180

Table 16—continued 383

a \ b	0·08	0·09	0·10	0·11
−0·20	0·9014 + j0·0387	0·9013 + j0·0435	0·9012 + j0·0483	0·9012 + j0·0532
	0·9669 + j0·0131	0·9669 + j0·0147	0·9669 + j0·0163	0·9669 + j0·0180
−0·21	0·8966 + j0·0386	0·8965 + j0·0434	0·8964 + j0·0483	0·8963 + j0·0531
	0·9653 + j0·0131	0·9653 + j0·0147	0·9653 + j0·0160	0·9653 + j0·0180
−0·22	0·8917 + j0·0385	0·8917 + j0·0434	0·8916 + j0·0482	0·8915 + j0·0530
	0·9637 + j0·0130	0·9637 + j0·0147	0·9637 + j0·0163	0·9636 + j0·0179
−0·23	0·8869 + j0·0385	0·8869 + j0·0433	0·8868 + j0·0481	0·8867 + j0·0529
	0·9621 + j0·0130	0·9620 + j0·0147	0·9620 + j0·0163	0·9620 + j0·0179
−0·24	0·8821 + j0·0384	0·8821 + j0·0432	0·8820 + j0·0480	0·8819 + j0·0528
	0·9604 + j0·0130	0·9604 + j0·0146	0·9604 + j0·0163	0·9604 + j0·0179
−0·25	0·8773 + j0·0384	0·8773 + j0·0431	0·8772 + j0·0479	0·8771 + j0·0527
	0·9588 + j0·0130	0·9588 + j0·0146	0·9588 + j0·0163	0·9588 + j0·0179
−0·26	0·8725 + j0·0383	0·8725 + j0·0431	0·8724 + j0·0479	0·8723 + j0·0526
	0·9572 + j0·0130	0·9572 + j0·0146	0·9571 + j0·0162	0·9571 + j0·0179
−0·27	0·8678 + j0·0382	0·8677 + j0·0430	0·8676 + j0·0478	0·8675 + j0·0526
	0·9556 + j0·0130	0·9555 + j0·0146	0·9555 + j0·0162	0·9555 + j0·0178
−0·28	0·8630 + j0·0382	0·8629 + j0·0429	0·8628 + j0·0477	0·8627 + j0·0525
	0·9539 + j0·0130	0·9539 + j0·0146	0·9539 + j0·0162	0·9539 + j0·0178
−0·29	0·8582 + j0·0381	0·8581 + j0·0429	0·8581 + j0·0476	0·8580 + j0·0524
	0·9523 + j0·0130	0·9523 + j0·0146	0·9523 + j0·0162	0·9523 + j0·0178
−0·30	0·8535 + j0·0380	0·8534 + j0·0428	0·8533 + j0·0475	0·8532 + j0·0523
	0·9507 + j0·0129	0·9507 + j0·0146	0·9507 + j0·0162	0·9506 + j0·0178
−0·31	0·8487 + j0·0380	0·8486 + j0·0427	0·8486 + j0·0475	0·8485 + j0·0522
	0·9491 + j0·0129	0·9491 + j0·0145	0·9490 + j0·0162	0·9490 + j0·0178
−0·32	0·8440 + j0·0379	0·8439 + j0·0426	0·8438 + j0·0474	0·8437 + j0·0521
	0·9475 + j0·0129	0·9474 + j0·0145	0·9474 + j0·0161	0·9474 + j0·0178
−0·33	0·8392 + j0·0378	0·8392 + j0·0426	0·8391 + j0·0473	0·8390 + j0·0520
	0·9458 + j0·0129	0·9458 + j0·0145	0·9458 + j0·0161	0·9458 + j0·0177
−0·34	0·8345 + j0·0378	0·8344 + j0·0425	0·8344 + j0·0472	0·8343 + j0·0519
	0·9442 + j0·0129	0·9442 + j0·0145	0·9442 + j0·0161	0·9442 + j0·0177
−0·35	0·8298 + j0·0377	0·8297 + j0·0424	0·8296 + j0·0471	0·8296 + j0·0518
	0·9426 + j0·0129	0·9426 + j0·0145	0·9426 + j0·0161	0·9426 + j0·0177

Table 16—continued

a \ b	0·12	0·13	0·14	0·15
0·00	0·9994 + j0·0600	0·9993 + j0·0650	0·9992 + j0·0700	0·9991 + j0·0750
	0·9999 + j0·0200	0·9999 + j0·0217	0·9998 + j0·0233	0·9998 + j0·0250
−0·01	0·9944 + j0·0599	0·9943 + j0·0649	0·9942 + j0·0699	0·9941 + j0·0749
	0·9982 + j0·0200	0·9982 + j0·0216	0·9982 + j0·0233	0·9981 + j0·0250
−0·02	0·9894 + j0·0598	0·9893 + j0·0648	0·9892 + j0·0698	0·9891 + j0·0747
	0·9966 + j0·0200	0·9965 + j0·0216	0·9965 + j0·0233	0·9965 + j0·0249
−0·03	0·9844 + j0·0597	0·9843 + j0·0647	0·9842 + j0·0696	0·9841 + j0·0746
	0·9949 + j0·0199	0·9949 + j0·0216	0·9948 + j0·0233	0·9948 + j0·0249
−0·04	0·9795 + j0·0596	0·9794 + j0·0646	0·9793 + j0·0695	0·9791 + j0·0745
	0·9932 + j0·0199	0·9932 + j0·0216	0·9932 + j0·0232	0·9932 + j0·0249
−0·05	0·9745 + j0·0595	0·9744 + j0·0645	0·9743 + j0·0694	0·9742 + j0·0744
	0·9916 + j0·0199	0·9915 + j0·0216	0·9915 + j0·0232	0·9915 + j0·0249
−0·06	0·9696 + j0·0594	0·9694 + j0·0643	0·9693 + j0·0693	0·9692 + j0·0742
	0·9899 + j0·0199	0·9899 + j0·0215	0·9899 + j0·0232	0·9898 + j0·0248
−0·07	0·9646 + j0·0593	0·9645 + j0·0642	0·9644 + j0·0692	0·9643 + j0·0741
	0·9883 + j0·0199	0·9882 + j0·0215	0·9882 + j0·0232	0·9882 + j0·0248
−0·08	0·9597 + j0·0592	0·9596 + j0·0641	0·9595 + j0·0691	0·9593 + j0·0740
	0·9866 + j0·0198	0·9866 + j0·0215	0·9866 + j0·0231	0·9865 + j0·0248
−0·09	0·9547 + j0·0591	0·9546 + j0·0640	0·9545 + j0·0690	0·9544 + j0·0739
	0·9849 + j0·0198	0·9849 + j0·0215	0·9849 + j0·0231	0·9849 + j0·0248
−0·10	0·9498 + j0·0590	0·9497 + j0·0639	0·9496 + j0·0688	0·9495 + j0·0738
	0·9833 + j0·0198	0·9833 + j0·0215	0·9833 + j0·0231	0·9832 + j0·0248
−0·11	0·9449 + j0·0589	0·9448 + j0·0638	0·9447 + j0·0687	0·9446 + j0·0736
	0·9816 + j0·0198	0·9816 + j0·0214	0·9816 + j0·0231	0·9816 + j0·0247
−0·12	0·9400 + j0·0588	0·9399 + j0·0637	0·9398 + j0·0686	0·9397 + j0·0735
	0·9800 + j0·0198	0·9800 + j0·0214	0·9800 + j0·0231	0·9799 + j0·0247
−0·13	0·9351 + j0·0587	0·9350 + j0·0636	0·9349 + j0·0685	0·9348 + j0·0734
	0·9784 + j0·0197	0·9783 + j0·0214	0·9783 + j0·0230	0·9783 + j0·0247
−0·14	0·9302 + j0·0586	0·9301 + j0·0635	0·9300 + j0·0684	0·9299 + j0·0733
	0·9767 + j0·0197	0·9767 + j0·0214	0·9767 + j0·0230	0·9766 + j0·0247
−0·15	0·9253 + j0·0585	0·9252 + j0·0634	0·9251 + j0·0683	0·9250 + j0·0731
	0·9751 + j0·0197	0·9750 + j0·0213	0·9750 + j0·0230	0·9750 + j0·0246
−0·16	0·9205 + j0·0584	0·9204 + j0·0633	0·9203 + j0·0681	0·9201 + j0·0730
	0·9734 + j0·0197	0·9734 + j0·0213	0·9734 + j0·0230	0·9734 + j0·0246
−0·17	0·9156 + j0·0583	0·9155 + j0·0632	0·9154 + j0·0680	0·9153 + j0·0729
	0·9718 + j0·0197	0·9718 + j0·0213	0·9717 + j0·0229	0·9717 + j0·0246
−0·18	0·9108 + j0·0582	0·9107 + j0·0631	0·9105 + j0·0679	0·9104 + j0·0728
	0·9702 + j0·0196	0·9701 + j0·0213	0·9701 + j0·0229	0·9701 + j0·0246
−0·19	0·9059 + j0·0581	0·9058 + j0·0630	0·9057 + j0·0678	0·9056 + j0·0726
	0·9685 + j0·0196	0·9685 + j0·0213	0·9685 + j0·0229	0·9684 + j0·0245

Table 16—continued 385

b a	0·12	0·13	0·14	0·15
−0·20	0·9011 + j0·0580	0·9010 + j0·0629	0·9009 + j0·0677	0·9007 + j0·0725
	0·9669 + j0·0196	0·9669 + j0·0212	0·9668 + j0·0229	0·9668 + j0·0245
−0·21	0·8962 + j0·0579	0·8961 + j0·0627	0·8960 + j0·0676	0·8959 + j0·0724
	0·9652 + j0·0196	0·9652 + j0·0212	0·9652 + j0·0228	0·9652 + j0·0245
−0·22	0·8914 + j0·0578	0·8913 + j0·0626	0·8912 + j0·0675	0·8911 + j0·0723
	0·9636 + j0·0196	0·9636 + j0·0212	0·9636 + j0·0228	0·9636 + j0·0245
−0·23	0·8866 + j0·0577	0·8865 + j0·0625	0·8864 + j0·0673	0·8863 + j0·0722
	0·9620 + j0·0195	0·9620 + j0·0212	0·9619 + j0·0228	0·9619 + j0·0244
−0·24	0·8818 + j0·0576	0·8817 + j0·0624	0·8816 + j0·0672	0·8815 + j0·0720
	0·9604 + j0·0195	0·9603 + j0·0212	0·9603 + j0·0228	0·9603 + j0·0244
−0·25	0·8770 + j0·0575	0·8769 + j0·0623	0·8768 + j0·0671	0·8767 + j0·0719
	0·9587 + j0·0195	0·9587 + j0·0211	0·9587 + j0·0228	0·9587 + j0·0244
−0·26	0·8722 + j0·0574	0·8721 + j0·0622	0·8720 + j0·0670	0·8719 + j0·0718
	0·9571 + j0·0195	0·9571 + j0·0211	0·9571 + j0·0227	0·9570 + j0·0244
−0·27	0·8674 + j0·0573	0·8673 + j0·0621	0·8672 + j0·0669	0·8671 + j0·0717
	0·9555 + j0·0195	0·9555 + j0·0211	0·9554 + j0·0227	0·9554 + j0·0243
−0·28	0·8627 + j0·0572	0·8626 + j0·0620	0·8624 + j0·0668	0·8623 + j0·0715
	0·9539 + j0·0194	0·9538 + j0·0211	0·9538 + j0·0227	0·9538 + j0·0243
−0·29	0·8579 + j0·0571	0·8578 + j0·0619	0·8577 + j0·0667	0·8576 + j0·0714
	0·9522 + j0·0194	0·9522 + j0·0210	0·9522 + j0·0227	0·9522 + j0·0243
−0·30	0·8531 + j0·0570	0·8530 + j0·0618	0·8529 + j0·0665	0·8528 + j0·0713
	0·9506 + j0·0194	0·9506 + j0·0210	0·9506 + j0·0226	0·9506 + j0·0243
−0·31	0·8484 + j0·0569	0·8483 + j0·0617	0·8482 + j0·0664	0·8481 + j0·0712
	0·9490 + j0·0194	0·9490 + j0·0210	0·9490 + j0·0226	0·9489 + j0·0242
−0·32	0·8436 + j0·0568	0·8435 + j0·0616	0·8434 + j0·0663	0·8433 + j0·0711
	0·9474 + j0·0194	0·9474 + j0·0210	0·9474 + j0·0226	0·9473 + j0·0242
−0·33	0·8389 + j0·0568	0·8388 + j0·0615	0·8387 + j0·0662	0·8386 + j0·0709
	0·9458 + j0·0193	0·9458 + j0·0210	0·9457 + j0·0226	0·9457 + j0·0242
−0·34	0·8342 + j0·0567	0·8341 + j0·0614	0·8340 + j0·0661	0·8339 + j0·0708
	0·9442 + j0·0193	0·9442 + j0·0209	0·9441 + j0·0225	0·9441 + j0·0242
−0·35	0·8295 + j0·0566	0·8294 + j0·0613	0·8293 + j0·0660	0·8291 + j0·0707
	0·9426 + j0·0193	0·9425 + j0·0209	0·9425 + j0·0225	0·9425 + j0·0241

Table 16—continued

b a	0·16	0·17	0·18	0·19
0·00	0·9989 + j0·0800	0·9988 + j0·0850	0·9987 + j0·0900	0·9985 + j0·0950
	0·9998 + j0·0267	0·9998 + j0·0283	0·9997 + j0·0300	0·9997 + j0·0317
−0·01	0·9939 + j0·0799	0·9938 + j0·0849	0·9937 + j0·0898	0·9935 + j0·0948
	0·9981 + j0·0266	0·9981 + j0·0283	0·9981 + j0·0300	0·9980 + j0·0316
−0·02	0·9890 + j0·0797	0·9888 + j0·0847	0·9887 + j0·0897	0·9885 + j0·0947
	0·9965 + j0·0266	0·9964 + j0·0283	0·9964 + j0·0299	0·9964 + j0·0316
−0·03	0·9840 + j0·0796	0·9838 + j0·0846	0·9837 + j0·0895	0·9835 + j0·0945
	0·9948 + j0·0266	0·9948 + j0·0282	0·9947 + j0·0299	0·9947 + j0·0316
−0·04	0·9790 + j0·0795	0·9789 + j0·0844	0·9787 + j0·0894	0·9786 + j0·0944
	0·9931 + j0·0266	0·9931 + j0·0282	0·9931 + j0·0299	0·9930 + j0·0315
−0·05	0·9740 + j0·0793	0·9739 + j0·0843	0·9738 + j0·0892	0·9736 + j0·0942
	0·9915 + j0·0265	0·9914 + j0·0282	0·9914 + j0·0298	0·9914 + j0·0315
−0·06	0·9691 + j0·0792	0·9690 + j0·0841	0·9688 + j0·0891	0·9687 + j0·0940
	0·9898 + j0·0265	0·9898 + j0·0282	0·9898 + j0·0298	0·9897 + j0·0315
−0·07	0·9641 + j0·0791	0·9640 + j0·0840	0·9639 + j0·0889	0·9637 + j0·0939
	0·9882 + j0·0265	0·9881 + j0·0281	0·9881 + j0·0298	0·9881 + j0·0314
−0·08	0·9592 + j0·0789	0·9591 + j0·0839	0·9589 + j0·0888	0·9588 + j0·0937
	0·9865 + j0·0265	0·9865 + j0·0281	0·9865 + j0·0298	0·9864 + j0·0314
−0·09	0·9543 + j0·0788	0·9541 + j0·0837	0·9540 + j0·0886	0·9538 + j0·0936
	0·9849 + j0·0264	0·9848 + j0·0281	0·9848 + j0·0297	0·9848 + j0·0314
−0·10	0·9494 + j0·0787	0·9492 + j0·0836	0·9491 + j0·0885	0·9489 + j0·0934
	0·9832 + j0·0264	0·9832 + j0·0281	0·9831 + j0·0297	0·9831 + j0·0313
−0·11	0·9444 + j0·0785	0·9443 + j0·0834	0·9442 + j0·0884	0·9440 + j0·0933
	0·9816 + j0·0264	0·9815 + j0·0280	0·9815 + j0·0297	0·9815 + j0·0313
−0·12	0·9395 + j0·0784	0·9394 + j0·0833	0·9393 + j0·0882	0·9391 + j0·0931
	0·9799 + j0·0263	0·9799 + j0·0280	0·9799 + j0·0296	0·9798 + j0·0313
−0·13	0·9346 + j0·0783	0·9345 + j0·0832	0·9344 + j0·0881	0·9342 + j0·0929
	0·9783 + j0·0263	0·9782 + j0·0280	0·9782 + j0·0296	0·9782 + j0·0313
−0·14	0·9298 + j0·0781	0·9296 + j0·0830	0·9295 + j0·0879	0·9293 + j0·0928
	0·9766 + j0·0263	0·9766 + j0·0279	0·9766 + j0·0296	0·9765 + j0·0312
−0·15	0·9249 + j0·0780	0·9247 + j0·0829	0·9246 + j0·0878	0·9245 + j0·0926
	0·9750 + j0·0263	0·9749 + j0·0279	0·9749 + j0·0296	0·9749 + j0·0312
−0·16	0·9200 + j0·0779	0·9199 + j0·0827	0·9197 + j0·0876	0·9196 + j0·0925
	0·9733 + j0·0262	0·9733 + j0·0279	0·9733 + j0·0295	0·9732 + j0·0312
−0·17	0·9150 + j0·0777	0·9150 + j0·0826	0·9149 + j0·0875	0·9147 + j0·0923
	0·9717 + j0·0262	0·9717 + j0·0279	0·9716 + j0·0295	0·9716 + j0·0311
−0·18	0·9103 + j0·0776	0·9102 + j0·0825	0·9100 + j0·0873	0·9099 + j0·0922
	0·9701 + j0·0262	0·9700 + j0·0278	0·9700 + j0·0295	0·9700 + j0·0311
−0·19	0·9054 + j0·0775	0·9053 + j0·0823	0·9052 + j0·0872	0·9050 + j0·0920
	0·9684 + j0·0262	0·9684 + j0·0278	0·9684 + j0·0294	0·9683 + j0·0311

Table 16—continued 387

b a	0·16	0·17	0·18	0·19
−0·20	0·9006 + j0·0774	0·9005 + j0·0822	0·9003 + j0·0870	0·9002 + j0·0919
	0·9668 + j0·0261	0·9668 + j0·0278	0·9667 + j0·0294	0·9667 + j0·0310
−0·21	0·8958 + j0·0772	0·8956 + j0·0820	0·8955 + j0·0869	0·8954 + j0·0917
	0·9652 + j0·0261	0·9651 + j0·0277	0·9651 + j0·0294	0·9651 + j0·0310
−0·22	0·8910 + j0·0771	0·8908 + j0·0819	0·8907 + j0·0867	0·8905 + j0·0915
	0·9635 + j0·0261	0·9635 + j0·0277	0·9635 + j0·0293	0·9634 + j0·0310
−0·23	0·8861 + j0·0770	0·8860 + j0·0818	0·8859 + j0·0866	0·8857 + j0·0914
	0·9619 + j0·0261	0·9619 + j0·0277	0·9618 + j0·0293	0·9618 + j0·0309
−0·24	0·8813 + j0·0768	0·8812 + j0·0816	0·8811 + j0·0864	0·8809 + j0·0912
	0·9603 + j0·0260	0·9602 + j0·0277	0·9602 + j0·0293	0·9602 + j0·0309
−0·25	0·8765 + j0·0767	0·8764 + j0·0815	0·8763 + j0·0863	0·8761 + j0·0911
	0·9586 + j0·0260	0·9586 + j0·0276	0·9586 + j0·0293	0·9586 + j0·0309
−0·26	0·8718 + j0·0766	0·8716 + j0·0814	0·8715 + j0·0861	0·8713 + j0·0909
	0·9570 + j0·0260	0·9570 + j0·0276	0·9570 + j0·0292	0·9569 + j0·0308
−0·27	0·8670 + j0·0764	0·8668 + j0·0812	0·8667 + j0·0860	0·8665 + j0·0908
	0·9554 + j0·0260	0·9554 + j0·0276	0·9553 + j0·0292	0·9553 + j0·0308
−0·28	0·8622 + j0·0763	0·8621 + j0·0811	0·8619 + j0·0859	0·8618 + j0·0906
	0·9538 + j0·0259	0·9537 + j0·0275	0·9537 + j0·0292	0·9537 + j0·0308
−0·29	0·8574 + j0·0762	0·8573 + j0·0809	0·8572 + j0·0857	0·8570 + j0·0905
	0·9522 + j0·0259	0·9521 + j0·0275	0·9521 + j0·0291	0·9521 + j0·0308
−0·30	0·8527 + j0·0761	0·8525 + j0·0808	0·8524 + j0·0856	0·8523 + j0·0903
	0·9505 + j0·0259	0·9505 + j0·0275	0·9505 + j0·0291	0·9505 + j0·0307
−0·31	0·8479 + j0·0759	0·8478 + j0·0807	0·8477 + j0·0854	0·8475 + j0·0902
	0·9489 + j0·0258	0·9489 + j0·0275	0·9489 + j0·0291	0·9488 + j0·0307
−0·32	0·8432 + j0·0758	0·8431 + j0·0805	0·8429 + j0·0853	0·8428 + j0·0900
	0·9473 + j0·0258	0·9473 + j0·0274	0·9472 + j0·0290	0·9472 + j0·0307
−0·33	0·8385 + j0·0757	0·8383 + j0·0804	0·8382 + j0·0851	0·8380 + j0·0899
	0·9457 + j0·0258	0·9457 + j0·0274	0·9456 + j0·0290	0·9456 + j0·0306
−0·34	0·8337 + j0·0755	0·8336 + j0·0803	0·8335 + j0·0850	0·8333 + j0·0897
	0·9441 + j0·0258	0·9441 + j0·0274	0·9440 + j0·0290	0·9440 + j0·0306
−0·35	0·8290 + j0·0754	0·8289 + j0·0801	0·8287 + j0·0848	0·8286 + j0·0895
	0·9425 + j0·0257	0·9424 + j0·0274	0·9424 + j0·0290	0·9424 + j0·0306

Table 16—continued

b a	0·20	0·21	0·22	0·23
0·00	0·9983 + j0·1000	0·9982 + j0·1050	0·9980 + j0·1100	0·9978 + j0·1150
	0·9997 + j0·0333	0·9996 + j0·0350	0·9996 + j0·0367	0·9996 + j0·0383
−0·01	0·9933 + j0·0998	0·9932 + j0·1048	0·9930 + j0·1098	0·9928 + j0·1148
	0·9980 + j0·0333	0·9980 + j0·0350	0·9979 + j0·0366	0·9979 + j0·0383
−0·02	0·9884 + j0·0997	0·9882 + j0·1046	0·9880 + j0·1096	0·9878 + j0·1146
	0·9963 + j0·0333	0·9963 + j0·0349	0·9963 + j0·0366	0·9962 + j0·0383
−0·03	0·9834 + j0·0995	0·9832 + j0·1045	0·9830 + j0·1094	0·9828 + j0·1144
	0·9947 + j0·0332	0·9946 + j0·0349	0·9946 + j0·0366	0·9946 + j0·0382
−0·04	0·9784 + j0·0993	0·9782 + j0·1043	0·9781 + j0·1093	0·9779 + j0·1142
	0·9930 + j0·0332	0·9930 + j0·0349	0·9929 + j0·0365	0·9929 + j0·0382
−0·05	0·9734 + j0·0992	0·9733 + j0·1041	0·9731 + j0·1091	0·9729 + j0·1140
	0·9914 + j0·0332	0·9913 + j0·0348	0·9913 + j0·0365	0·9912 + j0·0381
−0·06	0·9685 + j0·0990	0·9683 + j0·1039	0·9681 + j0·1089	0·9680 + j0·1138
	0·9897 + j0·0331	0·9897 + j0·0348	0·9896 + j0·0364	0·9896 + j0·0381
−0·07	0·9635 + j0·0988	0·9634 + j0·1038	0·9632 + j0·1087	0·9630 + j0·1136
	0·9880 + j0·0331	0·9880 + j0·0348	0·9880 + j0·0364	0·9879 + j0·0381
−0·08	0·9586 + j0·0987	0·9584 + j0·1036	0·9583 + j0·1085	0·9581 + j0·1135
	0·9864 + j0·0331	0·9864 + j0·0347	0·9863 + j0·0364	0·9863 + j0·0380
−0·09	0·9537 + j0·0985	0·9535 + j0·1034	0·9533 + j0·1083	0·9532 + j0·1133
	0·9847 + j0·0330	0·9847 + j0·0347	0·9847 + j0·0363	0·9846 + j0·0380
−0·10	0·9488 + j0·0983	0·9486 + j0·1032	0·9484 + j0·1082	0·9482 + j0·1131
	0·9831 + j0·0330	0·9831 + j0·0346	0·9830 + j0·0363	0·9830 + j0·0379
−0·11	0·9439 + j0·0982	0·9437 + j0·1031	0·9435 + j0·1080	0·9433 + j0·1129
	0·9814 + j0·0330	0·9814 + j0·0346	0·9814 + j0·0363	0·9813 + j0·0379
−0·12	0·9390 + j0·0980	0·9388 + j0·1029	0·9386 + j0·1078	0·9384 + j0·1127
	0·9798 + j0·0329	0·9798 + j0·0346	0·9797 + j0·0362	0·9797 + j0·0379
−0·13	0·9341 + j0·0978	0·9339 + j0·1027	0·9337 + j0·1076	0·9335 + j0·1125
	0·9781 + j0·0329	0·9781 + j0·0345	0·9781 + j0·0362	0·9780 + j0·0378
−0·14	0·9292 + j0·0977	0·9290 + j0·1026	0·9288 + j0·1074	0·9286 + j0·1123
	0·9765 + j0·0329	0·9765 + j0·0345	0·9764 + j0·0362	0·9764 + j0·0378
−0·15	0·9243 + j0·0975	0·9241 + j0·1024	0·9239 + j0·1073	0·9238 + j0·1121
	0·9749 + j0·0328	0·9748 + j0·0345	0·9748 + j0·0361	0·9748 + j0·0378
−0·16	0·9194 + j0·0973	0·9193 + j0·1022	0·9191 + j0·1071	0·9189 + j0·1119
	0·9732 + j0·0328	0·9732 + j0·0344	0·9731 + j0·0361	0·9731 + j0·0377
−0·17	0·9146 + j0·0972	0·9144 + j0·1020	0·9142 + j0·1069	0·9140 + j0·1118
	0·9716 + j0·0328	0·9715 + j0·0344	0·9715 + j0·0360	0·9715 + j0·0377
−0·18	0·9097 + j0·0970	0·9095 + j0·1019	0·9094 + j0·1067	0·9092 + j0·1116
	0·9699 + j0·0327	0·9699 + j0·0344	0·9699 + j0·0360	0·9698 + j0·0376
−0·19	0·9049 + j0·0969	0·9047 + j0·1017	0·9045 + j0·1065	0·9043 + j0·1114
	0·9683 + j0·0327	0·9683 + j0·0343	0·9682 + j0·0360	0·9682 + j0·0376

Table 16—continued 389

b a	0·20	0·21	0·22	0·23
−0·20	0·9000 + j0·0967	0·8999 + j0·1015	0·8997 + j0·1064	0·8995 + j0·1112
	0·9667 + j0·0327	0·9666 + j0·0343	0·9666 + j0·0359	0·9666 + j0·0376
−0·21	0·8952 + j0·0965	0·8950 + j0·1014	0·8949 + j0·1062	0·8947 + j0·1110
	0·9650 + j0·0326	0·9650 + j0·0343	0·9650 + j0·0359	0·9649 + j0·0375
−0·22	0·8904 + j0·0964	0·8902 + j0·1012	0·8900 + j0·1060	0·8898 + j0·1108
	0·9634 + j0·0326	0·9634 + j0·0342	0·9633 + j0·0359	0·9633 + j0·0375
−0·23	0·8856 + j0·0962	0·8854 + j0·1010	0·8852 + j0·1058	0·8850 + j0·1106
	0·9618 + j0·0326	0·9617 + j0·0342	0·9617 + j0·0358	0·9617 + j0·0375
−0·24	0·8808 + j0·0960	0·8806 + j0·1008	0·8804 + j0·1056	0·8802 + j0·1104
	0·9601 + j0·0325	0·9601 + j0·0342	0·9601 + j0·0358	0·9600 + j0·0374
−0·25	0·8760 + j0·0959	0·8758 + j0·1007	0·8756 + j0·1055	0·8754 + j0·1103
	0·9585 + j0·0325	0·9585 + j0·0341	0·9585 + j0·0358	0·9584 + j0·0374
−0·26	0·8712 + j0·0957	0·8710 + j0·1005	0·8708 + j0·1053	0·8706 + j0·1101
	0·9569 + j0·0325	0·9569 + j0·0341	0·9568 + j0·0357	0·9568 + j0·0373
−0·27	0·8664 + j0·0955	0·8662 + j0·1003	0·8660 + j0·1051	0·8659 + j0·1099
	0·9553 + j0·0324	0·9552 + j0·0341	0·9552 + j0·0357	0·9552 + j0·0373
−0·28	0·8616 + j0·0954	0·8615 + j0·1002	0·8613 + j0·1049	0·8611 + j0·1097
	0·9537 + j0·0324	0·9536 + j0·0340	0·9536 + j0·0356	0·9536 + j0·0373
−0·29	0·8569 + j0·0952	0·8567 + j0·1000	0·8565 + j0·1047	0·8563 + j0·1095
	0·9520 + j0·0324	0·9520 + j0·0340	0·9520 + j0·0356	0·9519 + j0·0372
−0·30	0·8521 + j0·0951	0·8519 + j0·0998	0·8518 + j0·1046	0·8516 + j0·1093
	0·9504 + j0·0323	0·9504 + j0·0340	0·9503 + j0·0356	0·9503 + j0·0372
−0·31	0·8473 + j0·0949	0·8472 + j0·0996	0·8470 + j0·1044	0·8468 + j0·1091
	0·9488 + j0·0323	0·9488 + j0·0339	0·9487 + j0·0355	0·9487 + j0·0372
−0·32	0·8426 + j0·0947	0·8424 + j0·0995	0·8423 + j0·1042	0·8421 + j0·1089
	0·9472 + j0·0323	0·9472 + j0·0339	0·9471 + j0·0355	0·9471 + j0·0371
−0·33	0·8379 + j0·0946	0·8377 + j0·0993	0·8375 + j0·1040	0·8374 + j0·1088
	0·9456 + j0·0322	0·9455 + j0·0339	0·9455 + j0·0355	0·9455 + j0·0371
−0·34	0·8332 + j0·0944	0·8330 + j0·0991	0·8328 + j0·1039	0·8326 + j0·1086
	0·9440 + j0·0322	0·9439 + j0·0338	0·9439 + j0·0354	0·9439 + j0·0370
−0·35	0·8284 + j0·0943	0·8283 + j0·0990	0·8281 + j0·1037	0·8279 + j0·1084
	0·9424 + j0·0322	0·9423 + j0·0338	0·9423 + j0·0354	0·9422 + j0·0370

Table 16—continued

b / a	0·24	0·25	0·26	0·27
0·00	0·9976 + j0·1200	0·9974 + j0·1250	0·9972 + j0·1300	0·9970 + j0·1350
	0·9995 + j0·0400	0·9995 + j0·0417	0·9994 + j0·0433	0·9994 + j0·0450
−0·01	0·9926 + j0·1198	0·9924 + j0·1248	0·9922 + j0·1298	0·9920 + j0·1347
	0·9979 + j0·0400	0·9978 + j0·0416	0·9978 + j0·0433	0·9977 + j0·0450
−0·02	0·9876 + j0·1196	0·9874 + j0·1246	0·9872 + j0·1295	0·9870 + j0·1345
	0·9962 + j0·0399	0·9961 + j0·0416	0·9961 + j0·0432	0·9961 + j0·0449
−0·03	0·9826 + j0·1194	0·9824 + j0·1244	0·9822 + j0·1293	0·9820 + j0·1343
	0·9945 + j0·0399	0·9945 + j0·0415	0·9944 + j0·0432	0·9944 + j0·0449
−0·04	0·9777 + j0·1192	0·9775 + j0·1241	0·9773 + j0·1291	0·9770 + j0·1341
	0·9929 + j0·0398	0·9928 + j0·0415	0·9928 + j0·0432	0·9927 + j0·0448
−0·05	0·9727 + j0·1190	0·9725 + j0·1239	0·9723 + j0·1289	0·9721 + j0·1339
	0·9912 + j0·0398	0·9912 + j0·0415	0·9911 + j0·0431	0·9911 + j0·0448
−0·06	0·9678 + j0·1188	0·9676 + j0·1237	0·9673 + j0·1287	0·9671 + j0·1336
	0·9896 + j0·0398	0·9895 + j0·0414	0·9895 + j0·0431	0·9894 + j0·0447
−0·07	0·9628 + j0·1186	0·9626 + j0·1235	0·9624 + j0·1285	0·9622 + j0·1334
	0·9879 + j0·0397	0·9879 + j0·0414	0·9878 + j0·0430	0·9878 + j0·0447
−0·08	0·9579 + j0·1184	0·9577 + j0·1233	0·9575 + j0·1282	0·9573 + j0·1332
	0·9862 + j0·0397	0·9862 + j0·0413	0·9862 + j0·0430	0·9861 + j0·0446
−0·09	0·9530 + j0·1182	0·9528 + j0·1231	0·9525 + j0·1280	0·9523 + j0·1330
	0·9846 + j0·0396	0·9845 + j0·0413	0·9845 + j0·0429	0·9845 + j0·0446
−0·10	0·9480 + j0·1180	0·9478 + j0·1229	0·9476 + j0·1278	0·9474 + j0·1327
	0·9829 + j0·0396	0·9829 + j0·0412	0·9829 + j0·0429	0·9828 + j0·0445
−0·11	0·9431 + j0·1178	0·9429 + j0·1227	0·9427 + j0·1276	0·9425 + j0·1325
	0·9813 + j0·0396	0·9813 + j0·0412	0·9812 + j0·0429	0·9812 + j0·0445
−0·12	0·9382 + j0·1176	0·9380 + j0·1225	0·9378 + j0·1274	0·9376 + j0·1323
	0·9796 + j0·0395	0·9796 + j0·0412	0·9796 + j0·0428	0·9795 + j0·0445
−0·13	0·9333 + j0·1174	0·9331 + j0·1223	0·9329 + j0·1272	0·9327 + j0·1321
	0·9780 + j0·0395	0·9780 + j0·0411	0·9779 + j0·0428	0·9779 + j0·0444
−0·14	0·9284 + j0·1172	0·9282 + j0·1221	0·9280 + j0·1270	0·9278 + j0·1318
	0·9764 + j0·0394	0·9763 + j0·0411	0·9763 + j0·0427	0·9762 + j0·0444
−0·15	0·9236 + j0·1170	0·9234 + j0·1219	0·9232 + j0·1267	0·9229 + j0·1316
	0·9747 + j0·0394	0·9747 + j0·0410	0·9746 + j0·0427	0·9746 + j0·0443
−0·16	0·9187 + j0·1168	0·9185 + j0·1217	0·9183 + j0·1265	0·9181 + j0·1314
	0·9731 + j0·0394	0·9730 + j0·0410	0·9730 + j0·0426	0·9729 + j0·0443
−0·17	0·9138 + j0·1166	0·9136 + j0·1215	0·9134 + j0·1263	0·9132 + j0·1312
	0·9714 + j0·0393	0·9714 + j0·0410	0·9714 + j0·0426	0·9713 + j0·0442
−0·18	0·9090 + j0·1164	0·9088 + j0·1213	0·9086 + j0·1261	0·9084 + j0·1310
	0·9698 + j0·0393	0·9698 + j0·0409	0·9697 + j0·0426	0·9797 + j0·0442
−0·19	0·9041 + j0·1162	0·9039 + j0·1211	0·9037 + j0·1259	0·9035 + j0·1307
	0·9682 + j0·0392	0·9681 + j0·0409	0·9681 + j0·0425	0·9680 + j0·0441

Table 16—continued   391

b	0·24	0·25	0·26	0·27
a				
−0·20	0·8993 + j0·1160	0·8991 + j0·1209	0·8989 + j0·1257	0·8987 + j0·1305
	0·9665 + j0·0392	0·9665 + j0·0408	0·9664 + j0·0425	0·9664 + j0·0441
−0·21	0·8945 + j0·1158	0·8943 + j0·1206	0·8941 + j0·1255	0·8939 + j0·1303
	0·9649 + j0·0392	0·9649 + j0·0408	0·9648 + j0·0424	0·9648 + j0·0441
−0·22	0·8897 + j0·1156	0·8895 + j0·1204	0·8892 + j0·1253	0·8890 + j0·1301
	0·9633 + j0·0391	0·9632 + j0·0408	0·9632 + j0·0424	0·9631 + j0·0440
−0·23	0·8848 + j0·1154	0·8846 + j0·1202	0·8844 + j0·1250	0·8842 + j0·1299
	0·9616 + j0·0391	0·9616 + j0·0407	0·9616 + j0·0423	0·9615 + j0·0440
−0·24	0·8800 + j0·1152	0·8798 + j0·1200	0·8796 + j0·1248	0·8794 + j0·1296
	0·9600 + j0·0390	0·9600 + j0·0407	0·9599 + j0·0423	0·9599 + j0·0439
−0·25	0·8752 + j0·1150	0·8750 + j0·1198	0·8748 + j0·1246	0·8746 + j0·1294
	0·9584 + j0·0390	0·9583 + j0·0406	0·9583 + j0·0423	0·9583 + j0·0439
−0·26	0·8705 + j0·1148	0·8703 + j0·1196	0·8700 + j0·1244	0·8698 + j0·1292
	0·9568 + j0·0390	0·9567 + j0·0406	0·9567 + j0·0422	0·9566 + j0·0438
−0·27	0·8657 + j0·1147	0·8655 + j0·1194	0·8563 + j0·1242	0·8651 + j0·1290
	0·9551 + j0·0389	0·9551 + j0·0405	0·9551 + j0·0422	0·9550 + j0·0438
−0·28	0·8609 + j0·1145	0·8607 + j0·1192	0·8605 + j0·1240	0·8603 + j0·1288
	0·9535 + j0·0389	0·9535 + j0·0405	0·9534 + j0·0421	0·9534 + j0·0437
−0·29	0·8561 + j0·1143	0·8559 + j0·1190	0·8557 + j0·1238	0·8555 + j0·1285
	0·9519 + j0·0388	0·9519 + j0·0405	0·9518 + j0·0421	0·9518 + j0·0437
−0·30	0·8514 + j0·1141	0·8512 + j0·1188	0·8510 + j0·1236	0·8508 + j0·1283
	0·9503 + j0·0388	0·9502 + j0·0404	0·9502 + j0·0420	0·9502 + j0·0437
−0·31	0·8466 + j0·1139	0·8464 + j0·1186	0·8462 + j0·1234	0·8460 + j0·1281
	0·9487 + j0·0388	0·9486 + j0·0404	0·9486 + j0·0420	0·9485 + j0·0436
−0·32	0·8419 + j0·1137	0·8417 + j0·1184	0·8415 + j0·1232	0·8413 + j0·1279
	0·9470 + j0·0387	0·9470 + j0·0403	0·9470 + j0·0420	0·9469 + j0·0436
−0·33	0·8372 + j0·1135	0·8370 + j0·1182	0·8368 + j0·1229	0·8366 + j0·1277
	0·9454 + j0·0387	0·9454 + j0·0403	0·9454 + j0·0419	0·9453 + j0·0435
−0·34	0·8324 + j0·1133	0·8322 + j0·1180	0·8320 + j0·1227	0·8318 + j0·1275
	0·9438 + j0·0387	0·9438 + j0·0403	0·9437 + j0·0419	0·9437 + j0·0435
−0·35	0·8277 + j0·1131	0·8275 + j0·1178	0·8273 + j0·1225	0·8271 + j0·1272
	0·9422 + j0·0386	0·9422 + j0·0402	0·9421 + j0·0418	0·9421 + j0·0434

INDEX